TRANSPORT RESEARCH

Automotive Braking

Collected papers from the 2nd International Seminar on
Automotive Braking – Recent Developments and Future Trends
held at Weetwood Hall, Leeds, UK, 14–15 May 1998.

The Seminar was organized by
The Yorkshire Centre, Automobile Division of
The Institution of Mechanical Engineers (IMechE)

In Association with
The University of Leeds
The Society of Automotive Engineers
The Institute of Road Transport Engineers

Seminar Organizing Committee

Dr David Barton (Joint Chairman)
University of Leeds

Mr Stuart Cowling
Vickers Defence Systems

Mr Steve Earle
Pioneer Weston

Dr Martin Haigh (Joint Chairman)
BBA Friction Limited

Mr Bert Smales
Consultant

Cover design/illustration by Terry Bambrook, Leeds, UK.
Tel: + 44 (0)113 236 2855

The Second International Seminar on

Automotive Braking

Recent Developments and Future Trends

Edited by

Dr David C Barton

and

Dr Martin J Haigh

Professional Engineering Publishing

Professional Engineering Publishing Limited
Bury St Edmunds and London, UK.

First Published 1998

This publication is copyright under the Berne Convention and the International Copyright Convention. All rights reserved. Apart from any fair dealing for the purpose of private study, research, criticism or review, as permitted under the Copyright, Designs and Patents Act, 1988, no part may be reproduced, stored in a retrieval system, or transmitted in any form or by any means, electronic, electrical, chemical, mechanical, photocopying, recording or otherwise, without the prior permission of the copyright owners. Reprographic reproduction is permitted only in accordance with the terms of licences issued by the Copyright Licensing Agency, 90 Tottenham Court Road, London W1P 9HE. *Unlicensed multiple copying of the contents of this publication is illegal.* Inquiries should be addressed to: The Publishing Editor, Professional Engineering Publishing Limited, Northgate Avenue, Bury St. Edmunds, Suffolk, IP32 6BW, UK. Fax: 01284 704006.

© With Authors 1998

ISBN 1 86058 131 5

A CIP catalogue record for this book is available from the British Library.

Printed and bound in Great Britain by Antony Rowe Limited, Chippenham, Wiltshire, UK.

The Publishers are not responsible for any statement made in this publication. Data, discussion, and conclusions developed by authors are for information only and are not intended for use without independent substantiating investigation on the part of potential users. Opinions expressed are those of the Author and are not necessarily those of the Institution of Mechanical Engineers or its Publishers.

Related Titles of Interest

Title	Editor/Author	ISBN
Brakes and Friction Materials – The history and development of the technologies	Graham A Harper	1 86058 127 7
Heat Treatment Processes for the Reduction of Wear in Vehicle Components	Alexander Schreiner	0 85298 773 0
Automotive Materials Technology	IMechE Seminar 1997–5	1 86058 117 X

For the full range of titles published by Professional Engineering Publishing contact:

Sales Department
Professional Engineering Publishing Limited
Northgate Avenue
Bury St Edmunds
Suffolk
IP32 6BW
UK

Tel: 01284 724384
Fax: 01284 718692

Contents

Foreword
D C Barton and M J Haigh — ix

Brake Refinement

On the stick–slip dynamics of an elastic slider on a vibrating disc
H Ouyang, J E Mottershead, D J Brookfield, S James, and M P Cartmell — 3

Noise insulators for brake squeal reduction – influence and selection of the damping material
P Heppes — 15

The value of noise fix shims – a case study
J D Fieldhouse and M Rennison — 27

Brake pull – a vehicle-based factorial experiment
R Bartholomew — 49

Rear disc brake moan – experimental investigation and ADAMS simulation
D Riesland, J Janevic, J Malosh, and W Stringham — 67

Brake vibration and disc thickness variation (DTV)
K Vikulov, B Tron, and P Buonfico — 77

Testing and Miscellaneous Topics

Prediction of whole vehicle braking performance using dual dynamometer differential effectiveness analysis (D^3EA^{TM})
C W Greening Jr — 87

Correlation of scale to full-size dynamometer testing
T Kuroda and J Abo — 99

Characterization of automotive friction behaviour using small specimens
J W Fash, T M Dalka, D L Hartsock, R L Hecht, and R Karthik — 113

Hayes' increased airflow rotor design
A R Daudi, W E Dickerson, and M Narain — 127

Determination of water content in brake fluid by refractive index
T E Ryan — 145

Automotive braking – a patent perspective
T H Lemon — 159

Materials and Commercial Vehicle Brakes

Cast iron – a brake disc material for the future?
M P Macnaughtan and J G Krosnar 177

The effect of cast iron disc brake metallurgy on friction and wear characteristics
K Ellis 191

Aluminium metal matrix composite rotors and drums – a future trend
M J Denholm 205

Production and selection of iron powder for the friction brake industry
C E Evans 215

Thermo-mechanical instability in braking and brake disc thermal judder: an experimental and finite element study
T K Kao, J W Richmond, and A Douarre 231

The brake disc – development challenge and factors for success in the pneumatic disc brake system for heavy trucks
V Kühne 265

The influence of heavy vehicle brake drum geometry on brake torque stability
T Varma and R McLellan 279

Authors' Index 293

Foreword

Following successful events at Leeds in 1994 and 1996, this two-day seminar brings together vehicle, brake, and friction material specialists from around the world. The speakers, drawn from academia and industry, cover both the theoretical bases and practical applications of their specialist topics. The aim is to provide a forum for technology transfer between academic establishments on the one hand and practising engineers in the automotive braking and related industries on the other, in order to enable real problems to be discussed and appropriate solutions sought.

The first session is concerned with brake refinement, i.e. the overall quality and safety of the braking system as perceived by that all-important consumer – the driver of the vehicle. Thus, the first three papers in this volume all address the issue of brake 'squeal', with the opening paper providing a theoretical basis for understanding the problem and the next two assessing practical solutions. The following three papers consider other aspects of brake refinement: brake 'pull', rear disc brake 'moan', and brake vibration ('judder') due to disc thickness variation.

The second session is concerned with testing techniques and other related topics. The first paper describes a new assessment technique based on dual dynamometer differential effectiveness analysis whilst the next two contributions consider the important topic of small-scale dynamometer testing. The next paper considers airflow through vented rotors and the following contribution discusses methods for determining water content in brake fluid. Finally the importance of proper protection and intelligent use of advances in braking technology as represented in the patent literature, is highlighted.

The third session considers developments in automotive braking materials. The opening two papers discuss advances in cast iron metallurgy for brake discs and the effect of these developments on the friction pair tribology, respectively. A potential replacement for cast iron in the form of aluminium metal matrix composite rotors is then considered, followed by a discussion of the different forms of iron powder now available for the friction brake industry. Attention is refocussed on the issue of brake 'judder' in the next paper whilst the final two contributions examine development in commercial vehicle brakes utilizing discs and drums, respectively.

The editors would like to thank all authors for their contributions to the Seminar. Acknowledgements are also due to the Seminar Planning Committee, the Secretariat and the staff of Weetwood Hall, Leeds for all their hard work in organizing the event. Finally the editors sincerely hope that the seminar delegates have an interesting and enjoyable two days and are able to take away with them something of value for their future involvement in this important section of the automotive industry. We look forward to seeing many of the delegates again at the next event in the series planned for late Spring 2000.

Dr D C Barton (Joint Chairman)
University of Leeds, UK

Dr M J Haigh (Joint Chairman)
BBA Friction Limited, UK

Brake Refinement

On the stick–slip dynamics of an elastic slider on a vibrating disc

H OUYANG, J E MOTTERSHEAD, D J BROOKFIELD, and S JAMES
Department of Mechanical Engineering, University of Liverpool, UK
M P CARTMELL
Department of Mechanical Engineering, University of Glasgow, UK

ABSTRACT

The in-plane vibration of a slider-mass which is driven around the surface of a flexible disc, and the transverse vibration of the disc, are investigated. The slider has flexibility and damping in the circumferential (in-plane) and transverse directions. The static friction coefficient is assumed to be higher than the kinetic friction. Numerical results show that initial braking pressure and the rotating speed of the drive point can drive the system into instability, while the rigidity and damping of the disc, and transverse stiffness and damping of the slider tend to suppress the vibrations. The motivation of this work is the understanding of instability and squeal in physical systems such as car brake discs where there are vibrations induced by non-smooth dry-friction forces.

NOMENCLATURE

a, b	inner and outer radii of the annular disc
c, c_p	damping coefficient of the slider in the transverse and in-plane directions
h	thickness of the disc
$i = \sqrt{-1}$	
k, k_p	transverse and in-plane stiffnesses of the slider system
m	mass of the slider
p	initial normal pressure applied by the slider onto the disc surface
q_{kl}	modal co-ordinate for k nodal circles and l nodal diameters
r	radial co-ordinate in cylindrical co-ordinate system
r_0	radial position of the slider
t	time
u, u_0	transverse displacement of the slider mass and its initial value

w, w_0	transverse displacement of the disc and its initial value
A	contact area of the slider-mass with the disc surface
D	flexural rigidity of the disc
D^*	Kelvin-type damping coefficient of the disc
E	Young's modulus of the disc material
N	initial normal load on the disc from the slider system ($N = A\,p$)
P	total normal force on the disc from the slider system
R_{kl}	combination of Bessel functions to represent the mode shape in the radial direction
θ	circumferential co-ordinate of cylindrical co-ordinate system
μ_k, μ_s	kinetic and static dry friction coefficient between the slider and the disc
ν	Poisson's ratio
ξ	damping ratio of the disc
ρ	mass-density of the disc
φ	absolute circumferential position of the slider
φ_{stick}	absolute circumferential position of the slider when it sticks to the disc
ψ	circumferential position of the slider relative to the drive point
ψ_{kl}	mode function for the transverse vibration of the disc corresponding to q_{kl}
ω_{kl}	natural (circular) frequency corresponding to q_{kl}
Ω	constant rotating speed of the drive point around the disc in radians per second

INTRODUCTION

There exists a whole class of mechanical systems which involve discs rotating relative to stationery parts, such as car brake discs, clutches, saws, computer discs and so on. In many of these systems, dry-friction induced vibration plays a crucial role in system performance. If the vibration becomes excessive, the system might fail, or cease to perform properly, or make offensive noises. In this paper we investigate the vibration of an in-plane slider system on an elastic disc, and the vibrations of the disc. The slider system consists of a transverse mass-spring-damper, attached through an in-plane spring and damper to a drive point which rotates at a constant speed. Dry friction acts between the slider system and the disc. The slider is assumed to be in perfect contact with the disc, which is taken to be a flat, thin, annular plate.

Dry-friction induced vibration has been studied extensively (1–7). For car brake vibration and squeal, see the review papers (8, 9). The stick-slip phenomenon of dry-friction induced vibration was studied in the context of chaotic vibration in (10–18). If sticking is absent at constant drive point speed, the problem is reduced to a linear parametric analysis which was carried out for a pin-on-disc system in (19, 20) and for a pad-on-disc system in (21).

IN-PLANE OSCILLATION OF THE SLIDER SYSTEM

As the drive point, which is connected to the slider-mass through an in-plane, elastic spring and damper, is rotated at constant angular speed around the disc, the driven slider will undergo stick-slip oscillations. The whole system of the slider and the disc is shown in Figure 1. The arrangement can be considered to represent an automotive disc brake where the resilience and dissipation of the pad material and the pad support structure (i.e. the calliper and its mounting) are modelled simply by the in-plane and transverse springs and dampers.

The equation of the in-plane motion of the slider system relative to the rotating drive point, in the sliding phase, is,

$$r_0(m\ddot{\psi} + c_p\dot{\psi} + k_p\psi) = \mu_k \text{sgn}(\dot{\varphi})P, \tag{1}$$

while in sticking, the equation of the motion becomes,

$$\psi = \Omega t - \varphi_{stick}. \tag{2}$$

The relationship between the relative motion of the slider system to the drive point and its absolute motion (relative to the stationary disc) is

$$\varphi = \Omega t - \psi, \quad \dot{\varphi} = \Omega - \dot{\psi}, \quad \ddot{\varphi} = -\ddot{\psi}. \tag{3}$$

We consider the following initial conditions. An initial pressure is exerted on the top surface of this disc, which produces an initial momentum and excites transverse vibration of the disc. At the same time, the slider is pulled around the disc surface by the drive point rotating at constant speed.

First, sliding from the initial sticking phase occurs when,

$$r_0(k_p\psi + c_p\dot{\psi}) \geq \mu_s P. \tag{4}$$

The slider will stick to the disc again when,

$$\dot{\psi} = \Omega, \quad r_0\left|(k_p\psi + c_p\dot{\psi})\right| \leq \mu_s P \quad \text{(during sliding)}, \tag{5}$$

or it will begin to slide again if,

$$r_0\left|(k_p\psi + c_p\dot{\psi})\right| \geq \mu_s P \quad \text{(during sticking)}. \tag{6}$$

Consequently, the slider system will stick and slide consecutively on the disc surface. This moving slider on the disc surface will input energy into and excite the disc.

TRANSVERSE VIBRATION OF THE ANNULAR DISC

The equation of motion of the disc under the slider system is,

$$\rho h \frac{\partial^2 w}{\partial t^2} + D^* \nabla^4 \dot{w} + D \nabla^4 w = -\frac{1}{r}\delta(r - r_0)\delta(\theta - \varphi)P. \tag{7}$$

The total force P is the summation of initial normal load N and the resultant force induced by transverse motion u of the slider. It may be expressed as,

$$P = N + m\ddot{u} + c\dot{u} + k(u - u_0). \tag{8}$$

Since it is assumed that the slider system is always in perfect contact with the disc, then,

$$u(t) = w(r_0, \varphi(t), t). \tag{9}$$

Substitution of equations (8) and (9) into (7) leads to,

$$\rho h \frac{\partial^2 w}{\partial t^2} + D^{\bullet}\nabla^4 \frac{\partial w}{\partial t} + D\nabla^4 w = -\frac{1}{r}\delta(r - r_0)\delta(\theta - \varphi)[N + $$
$$m(\ddot{\varphi}\frac{\partial w}{\partial \theta} + \dot{\varphi}^2 \frac{\partial^2 w}{\partial \theta^2} + 2\dot{\varphi}\frac{\partial^2 w}{\partial \theta \partial t} + \frac{\partial^2 w}{\partial t^2}) + c(\dot{\varphi}\frac{\partial w}{\partial \theta} + \frac{\partial w}{\partial t}) + \tag{10}$$
$$k(w - w_0)].$$

It should be noted that equation (10) is valid whether the slider system is sticking or sliding. When the slider sticks to the disc, equation (10) reduces to,

$$\rho h \frac{\partial^2 w}{\partial t^2} + D^{\bullet}\nabla^4 \frac{\partial w}{\partial t} + D\nabla^4 w = -\frac{1}{r}\delta(r - r_0)\delta(\theta - \varphi) \times$$
$$[N + m\frac{\partial^2 w}{\partial t^2} + c\frac{\partial w}{\partial t} + k(w - w_0)]. \tag{11}$$

COUPLED VIBRATIONS OF THE SLIDER AND THE DISC

The transverse motion of the disc can be expressed in modal co-ordinates such that,

$$w(r, \theta, t) = \sum_{k=0}^{\infty}\sum_{l=-\infty}^{\infty} \psi_{kl}(r, \theta) q_{kl}(t), \tag{12}$$

where,

$$\psi_{kl}(r, \theta) = \frac{1}{\sqrt{\rho h b^2}} R_{kl}(r) \exp(il\theta), \tag{13}$$

The modal functions satisfy the ortho-normality conditions,

$$\int_a^b \rho h \overline{\psi}_{kl} \psi_{rs} r dr d\theta = \delta_{kr}\delta_{ls},$$
$$\int_a^b D\overline{\psi}_{kl}\nabla^4 \psi_{rs} r dr d\theta = \omega_{rs}^2 \delta_{kr}\delta_{ls}. \tag{14}$$

where the overbar denotes complex conjugation.

Equations (10) and (11) are simplified when they are written in terms of the modal coordinates from equation (12).

During sticking, the motion of the whole system of the slider and the disc can be determined from,

$$\ddot{q}_{kl} + 2\xi\omega_{kl}\dot{q}_{kl} + \omega_{kl}^2 q_{kl} = -\frac{N}{\sqrt{\rho h b^2}} R_{kl}(r_0)\exp(-il\varphi) -$$
$$\frac{1}{\rho h b^2}\sum_{r=0}^{\infty}\sum_{s=-\infty}^{\infty} R_{rs}(r_0)R_{kl}(r_0)\exp[i(s-l)\varphi] \times \quad (15)$$
$$\{m\ddot{q}_{rs} + c\dot{q}_{rs} + k[q_{rs} - q_{rs}(0)]\},$$

where,

$$\varphi = \varphi_{stick}, \quad \psi = \Omega t - \varphi_{stick}, \quad (16)$$

and the sticking phase will be maintained when,

$$r_0\left|(k_p\psi + c_p\dot{\psi})\right| < \mu_s\left[N + \frac{1}{\sqrt{\rho h b^2}}\sum_{k=0}^{\infty}\sum_{l=-\infty}^{\infty} R_{kl}(r) \times \right.$$
$$\left.\exp(il\varphi)\{m\ddot{q}_{rs} + c\dot{q}_{rs} + k[q_{rs} - q_{rs}(0)]\}\right]. \quad (17)$$

In the sliding phase, the motion of the whole system can be represented by the coupled system of equations,

$$\ddot{q}_{kl} + 2\xi\omega_{kl}\dot{q}_{kl} + \omega_{kl}^2 q_{kl} = -\frac{N}{\sqrt{\rho h b^2}} R_{kl}(r_0)\exp(-il\varphi) - \frac{1}{\rho h b^2}\sum_{r=0}^{\infty}\sum_{s=-\infty}^{\infty} R_{rs}(r_0) \times$$
$$R_{kl}(r_0)\exp[i(s-l)\varphi]\{m[\ddot{q}_{rs} + i2s\dot{\varphi}\dot{q}_{rs} + (is\ddot{\varphi} - s^2\dot{\varphi}^2)q_{rs}] + \quad (18)$$
$$c(\dot{q}_{rs} + is\dot{\varphi}q_{rs}) + k[q_{rs} - q_{rs}(0)]\},$$

and,

$$r_0(m\ddot{\psi} + c_p\dot{\psi} + k_p\psi) = \mu_k\text{sgn}(\dot{\varphi})[N + \frac{1}{\sqrt{\rho h b^2}} \times$$
$$\sum_{r=0}^{\infty}\sum_{s=-\infty}^{\infty} R_{rs}(r_0)\exp(is\varphi)\{m[\ddot{q}_{rs} + i2s\dot{\varphi}\dot{q}_{rs} + (is\ddot{\varphi} - \quad (19)$$
$$s^2\dot{\varphi}^2)q_{rs}] + c(\dot{q}_{rs} + is\dot{\varphi}q_{rs}) + k[q_{rs} - q_{rs}(0)]\}].$$

The sliding phase will be maintained when either

$$r_0\left|(k_p\psi + c_p\dot{\psi})\right| > \mu_s[N + \frac{1}{\sqrt{\rho h b^2}}\sum_{r=0}^{\infty}\sum_{s=-\infty}^{\infty} R_{rs}(r_0) \times$$
$$\exp(is\varphi)\{m[\ddot{q}_{rs} + i2s\dot{\varphi}\dot{q}_{rs} + (is\ddot{\varphi} - s^2\dot{\varphi}^2)q_{rs}] + \quad , \quad (20)$$
$$c(\dot{q}_{rs} + is\dot{\varphi}q_{rs}) + k[q_{rs} - q_{rs}(0)]\}],$$

or,

$$\dot{\psi} \neq \Omega. \tag{21}$$

NUMERICAL ANALYSIS

As the slider system sticks and slides consecutively, the governing equations of the coupled motions of the whole system switch repeatedly from equations (15), (16) and (17) to equations (18), (19) and (20) or (21). The system is not smooth. Since the condition which controls the phases of the slider system itself depends on the motions, it is also a nonlinear system, whether μ_k is a constant or a function of relative speed $\dot{\varphi}$. In order to get modal co-ordinates, we have to truncate the infinite series in equation (12) to finite terms, as was also done in (18). Then numerical integration is used to solve equations (15), (18) and (19). Here a fourth order Runge-Kutta method (22), combined with a time step prediction criterion, is used for second order simultaneous ordinary differential equations. Since equation (18) has time-dependent coefficients, the time step length has to be very small. Constant time step lengths are chosen when the in-plane slider motion is well within the sticking phase or the sliding phase in the numerical integration. As it is imperative that the time step should be chosen such that at the end of some time intervals the slider happens to be on the sticking-sliding interfaces, we use a prediction criterion at the beginning of each time step in order to determine the next time step length when approaching these interfaces. Therefore, at the sticking-sliding interfaces, the time step length is variable (actually smaller than it is while well within the sticking or sliding regimes) so that the transition between the two regimes is accurately located. A few trial runs are needed to determine the constant time step length to be used well within the two regimes in order that both accuracy and efficiency can be achieved at the same time.

When transverse motion of the slider system becomes so violent that the total normal force P becomes negative or becomes several times larger than the initial normal load N, we describe the system as being unstable. Then the motion tends to diverge. But this instability should be distinguished from a chaotic motion which is bounded but never converges to a few points or a closed curve on the Poincare map.

The following data are used in the computation of numerical examples: $a = 0.044$m, $b = 0.12$m, $h = 0.002$m, $E = 150$GPa, $v = 0.211$, $\rho = 7200$kg/m^3; $r_0 = 0.1$m, $\Omega = 15$rad/sec, $D^* = 10^{-5}$Nms (very small damping factor of about 7.2×10^{-5} for the first mode), $c = 0$, $c_p = 0$, $\mu_s = 0.4$, $\mu_k = 0.24$, $k = 5 \times 10^4$N/m, $k_p = 2 \times 10^4$N/m, $m = 0.1$kg, $A = 0.0035$m^2. The disc is clamped at inner radius and free at outer radius. Note that in these numerical examples, the disc thickness is deliberately taken to be very small in order to reduce the amount of computing work. However, this should not affect the qualitative features of the results or conclusions drawn from the results thus obtained. The first five natural frequencies (in radians per second) of the disc are 1507, 1520, 1520, 1866, 1866. We will study vibration solutions at different values of all significant parameters, proceeding by varying one parameter while keeping other parameters constant. Poincare maps (the left-hand side ones are for the in-plane vibration of the slider whereas the right-hand side ones the transverse vibration of the disc at $r_0 = 0.1$ and $\theta = 0$) are given to show the dynamic behaviour.

First of all, we briefly study the effect of the initial normal pressure p (or the initial normal load N) and rotating speed Ω. When p is very small, the in-plane vibration of the slider system is periodic and the sticking interval is very short, as the transverse vibration of the disc is too small to affect the total normal force P. One of such cases is illustrated in Figure 2. As p increases, the sticking interval gets longer and more harmonics of the transverse vibration of the disc are involved, so that the in-plane stick-slip vibration becomes quasi-periodic. At $p=110$kPa, the whole system is driven into unbounded instability as shown in Figure 3. If a constant normal pressure is maintained but speed Ω is reduced, the amplitude of vibration becomes smaller. On the other hand, at certain critical value of Ω, the system again goes unstable, as shown in Figure 4.

Secondly, we look at the influence of the various damping terms. If disc damping D^* is increased, vibrations will become more regular and the amplitude of the transverse vibration will diminish. Figure 5 shows (when compared to Figure 3) that an unstable system can be stabilised by a sufficiently large disc damping term. Conversely the system will become unstable at $p=100$kPa, $\Omega=15$rad/sec when $D^*=8\times10^{-6}$Nms. If a smaller D^* were used, the system would go unstable at a lower value of p. Figure 6 shows what happens when D^* is absent. Transverse damping of the slider is also stabilising, as seen in Figure 7 in which an otherwise unstable system is stabilised by the introduction $c=0.707$Ns/m (damping ratio of 0.5%). As is already known, the in-plane damping of a slider (under a constant normal load on a rigid surface that moves at constant speed) makes the Poincare map of the slider spiral towards a center and finally settles down on a point. Here the in-plane damping has a similar but more complex effect because the normal load, P, changes with time instead of remaining constant. Consequently, the Poincare map of the slider first vibrates in stick-slip mode for a while and then after spiralling inward settles down on its equilibrium point represented by the solid area of spots. This feature is illustrated in Figure 8 in the case of $p=80$kPa with $c_p=0.447$Ns/m (damping ratio of 0.5%), which seems to indicate in-plane damping is also stabilising. The reduction in vibration amplitude is even greater at smaller pressures. It should also be noted that with this value of c_p sticking is absent in the later stage of vibration so a regime of persistent sliding is present when transient dies out. However, at $p=110$kPa, the large in-plane damping (damping factor as large as 5%) does not stabilise this unstable system. The reason can be clear by looking at Figure 8 where it has taken a large number of cycles (hence rather long time) for c_p to settle the vibrations down to the equilibrium points, so that in the present unstable case, instability happens before c_p is able to sufficiently reduce vibration. This will have a bearing on its effectiveness of stabilising effect. Therefore, the role of c_p is less effective than c or D^* in curbing vibrations, and although it can reduce vibration it may not stabilise an unstable system.

Thirdly, we examine the effect of stiffening the disc and the slider in its transverse and in-plane directions. When E is increased, the vibrations become smaller. As a result, an unstable system can be stabilised. The transverse stiffness k of the slider provides a restraint to the transverse vibration of the disc, and hence has a similar effect to E in curbing the vibrations. On the other hand, reduction of either E or k can destabilise the system. For instance, at $p=80$kPa, the vibrations become unstable at $k=35000$N/m. Numerical results indicate that the role of the in-plane stiffness k_p is more subtle. The whole range of k_p up to its critical value (where the system goes unstable) can be divided into several similar intervals. Within each interval, smaller values of k_p are stabilising while larger ones are not. At a sufficient large value of k_p the system will go unstable, and this case is illustrated in Figure 9 when $k_p = 4.5 \times 10^4$ N/m and $p = 80$ kPa. The critical values of k_p depend on the normal pressures. At smaller pressures the critical stiffnesses are found to be larger. Typically the system will become unstable at $p = 60$ kPa if $k_p = 7.4 \times 10^4$ N/m. Thus, k_p should not be considered as a stabilising factor and its exact role needs further investigation.

CONCLUSIONS

In this paper the stick-slip slider vibration and transverse vibration of an elastic disc have been studied. The whole system has been reduced to six degrees-of-freedom and we are able to make the following conclusions.

1. The stick-slip vibration of the slider and the transverse vibrations of the disc are both complex as one might expect with a multi-degree of freedom, non-smooth system.
2. For the normal pressure parameter, smaller values allow periodic solutions. Greater pressures result in quasi-periodic motions. At certain large pressures, the vibrations become unstable.
3. When the drive point is rotated at a low speed the vibrations are regular. An increase in the rotating speed can make the vibrations larger or more unstable.
4. Damping from the disc or slider in transverse direction can effectively reduce the magnitude of vibrations and can stabilise otherwise unstable vibrations. Damping from the slider in the in-plane direction can also reduce the magnitude of vibrations but fails to stabilise unstable vibrations.
5. Stiffness of the disc or the slider in transverse direction is stabilising. However, the in-plane stiffness from the slider can reduce the magnitude of transverse vibration at some values while increasing the transverse vibration at other values. The in-plane stiffness has a critical value at which the system will become unstable.

ACKNOWLEDGEMENT

This research is supported by the Engineering and Physical Sciences Research Council (grant numbers J35177 and L91061), BBA Friction Ltd and Lucas Varity GmbH.

REFERENCES

1. Marui, E. and Kato, S., Forced vibration of a base-excited single-degree-of freedom system with Coulomb friction. *J. Dyn. Sys., Measurement & Control, Trans. ASME*, 1984, **106**, pp.280-285
2. Shaw, S.W., On the dynamic response of a system with dry friction. *J. Sound Vib.*, 1986, **108**(2), pp.305-325
3. Lin, Y-Q and Wang Y-H, Stick-slip vibration of drill strings. *J. Eng. Ind., Trans. ASME*, 1991, **113**, pp.38-43

4. Ferri, A.A. and Bindemann, A.C., Damping and vibration of beams with various types of frictional support conditions. *J. Vib. Acoust., Trans. ASME*, 1992, **114**, pp.289-96
5. Lee, A.C., Study of disc brake noise using multi-body mechanism with friction interface. In *Friction-Induced Vibration, Chatter, Squeal, and Chaos*, Ed. Ibrahim, R.A. and Soom, A., DE-Vol. 49, ASME 1992, pp.99-105
6. Chen, J.H. and Chen, C.C., A study of stick slip motion and its influence on the cutting process. *Int. J. Mech. Sci.*, 1993, **35**(5), pp.353-370
7. Gao, C., Kuhlmann-Wilsdorf, D. and Makel, D.D., The dynamic analysis of stick-slip motion. *Wear*, 1994, **173**(1-2), pp.1-12
8. Ibrahim, R.A., Friction-induced vibration, chatter, squeal, and chaos. In *Friction-Induced Vibration, Chatter, Squeal, and Chaos*, Ed. Ibrahim, R.A. and Soom, A., DE-Vol. 49, ASME 1992, pp.107-38
9. Yang, S. and Gibson, R.F., Brake vibration and noise: reviews, comments, and proposed considerations. *Proceedings of the 14th Modal Analysis Conference*, The Society of Experimental Mechanics, Inc., 1996, pp.1342-9
10. Popp, K. and Stelter, P., Stick-slip vibration and chaos. *Phil. Trans. R. Soc. Lond.* A(1990), **332**, pp.89-106
11. Pfeiffer, F. and Majek, M., Stick-slip motion of turbine blade dampers. *Phil. Trans. R. Soc. Lond.* A(1992), **338**, pp.503-18
12. Knudsen, C., Feldberg, R. and True, H., Bifurcation and chaos in a model of a rolling railway wheelset. *Phil. Trans. R. Soc. Lond.* A(1992), **338**, pp.455-469
13. Wojewoda, J., Kapitaniak, T., Barron, R. and Brindley, J., Complex behaviour of a quasiperiodically forced experimental system with dry friction. *Chaos, Solitons and Fractals*, 1993, **3**, pp.35-46
14. Wiercigroch, M., A note on the switch function for the stick-slip phenomenon. *J. Sound. Vib.*, 1994, **175**(5), pp.700-4
15. Galvanetto, U. and Bishop, R.S., Stick-slip vibration of a two degree-of-freedom geophysical fault model. *Int. J. Mech. Sci.*, 1994, **36**, pp.683-698
16. Galvanetto, U. and Bishop, R.S., Characterization of the dynamics of a 4-dimensional stick-slip system by a scalar variable. *Chaos, Solitons and Fractals*, 1995, **5**(11), pp.2171-2179
17. Feeny, B.F. and Liang, J.W., Phase space reconstructions of stick-slip systems. *1995 Design Engineering Technical Conferences Proceedings*, Vol.3, Part A, 1995, pp.1049-1059
18. Whiteman, W.E. and Ferri, A.A., Analysis of beam-like structures with displacement-dependent friction forces: Part I: Single-degree-of-freedom model; Part II: Multi-degree-of-freedom model. *1995 Design Engineering Technical Conferences Proceedings*, Vol.3, Part A, 1995, pp.1093-1108
19. Chan, S.N., Mottershead, J.E. and Cartmell, M.P., Parametric resonances at subcritical speeds in discs with rotating frictional loads. *Proc. Instn. Mech. Engrs*, 1994, **208**, Part C, pp.417-25
20. Mottershead, J.E., Ouyang, H., Cartmell, M.P. and Friswell, M.I., Friction-induced parametric resonances in discs: effects of a negative-velocity relationship. *J. Sound Vib.*, 1998, **209**(2): 251-264
21. Mottershead, J.E., Ouyang, H., Cartmell, M.P. and Friswell, M.I., Parametric Resonances in an annular disc, with a rotating system of distributed mass and elasticity; and the effects of friction and damping. *Proc. Royal Soc. Lond. A.*, 1997, **453**(1), pp.1-19
22. M Tenebaum and H Pollard. Ordinary Differential Equations. Harper & Row, New York, 1963, pp.407-420

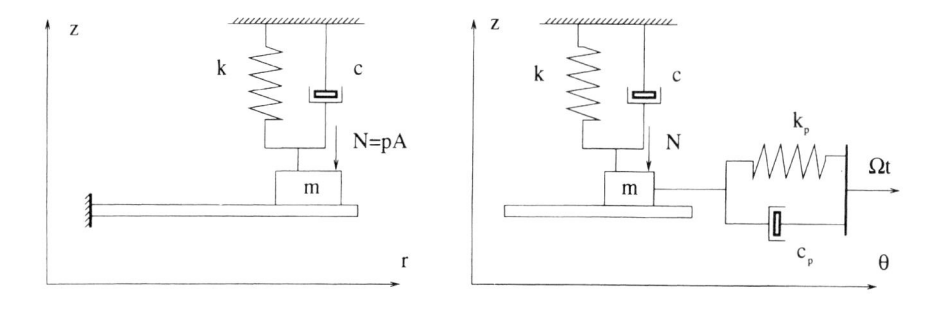

Figure 1. Slider system and disc in cylindrical co-ordinate system

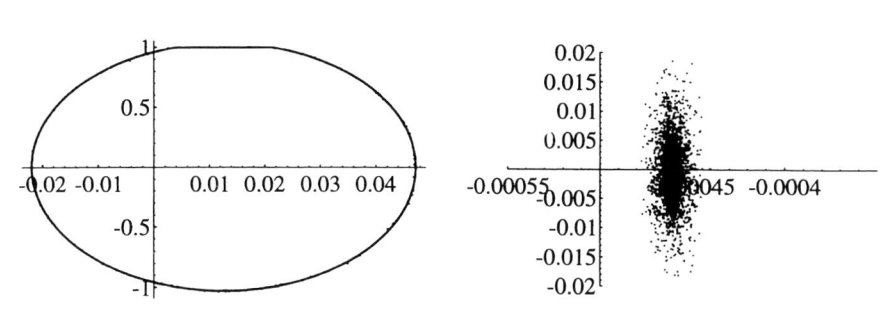

Figure 2. Poincare maps for the slider and the disc at p=30kPa

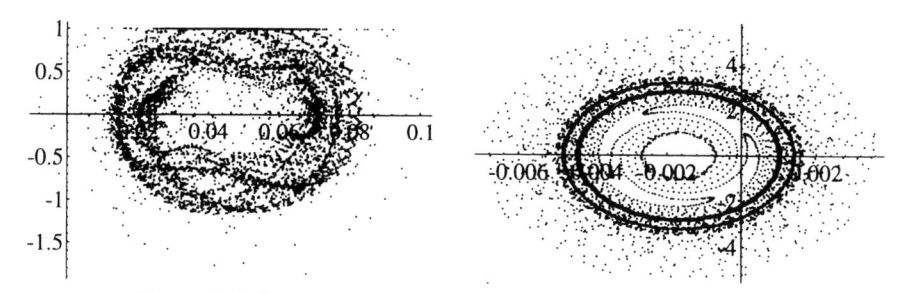

Figure 3. Poincare maps for the slider and the disc at p=110kPa

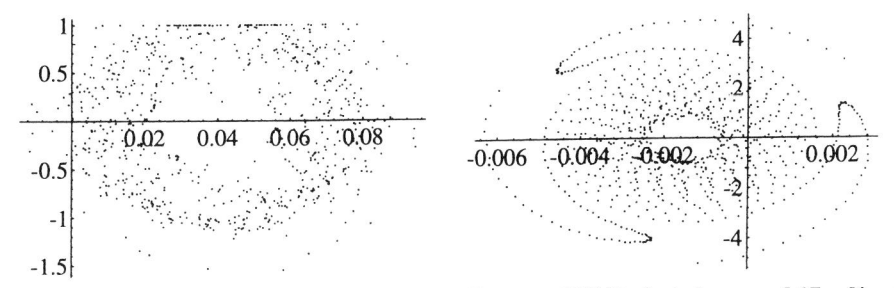

Figure 4. Poincare maps for the slider and the disc at p=100kPa (rotating speed 17rad/sec)

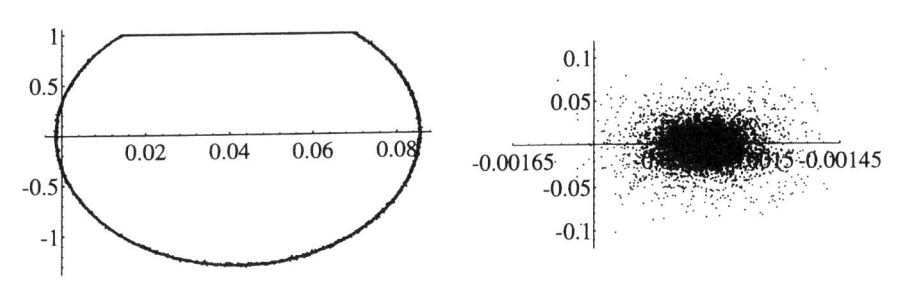

Figure 5. Poincare maps for the slider and the disc at p=100kPa (disc damping 0.00002)

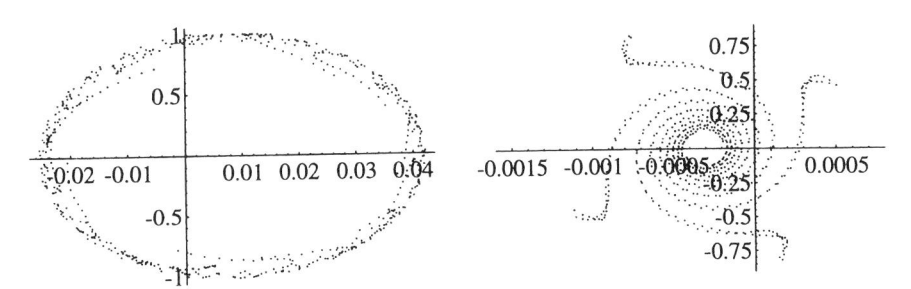

Figure 6. Poincare maps for the slider and the disc at p=20kPa (no damping)

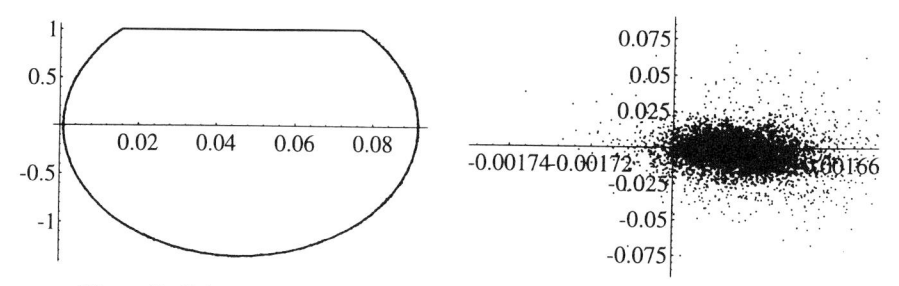

Figure 7. Poincare maps for the slider and the disc at p=110kpa (c=0.707)

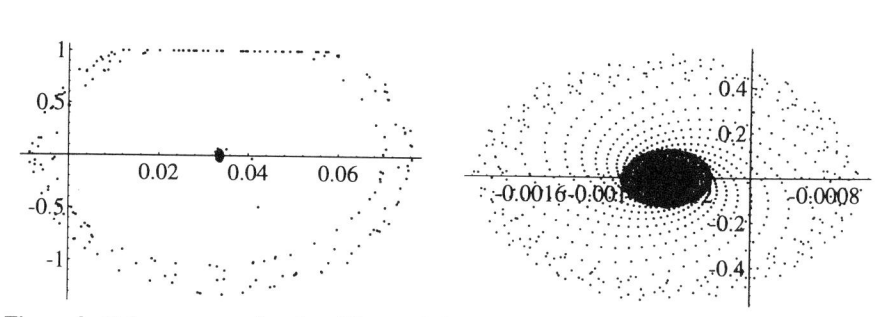

Figure 8. Poincare maps for the slider and the disc at p=80kPa (in-plane damping 0.447)

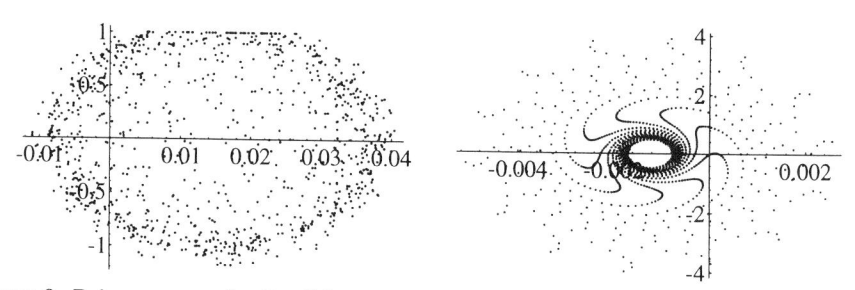

Figure 9. Poincare maps for the slider and the disc at p=80kPa (in-plane stiffness 45000N/m)

Noise insulators for brake squeal reduction – influence and selection of the damping material

P HEPPES
Eagle-Picher Wolverine® GmbH, Ohringen, Germany

ABSTRACT

Considering the action principles and functional requirements of noise insulators, a look at the development process shows, that the choice of the damping material is the most important parameter in reducing brake squeal. To evaluate the noise reducing function, different kinds of test runs on a brake noise dynamometer are used because of the good reproducibility. With the example of a completed project the selection of the damping material is explained by showing the influence of different material-parameters and considering other restrictions given in this project. Finally the possibility to transfer these results to similar applications is discussed.

1 INTRODUCTION

Brake noise insulators or noise-damping-shims generally can be characterized as follows: they are additional parts within the wheel brake which are not substantially necessary for the main function of braking. The only purpose of these additional parts is to reduce the squealing of the brake. In addition to the well known rubber coated Wolverine shims ((5), see fig. 1), which are bonded or clipped onto a backplate, the term "noise insulators" also includes washers, tubes or damping rings which will not be discussed further in this paper because they are special applications which are seldom used within brakes.

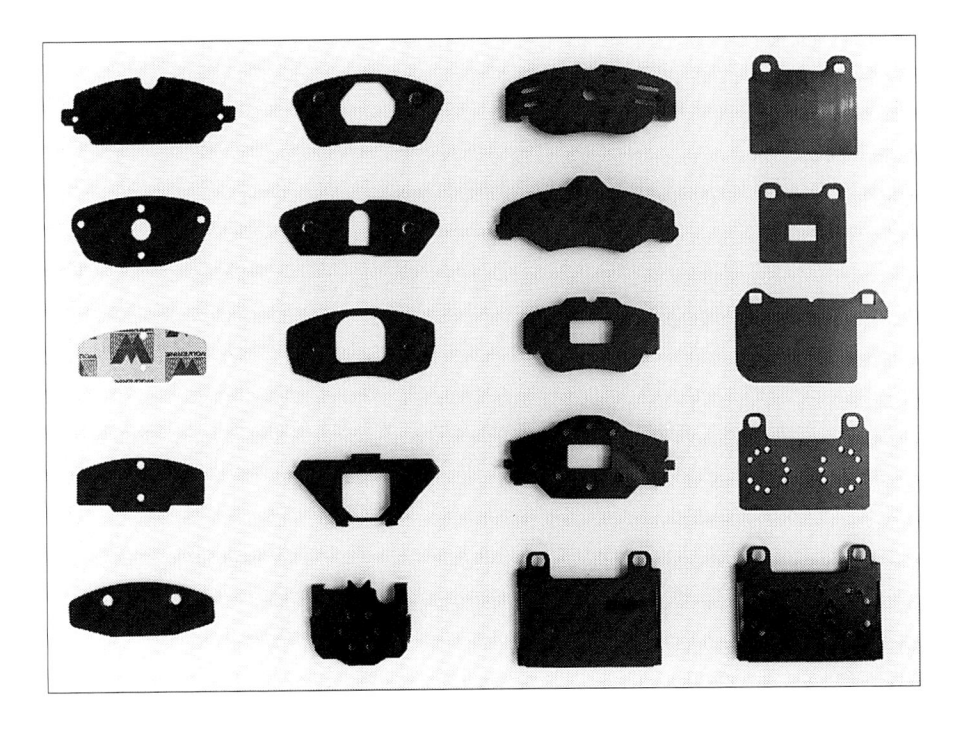

Figure 1: Some Examples of Noise Insulators

1.1 Functional Principles

The generally used terms "noise-damping-shims" respectively "noise insulators" express the two functional principles of vibration damping and vibration isolation, which are utilized with the application of such parts. With actual applications mostly both functional principles are combined – this makes development work quite demanding.

1.2 Requirements on a Noise Insulator

The most important requirement on a noise-damping-shim is to reduce brake squeal or to avoid it completely. In more detail this means that vibrations, which cause audible noise in a frequency range from 1500 to 15000 Hz, have to be reduced to an unnoticeable level with regards to their amplitude as well as their frequency of occurrence.

Almost as important is the fatigue strength of the elaborated solution. In particular the rubber coating (regarding abrasive strength) and bonding of the noise-damping-shims have to withstand a high mechanical and thermic strain, if the noise reduction should be reliable and efficient for the lifetime of the friction pads.

Functional requirements also include an easy integration into the production process of the pad manufacturer.

In order to fulfil these requirements high quality standards are applied to noise insulators, which consist of constructional and production engineering features (flatness, dimensions, tolerances) as well as material features (chemical resistance, mechanical values).

2 NOISE INSULATOR MANUFACTURING AND DEVELOPMENT PROCESS

In the past the development of noise insulators has been carried out mainly by pad and brake manufacturers and in accordance to their test results shape and material of the noise insulators have been determined. Due to the long test periods material variations were restricted to a few standard materials. The only way for a shim manufacturer to get involved into the development process was to supply prototype parts.

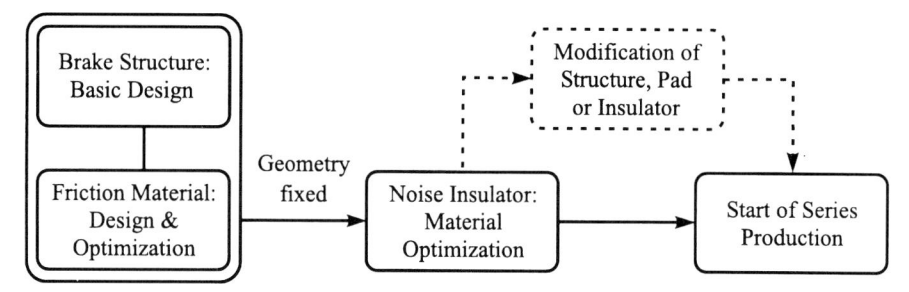

Figure 2: Noise Insulators within an usual Development Process

Shim manufactures only were involved into the development process in case of extreme noise problems occurred shortly before or even after series production started. At this stage the only possibility to solve a noise problem is to change the damping material because the shape of pads and shims cannot be modified on an economic level. Figure 2 explains the chronological order of this development scheme. As a result of this situation the shim manufacturers developed various damping materials in order to be able to provide the best solution for any individual application.

Lately cars have got more silent – therefore the squealing noise of a brake becomes more and more obvious. System suppliers make much efforts to avoid noise problems right from the start. For example: When constructing a new brake the space needed for a noise-damping-shim is always included in the layout. Noise evaluations are carried out as a matter of routine. In addition shim manufacturers are involved in the development process earlier and application specific noise problems become obvious in time so corresponding countermeasures can be introduced. Nevertheless it is still essential to choose the right damping material due to the fact that the shape of the pads and other components cannot always be changed for economic or technical reasons. (see fig. 2 broken line loop).

3 MATERIAL STRUCTURE OF DAMPING MATERIALS

Figure 3 shows the structure of two damping materials (4). The material on the left is a widely used standard material for European applications. The cross-section on the right makes clear which complex damping material structure can be achieved by combining several material layers.

A wide range of different materials can be derived from the standard material structure. The following variations are possible (in square brackets the number of variations currently used by Wolverine):

- Steel base: different thicknesses [4], different steel qualities [4 to 5].
- Rubber coating: different thicknesses and shore hardnesses [3], coated on one or both sides [3].

- Adhesives: different pressure sensitive and thermoset adhesives, coated on one or both sides [15].
- Additional layers: fiberglass, other laminates [3].

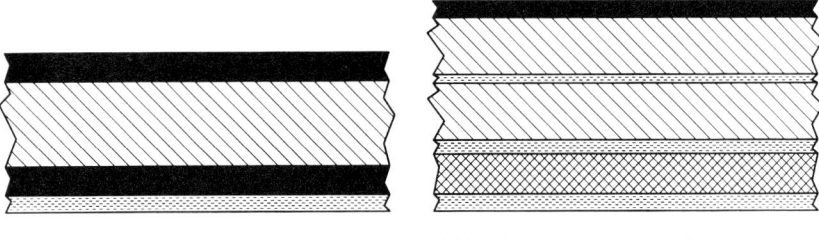

MS 18035 BH 1510-2   MS 18818 Grade 03-10-27-10-27-07-27

Figure 3: Different Damping Materials (Examples)

Theoretically more than 5.000 different damping materials could be composed with the standard structure. Approximately 40 compositions are included in the production range of Wolverine and 100 to 200 further experimental materials have been tested. When including the more complex sandwich materials the number of variations possible increases immensely.

4 METHODS TO CHOOSE AND EVALUATE DAMPING MATERIALS

With many applications a noise-damping-shim made out of a standard damping material already reduces brake noise remarkably (e. g. compare fig. 6 to fig. 7). Usually a standard material is chosen with regards to production engineering and process specific requirements.

In case a noise problem occurs in a certain application whose relationship to parameters such as "disc brake temperature" or "brake pressure" can be determined, a damping material adjusted to this problem parameters can provide a solution. An aid to choose a suitable material might be the temperature curve of the damping degree. Damping characteristics like this are determined either with the help of standardized tests (2) or by measuring the damping degree of a damping material test specimen assembled on a backplate in its natural frequencies. Unfortunately it is not always successful to use these damping characteristics. One aspect can be found in the restrictions which have to be considered when transferring the test results to an application (nonlinearity of the vibration system, different boundary conditions, etc.). Another reason is that to increase the damping might be disadvantageous if a problem has to be solved by a vibration isolation (3).

Due to the above the method mostly applied to select a damping material is "try and error". Certain strategies can be used with this method which depend strongly on the individual application as well as on specific experiences made by every company.

The methods used to evaluate the function of a damping material are illustrated in figure 4. The most relevant result is obtained from the road test. The prototypes necessary for those tests are available quite late during the development process of the car and to carry out driving tests is time intensive. Due to climatic influences it is difficult to reproduce the test conditions.

As a result more and more brake noise dynamometers have been established for noise evaluation within the past years. The idea of a test rig is to pick out those components from the "whole car" which cause the noise. The main problem in reproducing a noise problem is to choose the correct interfaces. Experience shows that it is necessary to assemble the axle suspension together with the shock absorbing strut as part of the chassis and to find the corresponding

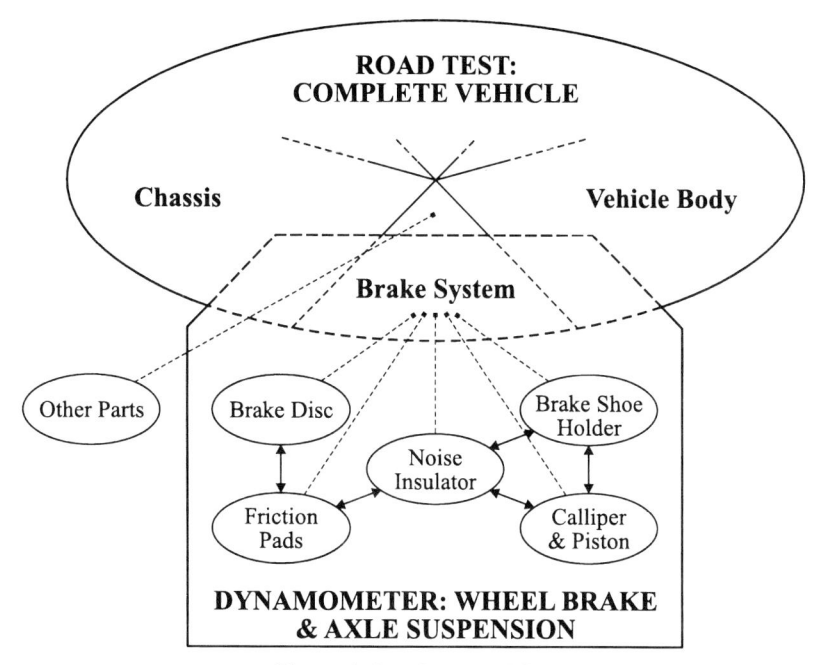

Figure 4: Development Methods

fixing positions at the vehicle body. This kind of assembly also shows some weak points of test rig work: brake squeal which only appears when braking in a road bend cannot be reproduced with the currently used test rig assemblies. Nevertheless most noise problems can be solved with the assembly illustrated in figure 4 which includes the components mainly involved in the generating of noise: brake disc, friction pads, calliper and brake shoe holder (1), (6).

5 DEVELOPMENT EXAMPLE

5.1 General Approach

Figure 5 shows the principal process of development work on a brake noise dynamometer. The pre-selection of the damping materials takes place in accordance to the edge conditions received and the noise problems which have to be solved.

Some examples for these edge conditions are temperature load on the adhesive caused by heat development when braking or restrictions of the material selection due to the manufacturing process. Another condition is the margin of possible changes to the shape of the shims, for example maximum material thickness, geometric variations or simply if the same shim has to be applied on both backplates of a set.

To ensure to work on a fixed problem and to avoid that influences due to the specimen assembly spoil test results it is important to have a description of the noise problem by the way of the frequency of occurrence, noise frequencies and levels in relationship to the parameters brake pressure, disc temperature, rotational speed and if necessary the climatic conditions. In case of this problem description is not known from road tests or other test work it has to be determined during the first test run.

PICK UP EDGE CONDITIONS, E. G.:
○ Backplate Temperature Level
○ Limiting Conditions with respect to Manufacturing Process
○ Maximum/minimum Thickness of Damping Material
○ Geometric Modifications possible
○ Identical Noise Insulators for inner and outer Friction Pad

DESCRIPTION OF THE NOISE PROBLEM:
○ Frequencies?   ○ Brake Pressure
○ Occurrence? at: ○ Disc Temperature
○ Level?   ○ Speed

PRE-SELECTION OF WOLVERINE-MATERIAL

BRAKE NOISE DYNAMOMETER: SERIES OF TESTS

Identification of a Structural Problem ← → Material Development

RECOMMENDATION OF NOISE INSULATOR

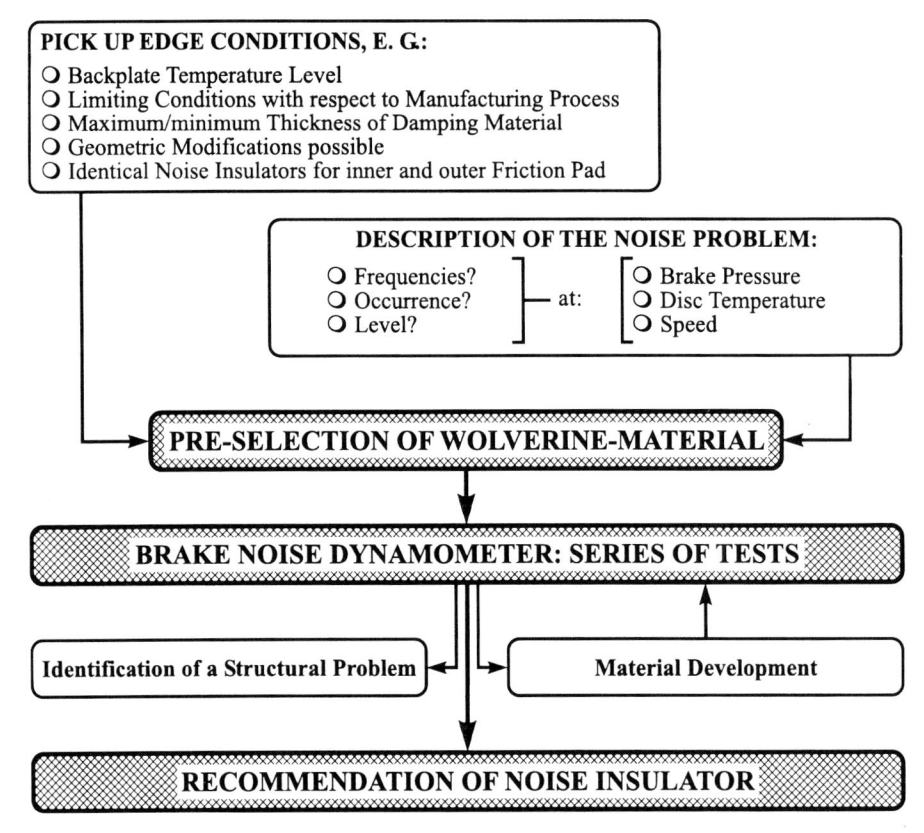

Figure 5: Practical Development Flow Chart

To find the most important factors of influence of an individual application series of short test programs (approximately 1.000 brakings including auxiliary and conditioning brakings) are run. These programs cause as less wear of the friction pads as possible and have only a moderate temperature load (325 °C maximum) on the friction lining. This is important for a direct comparison of the test results also because usually only a limited quantity of test material is available. The damping material which comes off best is submitted to several long term tests in accordance to a standard test program which has been developed by a team consisting of members of the European car, brake and pad industry. If – resulting from these tests – noise occurrence and level remain on an acceptable level the elaborated solution is recommended for the final road tests.

Another result from the test programs may be the identification of a structural problem which cannot be solved by using merely a noise-damping-shim. Also the first series of tests may reveal tendencies to further reduce noise by applying a specific new damping material. After composing this new material it will be involved into the tests.

5.2 Case Study
This procedure has been employed by using a typical brake of a leading manufacturer. The edge conditions have been determined as follows:

- bonding with pressure sensitive adhesive
- identical shims on outer and inner friction pad
- total thickness of the shim not higher than thickness of the standard damping material MS 18035 BH 1510-2

The noise problem was characterized as follows: Without noise insulators as well as with a shim made out of Wolverine standard damping material MS 18035 BH 1510-2 noise at approximately 8 kHz occurred with a high level of frequency. The noise level was particularly high when changing from front to reverse gear. Another noise problem within a frequency range from 3 to 4 kHz should be solved later by structural modifications. The brake squeal mainly occurred at disc temperatures from 100°C to 150°C and with brake pressures between 8 and 12 bar. This situation is illustrated in figures 6 and 7, which show the frequencies of noisy brakings related to all brakings within the different frequency ranges (top diagram) and the chronological sequence of the noisy brakings marked by the maximum level of the dominant squeal frequency (lower diagram).

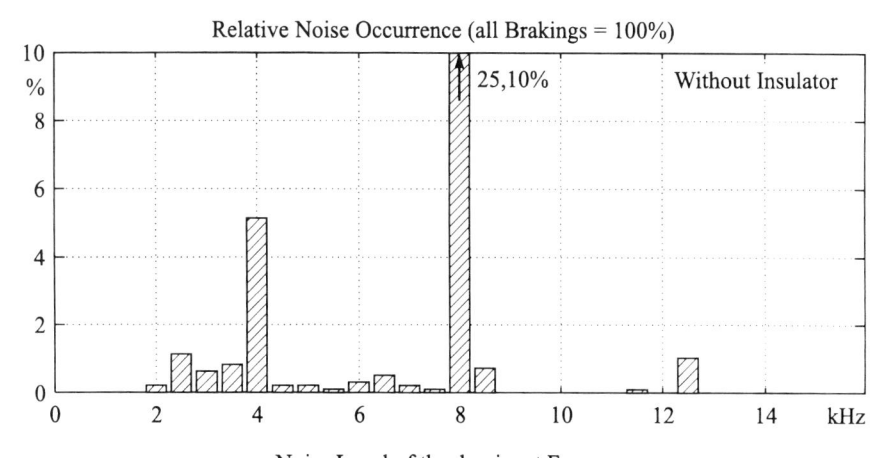

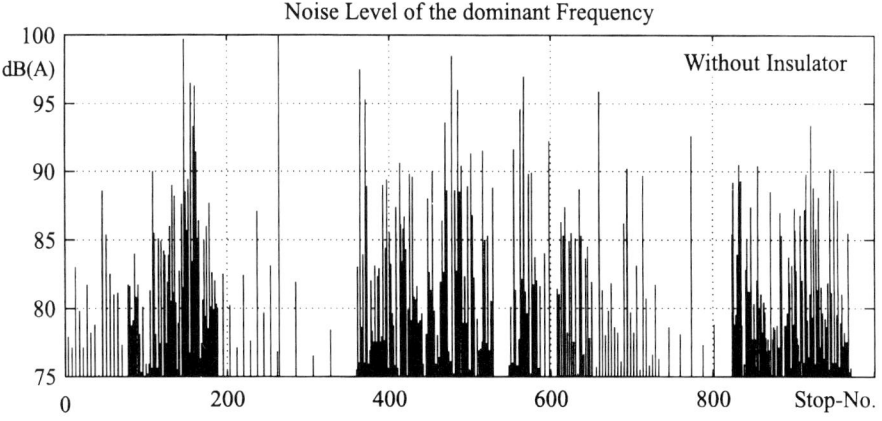

Figure 6: Noise Occurrence without Insulator

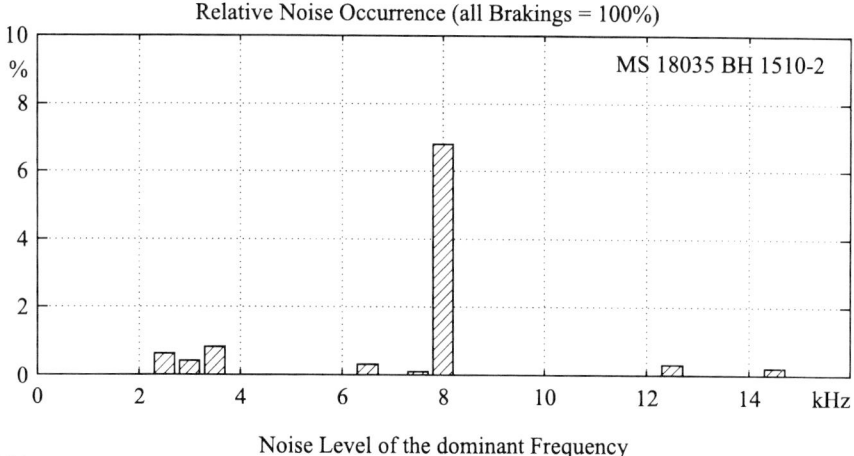

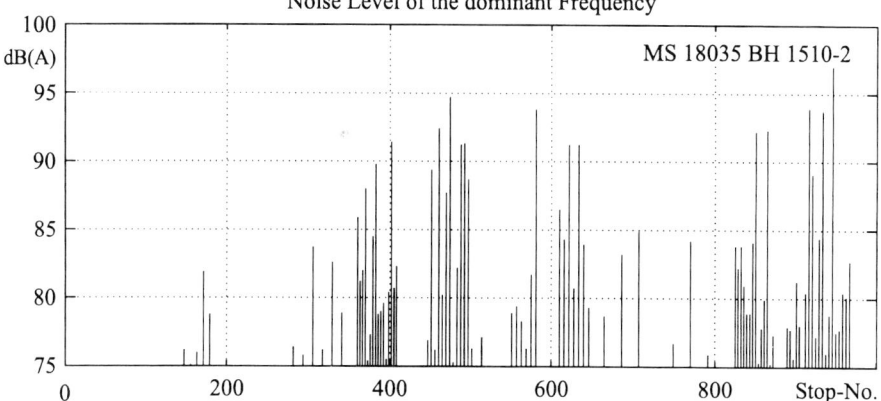

Figure 7: Noise Occurrence with Insulator of Standard-Material

5.3 Influence of the Damping Material

The noise reducing effect of the standard damping material becomes obvious by comparing figures 6 and 7. As illustrated in the development flow chart (see figure 5) with a series of noise-damping-shims out of various Wolverine materials the influence of different components was evaluated on the brake noise dynamometer. The effect of these components cannot be evaluated separately due to the fact that only the effect of a complete shim can be evaluated. In order to get some information on the effect of single components it is necessary to carry out a series of tests in which the structure of the damping material remains unchanged except for the interesting component.

To make the comparison of the variations easier, the information included of the two diagrams of figure 6 or 7 is combined into one characteristic value: The "relative weighted noise occurrence" used in figures 8 to 10 means the sum of all noise within the various frequency ranges where every event recorded is weighted by its level. This sum is set into relation with the result of the fundamental test without noise insulator which is used as reference value (equal 100%).

Figure 8 shows the effect of different adhesives: a significant noise reduction is achieved by using the pressure sensitive adhesive "-27" which has an improved thermal resistance compared to the standard pressure sensitive adhesive "-2". A further improvement is obtained by applying the thermoset adhesive "-29" which unfortunately could not be used in the series considering the edge conditions. Nevertheless this first series of tests shows the tendency to use an adhesive which withstands thermal load best (remark: the pressure sensitive adhesive "-24" selected for the later series application cannot be coated on elastomers, so it could not be included in this first series of tests because the material structure would have been changed in more than one component).

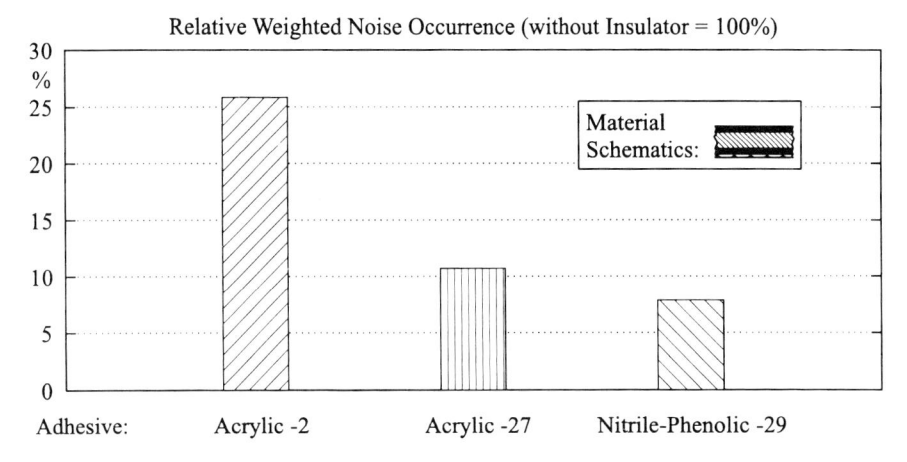

Figure 8: Influence of Adhesives

The influence of the rubber coating was examined in a second series of tests. This series includes two comparisons: the first one was made between an uncoated noise-damping-shim and an insulator coated on both sides of the steel base. The second comparison was carried out between noise-damping-shims coated on one respectively both sides of the steel base. From the results illustrated on the left hand in figure 9 it becomes obvious, that the insulator coated on both sides reduces noise better than the uncoated noise-damping-shim. The second comparison shows that the insulator coated only on one side of the steel base leads to better results than the one coated on both sides (right hand in figure 9). The different level of the two shims coated on both sides makes plain the high influence of the different material structures used for the two comparisons. For the first comparison a very stiff damping material was used (no coating, "-29"-adhesive), for the second one the damping material was softer because of an additional primed fiberglass cloth and an additional thin rubber coating applied on this cloth. From this second series of tests two trends can be derived: to obtain low noise levels, the damping material should only have a rubber coating on one side of the steel base and the damping material should have a high compressive strength.

Due to the results of those first two series of tests a damping material out of the following structure was tested in a third series: on the one side of the steel base a rubber coating is applied, on the other one a silicon based pressure sensitive adhesive. The parameter varied in this third series was the steel base thickness. Figure 10 shows on the right hand the trend to use a steel base thickness as thin as possible. Another comparison in that way which was carried out earlier using the standard damping material structure confirmed this trend (left hand in figure 10).

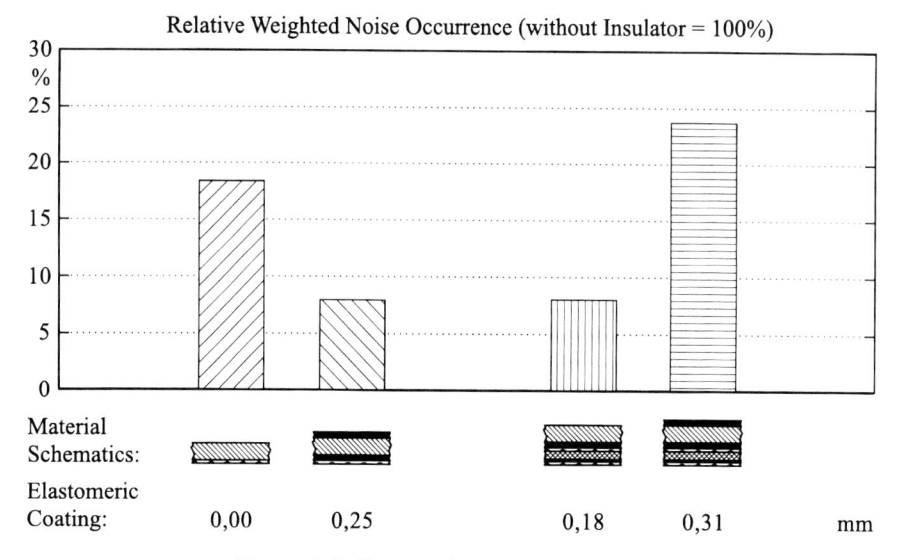

Figure 9: Influence of the Rubber Coating

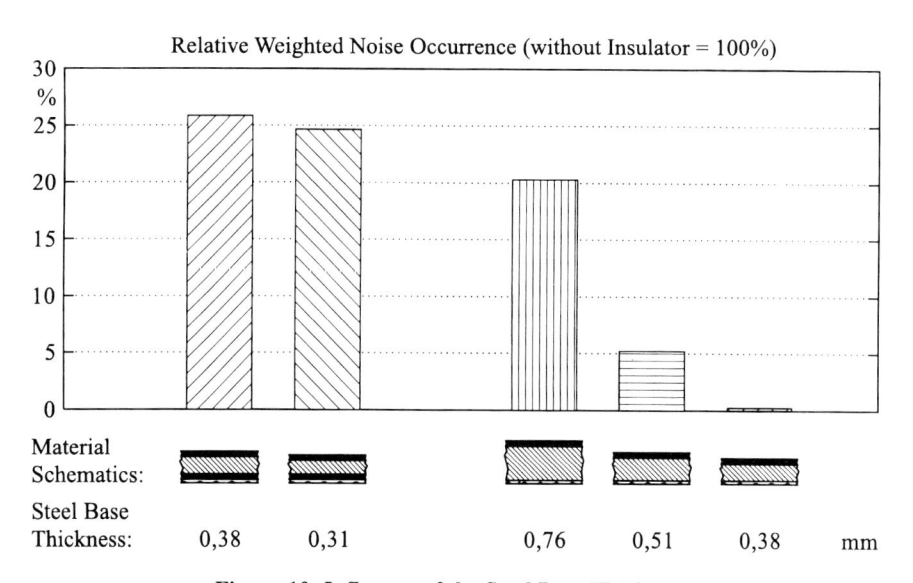

Figure 10: Influence of the Steel Base Thickness

5.4 Recommendation

In spite of this last tendency for later series application the damping material with the thinnest steel carrier has not been selected but the material with a steel base of 0,51 mm thickness due to the following reasons:
- The 8 kHz noise problem could be eliminated completely (see figure 11), this could not be achieved with the thinner steel base.
- The average level of the noise remaining was not as high as with the thinner material.

Of less importance also was:
- The thickness of the shim recommended was the same as the thickness of the standard material scheduled initially.
- As a production material the damping material selected was available immediately for production of parts, the material with the thinner steel base was an experimental material.

Figure 11 gives the detailed illustration of the test results with the damping material selected. The improvement compared to the initial situation is obvious (see figure 6 and 7). These excellent results have been confirmed in long running tests carried out afterwards.

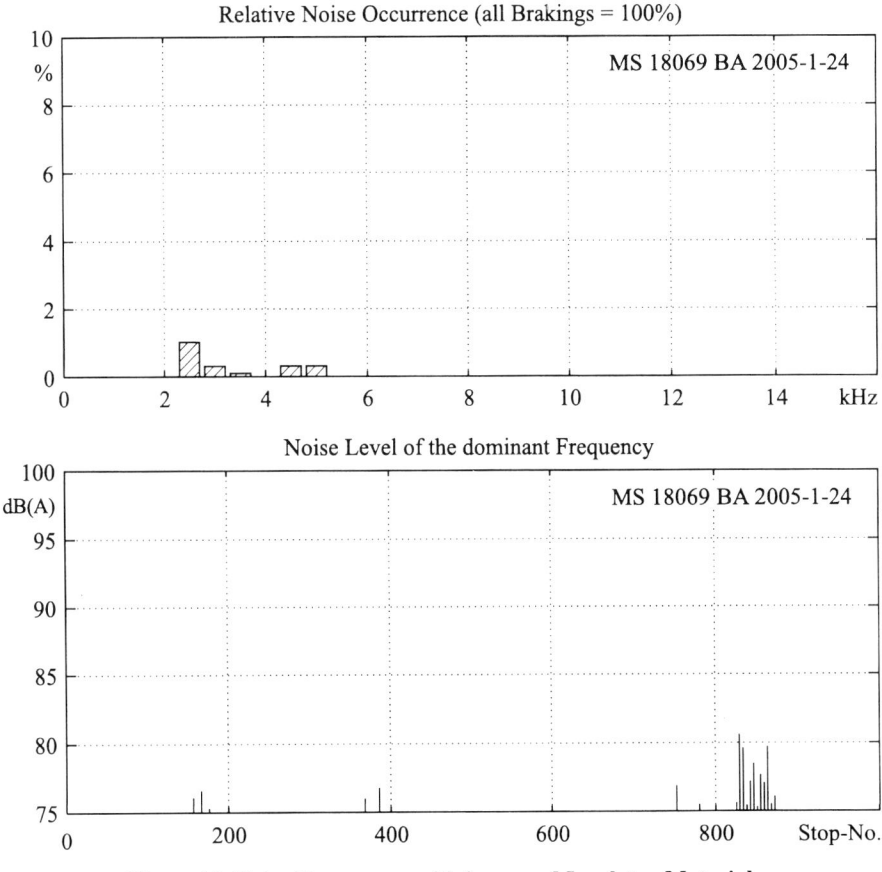

Figure 11: Noise Occurrence with improved Insulator-Material

6 TRANSFERABILITY – PROSPECT

From the above illustration it is possible to prove that different components and compositions of damping materials have a substantial influence on the brake noise occurring. Unfortunately it is not possible to transfer this knowledge of the damping material suited best in that examined application to arbitrary brake designs. The knowledge gained from systematic research made in various projects is enlarged constantly and it is planned to summarize this knowledge within a data base. For future applications it is expected that this data base will facilitate to select the best suitable damping material with respect to the special demands of any application. Thus the series of test runs necessary may be reduced to the optimization of one or two parameters.

The damping characteristics mentioned above used for the pre-selection of a suitable Wolverine material may be a further starting point for faster optimization. In this field currently basic research is carried out with the aim to achieve a better correlation of the values measured with the efficiency in a special application. Besides more efficient measurement methods the definition and determination of further characteristics, e. g. of the dynamic stiffness, may be a basis for a further reduction of testing effort in the future.

7 REFERENCES

(1) Rinsdorf, A: Theoretische und experimentelle Untersuchungen zur Komfortoptimierung von Scheibenbremsen. Publishing house Höpner und Göttert, Siegen, 1996, p. 95ff.
(2) ASTM E756-93: Standard Test Method for Measuring Vibration-Damping Properties of Materials. American Society for Testing and Material, 1993
(3) Krämer, E.: Maschinendynamik. Springer publishing house, Berlin, Heidelberg, New York, Tokyo, 1984
(4) Wolverine Gasket Division: Design, Specification & Test Manual for Disc Brake Noise Insulators. Company Booklet, Inkster, 1997
(5) Burkhardt, M.: Fahrwerktechnik: Bremsdynamik und Pkw-Bremsanlagen. Vogel publishing house, Würzburg, 1991, p. 243
(6) Hoffrichter, W./Lange, J./Rinsdorf, A.: Analsyetechniken zur Beeinflussung von Bremsgeräuschen (a systematic approach to analyse brake-noise). ATZ 96 (1994) 10, p. 606-611

The value of noise fix shims – a case study

J D FIELDHOUSE and **M RENNISON**
The University of Huddersfield, UK

ABSTRACT

A noisy sliding fist type brake is used to evaluate two different types of shim when they are applied to the brake pads in a systematic manner and in a variety of ways. For comparisons of noise propensities to be made the noise signature of the basic brake is measured and its value represented as a single value "equivalent noise dose rating". The same process is used to establish the resulting noise for each of the shim arrangements so that the same standard is used throughout the tests, the basic unshimmed pads being used as a "benchmark". It is seen that in some instances the application of noise fix shims has an overall detrimental effect on the noise propensity and in other cases the elevated temperatures result in certain types of shim detaching from the surface of the pad. It is also seen that the application of noise fix shims affects the predominant frequencies the brake generates.

The analysis allowed the shim supplier to identify the most cost effective way of applying the shims in addition to identifying the most overall effective arrangement in order to reduce the propensity of the brake to generate noise.

1.0 INTRODUCTION

Probably the most common noise fix for disc brakes is the "noise fix shim", a metallic strip with an elastomer bonded to each surface. This laminate has adhesive applied to both faces and is then sandwiched between the pad and the pad loading points - the calliper piston or calliper fingers. Fixing may be by cold bonding whereby a pressure of around 10 bar would need to be applied for 5 seconds at room temperature. Thermal bonding is also undertaken with some shim types. The shim acts as a damper where the pad displacement is used to create a loss of energy in the excited system. This is achieved through shearing of the elastomer and the creation of hysterisis loss - generally termed the visco-elastic effect. For the shim to be effective the pad must cause the elastomer to shear in some manner which means the pad has to either bend or twist. The adhesive ensures the shim remains in contact with the pad and so replicates the movement of the pad. The attachment to the pad loading points (calliper fingers or piston) maximises damping effect if the pad moves relative to these points either in-plane or out of plane. This type of attachment to the pad and calliper loading points allows ready removal and replacement of the pads without too much damage to the effectiveness on the shim on re-fitting to the calliper. Such a shim construction is generally shown in Figure 1 following.

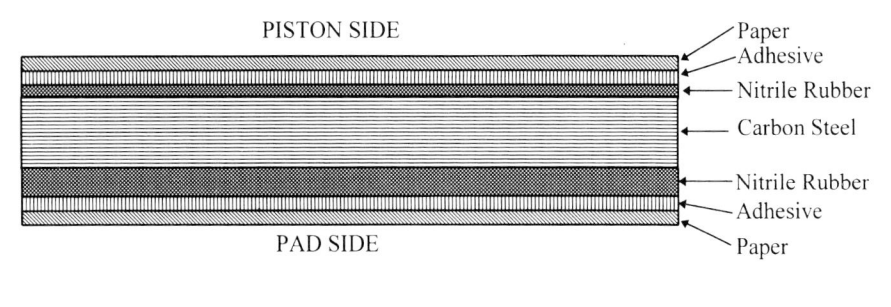

```
                       PISTON SIDE                           Paper
                                                             Adhesive
                                                             Nitrile Rubber
                                                             Carbon Steel
                                                             Nitrile Rubber
                                                             Adhesive
                       PAD SIDE                              Paper
```

Typical Dimensions (Piston to Pad Side):
Adhesive	0.10 mm
Nitrile Rubber	0.15 mm
Carbon Steel	0.40 mm
Nitrile Rubber	0.25 mm
Adhesive	0.10 mm
Total Working Thickness	1.00 mm
Paper (for protection)	0.15 mm (removed before application)

Figure 1 - Typical construction and dimensions of a noise fix shim.

It will be noticed that for a given total thickness of elastomer the distribution is such that some 60% is applied between the metal shim and the pad - where the majority of shear would be experienced. Other more economical shim designs may have 100% of the elastomer on the pad side, the piston side being adhesive directly onto the steel shim.

In addition to restraining the shim using adhesives other dedicated designs use spring clips which locate within the bore of the calliper piston. This type of fixing is not only expensive, and sometimes more costly than the pad, but they also may need to be destroyed in order to remove the pads from the calliper.

If the pad does not exhibit deformation to cause the elastomer to flex and shear then the shim "visco-elastic effect" will not take place and the shim will not be allowed to work. Similarly if the shim is not in constant contact with a deflecting pad then again the shim will not prove to be effective since again shim deflection will not be encouraged.

In general noise "fixes" are not perceived as a permanent solution to a brake noise problem but if they are effective they are accepted as a viable option. Regardless, shims are not inexpensive and their application needs to reflect a measurable reduction in noise generation. In order to demonstrate the value of applying noise fix shims it is necessary to demonstrate a marked improvement in noise reduction.

It is the purpose of this paper to evaluate two different types of shim, referred to as a single layer type and a multi-layer type, each arranged on the brake pads in a variety of ways in order to determine their most effective application in terms of noise reduction and cost. In all comparisons the "equivalent noise dose rating" of the standard pad is used as a "benchmark".

2.0 MEASURABLE PARAMETERS OF A NOISY BRAKE

Current methods of classifying a noisy brake are subjective and liable to variation from one investigator to another. The following investigation avoids subjectivity by reducing the noise variables over the test range to a single value referred to as the "equivalent noise dose rating".

2.1 "Equivalent Noise Dose Rating"

To calculate a meaningful single value on which to categorise a noisy brake the normal measurable parameters are identified and discussed. The approach to calculating a single "equivalent noise dose rating" is to reduce the variations over each test range to an acceptable average value. Because it is difficult to have an international "standard" test for all brakes it is proposed that it is only necessary to establish the brake's "noise signature" under a range of conditions for meaningful comparisons to be made. This may be carried out either on-the-road or under laboratory conditions - provided it is made clear at the outset and the same test is applied to all investigations being subject to comparison. The tests described herein are undertaken under laboratory conditions.

The variables considered were: Noise amplitude, noise duration over one revolution, noise frequencies and associated harmonics, rotor temperatures and brake pressures.

Rotor speed does not influence the propensity of a brake to generate noise and as such is not considered. A noisy brake will have a varying degree of noise levels measured in decibels (dB) - a logarithmic scale. The noise may be continuous over one revolution of the wheel or intermittent. The frequencies generated may differ dependent on the condition of the brake and may be either a pure tone or be comprised of several harmonics. Two frequencies may occur simultaneously or separately over one revolution of the wheel. Additionally, temperature plays a part in establishing and defining the condition of the brake as does the brake pressure. It is these parameters which need to be evaluated and consolidated to form an acceptable "equivalent noise dose rating".

2.2 Test Procedure:

All brake testing has been undertaken on a test rig which allows the brake to be arranged on its mountings as in the vehicle but without the suspension members. A variable speed motor is used to bed the brake at an elevated speed and then run at a reduced speed for measurement purposes - generally 10r/min. A normal process of conditioning the brake takes place as follows. The brake is driven, under varying random loads and at an elevated speed until the brake rotor temperature exceeds 250°C. The rotor speed is then reduced to 10r/min and as the temperature falls to around 200°C the system pressure is raised to around 25 bar and then progressively reduced to zero. At discrete measuring points the pressure is held and the system allowed to settle, normally for around 10 revolutions of the rotor. At these points the noise frequency, magnitude and duration over 360° of rotor rotation is recorded. The procedure is repeated at the same temperature setting as the pressure is progressively increased from zero to 25 bar.

Temperature is then allowed to fall 10°C and the tests repeated. This allows the brake's "noise signature" to be established and should be repeatable if it is to be meaningful. The recorded parameters are considered in turn.

Noise Frequency - The current calculation of an "equivalent noise dose rating" does not account for different frequencies. The initial concern is that the noise falls within the audible range of up to 16 kHz. The brake is unstable and generating audible noise and that is fundamentally unacceptable. The inclusion of frequency is discussed (1) where it is indicated that the importance of its inclusion very much depends on whether the frequency value falls within the sensitivity range of the ear.

Noise Level - The noise level clearly forms the basis on which a brake is classified as noisy. The intermittent noise needs to be reduced to an equivalent noise level over one revolution of the wheel for subsequent presentation and analysis.

For intermittent noise the noise level equivalent ($L_{pi(eq)}$) over 360° is calculated using the equation:

$$L_{pi(eq)} = 10 \log_{10} \left[\frac{1}{T} \Sigma t_n 10^{0.1 L_{pi}} \right] \quad (1)$$

where:

$L_{pi(eq)}$ = Noise level equivalent.

L_{pi} = Unweighted noise level of the particular noise.

t_n = Duration of the particular noise level.

T = Time for one revolution.

This may be used to sum several differing noise levels over one revolution.

Temperature - Since this will vary as the tests are recorded a mean temperature is normally used. All measured mean temperatures are generally ± 5°C.

Pressure - A set level from zero up to around 25 bar.

The resulting noise signatures are plotted on a 3-axis graph with the noise level equivalent the y axis with pressure and temperature the horizontal axes. The appropriate values are also be supplied in tabular form.

Such a typical noise signature is shown in Figure 2 following with the accompanying noise frequency plot for the noise signature in Figure 3.

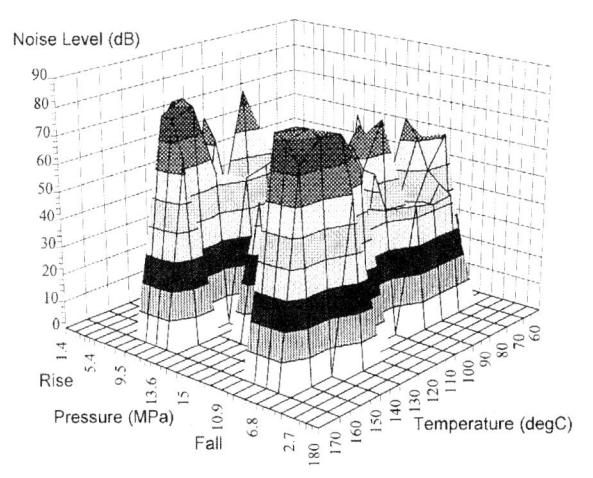

Figure 2 - Typical Noise Signature of a Disc Brake

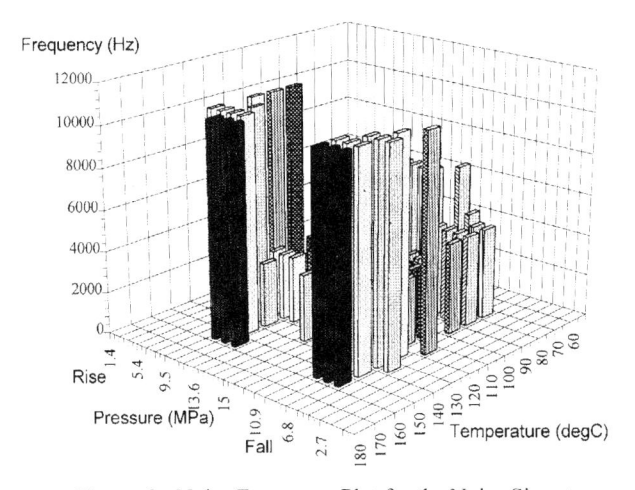

Figure 3 - Noise Frequency Plot for the Noise Signature

Noise Dose - Although the graphical representations allow useful superficial comparisons to be made it is necessary to make reference to the z-axes scale in order to appreciate the level of the problem. To determine whether there is a significant difference between any two systems it will prove necessary to also note the area covered on the x-y plane. The graphical representations need to be reduced to an acceptable value before sensible decisions may be taken. This value is referred to as the "equivalent noise dose rating". The "equivalent noise dose rating" over the test range considered is calculated for each brake to provide three values, a logarithmic average, a log/linear average and a linear average as follows.

Logarithmic Average = antilog of decibel readings. These antilog values are summed and averaged over the test area and then the logarithmic average value recalculated.

$$= 10 \log_{10} \left[\frac{1}{A} \sum 10^{0.1 L_{pi(eq)}} \right] \quad (2)$$

Log/Linear Average = the logarithmic average value is calculated for the number of noise occurrences and then this logarithmic value averaged linearly over the matrix area

$$= \frac{n}{A} 10 \log_{10} \left[\frac{1}{n} \sum_{1}^{n} 10^{0.1 L_{pi(eq)_i}} \right] \quad (3)$$

Linear Average = average of decibel readings over test area.

$$= \sum L_{pi(eq)} \, / \text{ test matrix area (A)} \quad (4)$$

where n = number of occurrences and test matrix area "A" = 312 samples (24x13 matrix).

3.0 CASE STUDY:

A company wished to undertake a series of tests to evaluate and compare the effectiveness of fitting two types of noise fix shims to a sliding fist type calliper. One shim was a single layer type (Type 1) whereas the second was a multi-layered type (Type 2). The company needed to know the shim arrangement to achieve the most effective noise reduction and the arrangement to achieve a significant reduction in noise yet be a cost effective "fix".

To ensure comparable results the same set of pads were used throughout.

3.1 Test Procedure - The procedure was as follows:

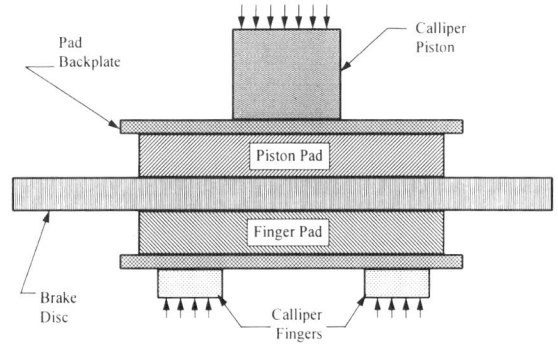

Standard Pad - (No noise fix shims applied) Determine the noise signature and noise frequencies for a standard pad arrangement in order to establish a "benchmark", Figure 4.

Figure 4 - Standard Pad Arrangement

For each shim type the following procedure was followed.

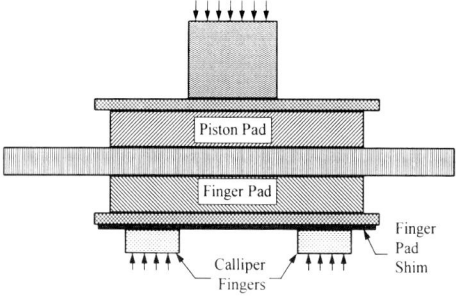

Figure 5 - Brake with Shim on Finger Pad Only

1) Determine the noise signature and frequencies for the standard pads with a noise fix shim fitted to the calliper finger pad only, Figure 5.

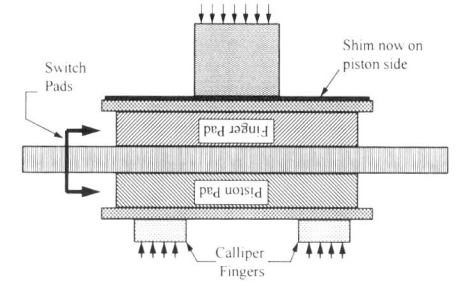

Figure 6 - Reverse Pads to Relocate Shim On Piston Side.

2) Reverse the pads to relocate the shimmed pad onto the piston side, Figure 6.

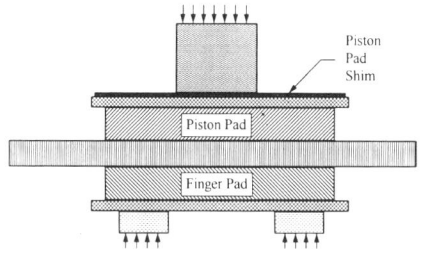

Figure 7 - Pads Re-bedded to Establish New Arrangement

3) Re-bed to establish the pad re-arrangement and determine the noise signature with the shim now relocated on the piston pad.

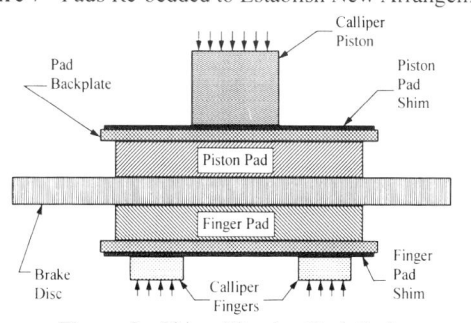

Figure 8 - Shims Fitted to Both Pads

4) Fit shim to finger pad, re-bed and determine the noise signature for shims now fitted to both pads.

3.2 Test Results:

The noise signatures and accompanying noise frequency plots are shown in Figures 9a &b for the Standard Pad, Figures 10, 11 & 12 a & b for Type 1 Shim and 13, 14 & 15 a & b for the Type 2 Shim.

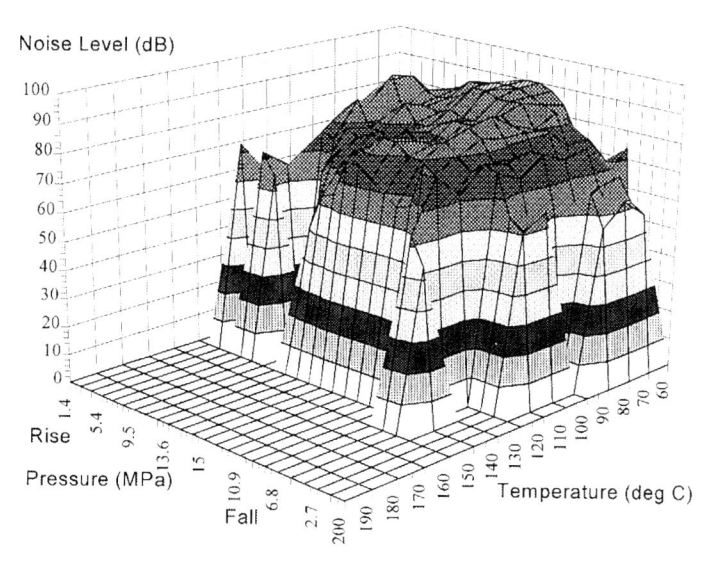

Figure 9a - Standard Pads, Noise Signature

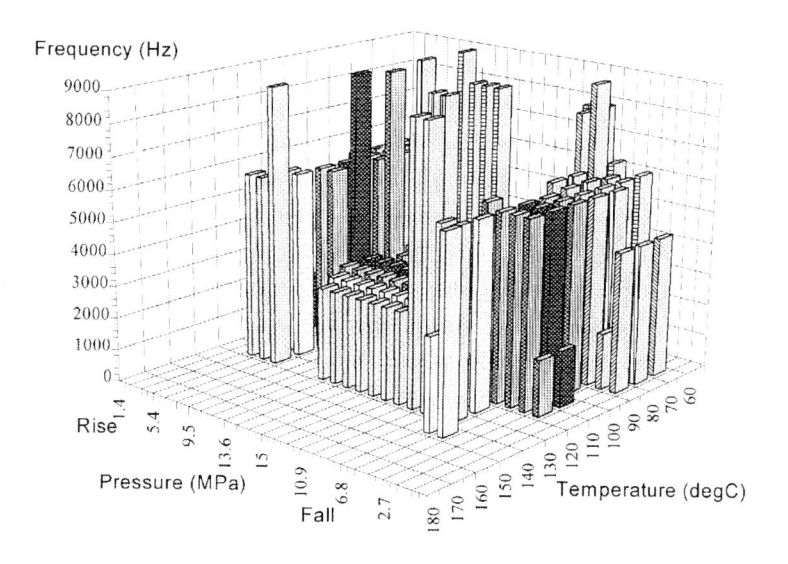

Figure 9b - Standard Pads, Frequency Plot, Frequency Plot

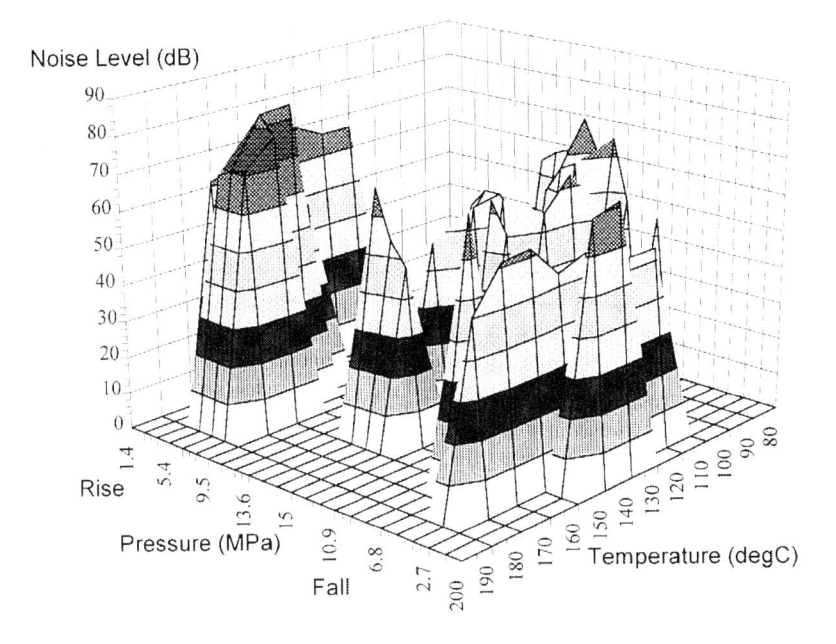

Figure 10a - Type 1 Shim on Finger Pad, Noise Signature

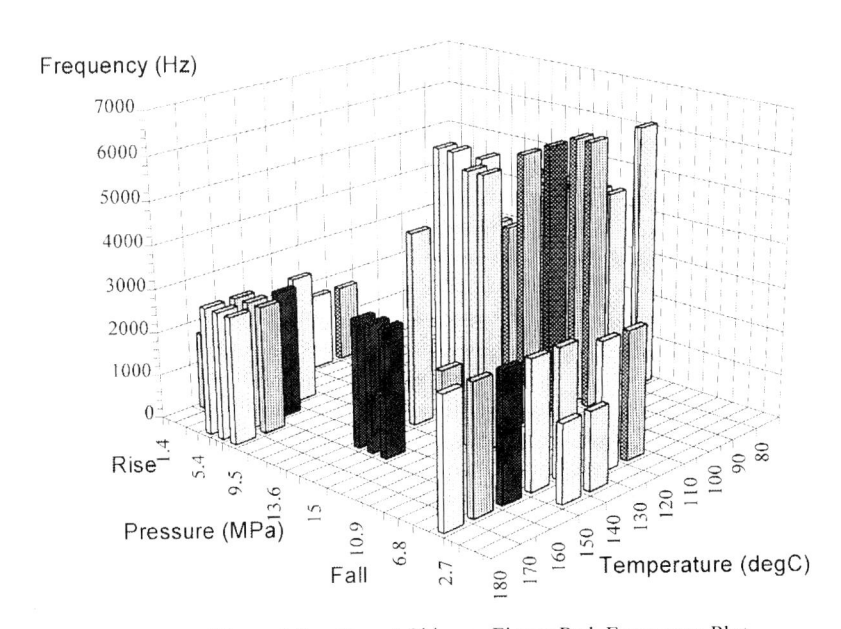

Figure 10b - Type 1 Shim on Finger Pad, Frequency Plot

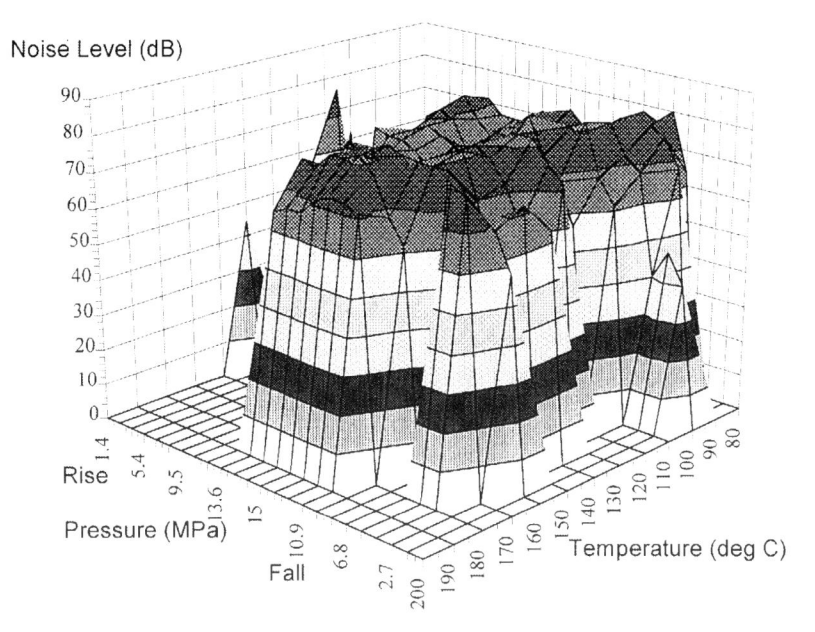

Figure 11a - Type 1 Shim on Piston Pad, Noise Signature

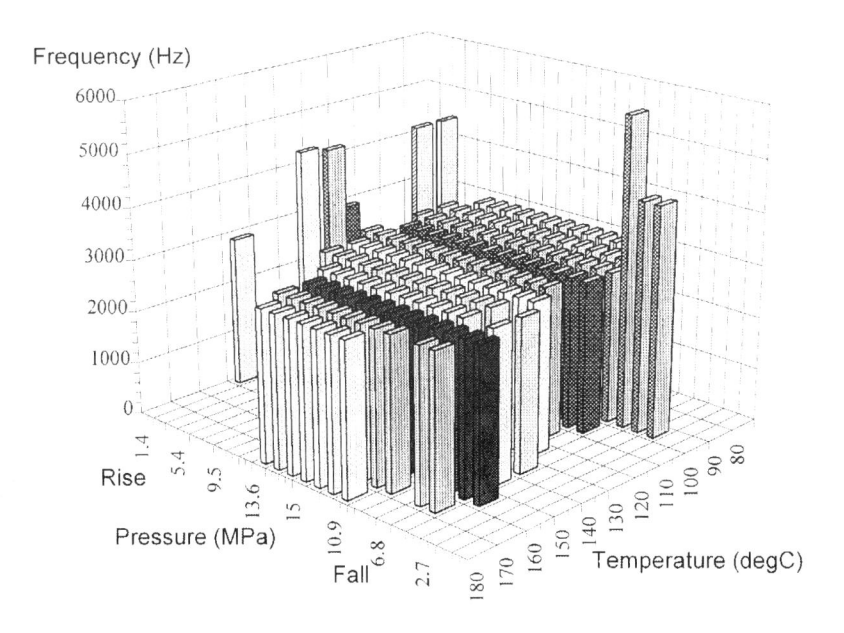

Figure 11b - Type 1 Shim on Piston Pad, Frequency Plot

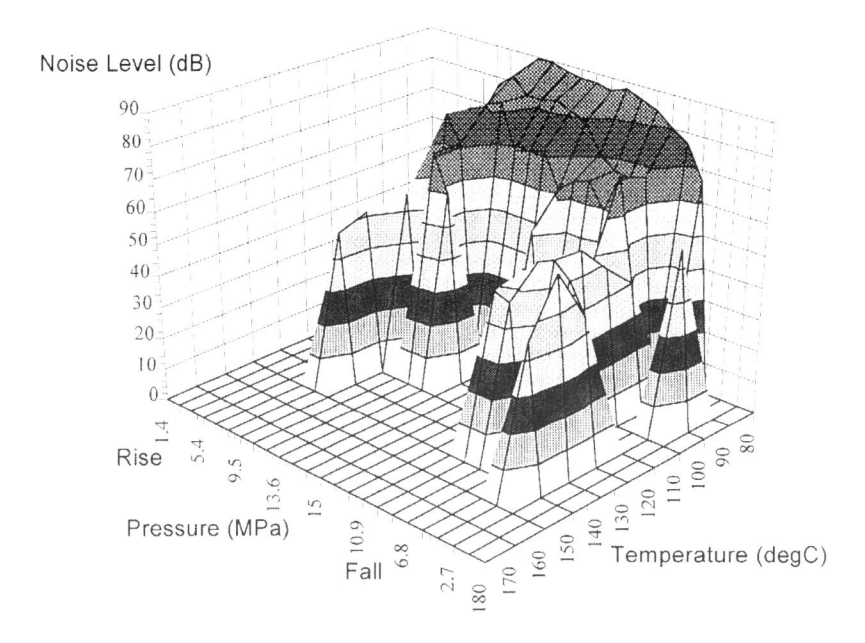

Figure 12a - Type 1 Shim on Both Pads, Noise Signature

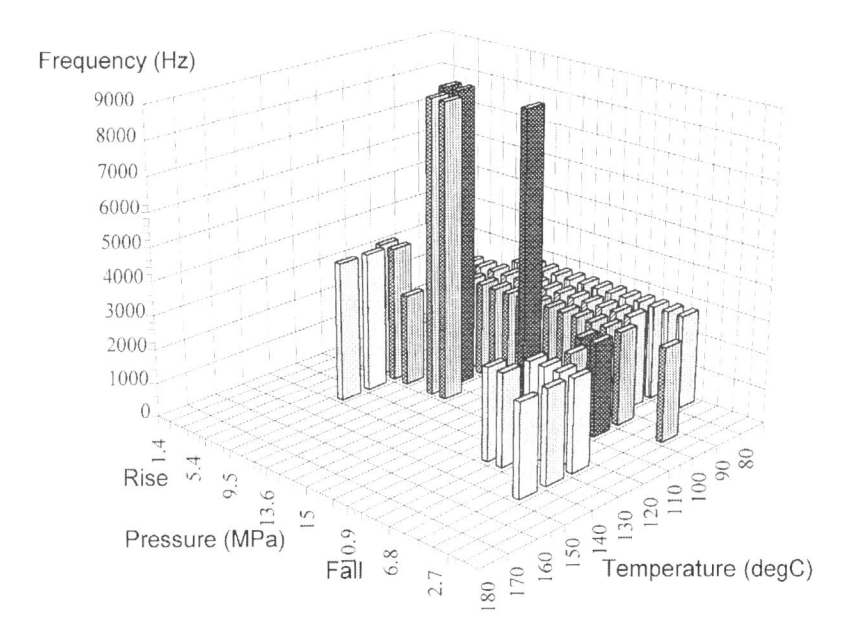

Figure 12b - Type 1 Shim on Both Pads, Frequency Plot

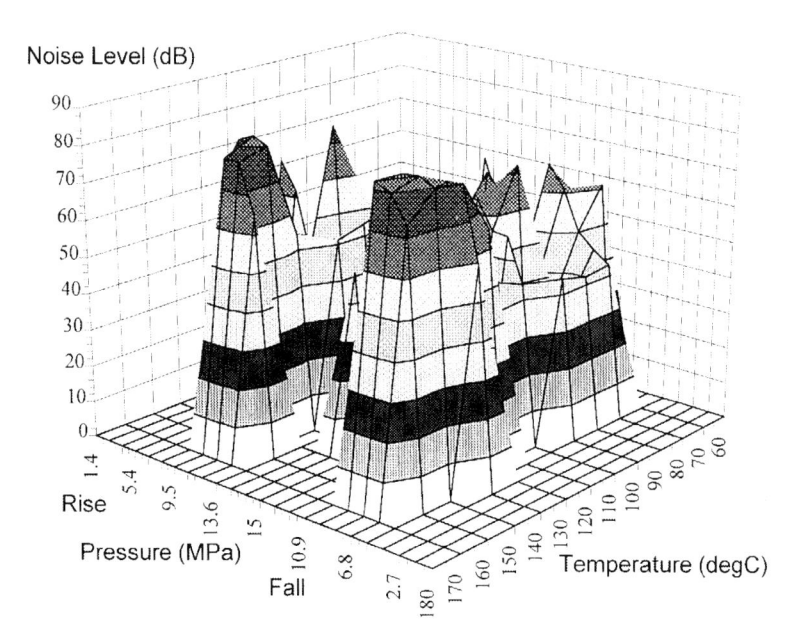

Figure 13a - Type 2 Shim on Finger Pad, Noise Signature

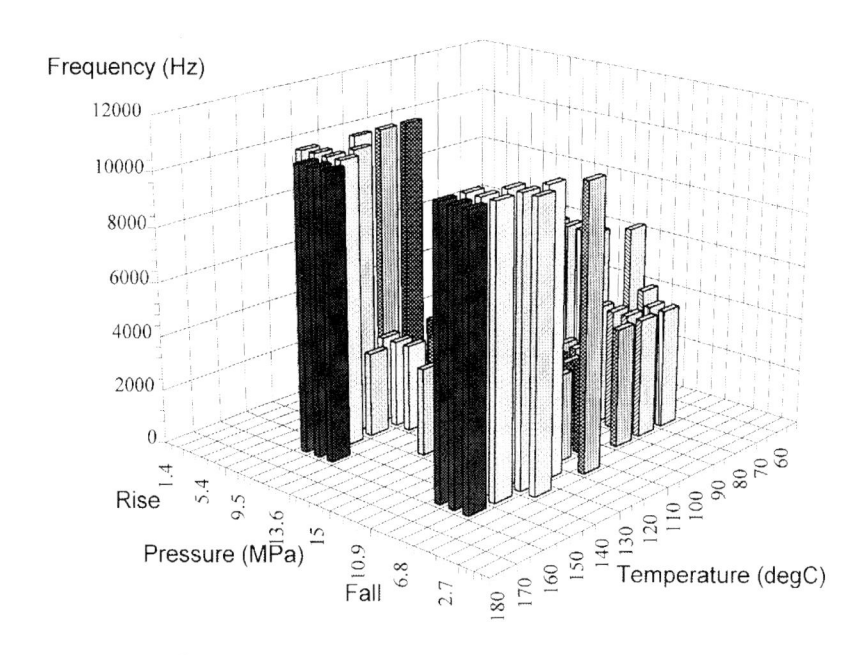

Figure 13b - Type 2 Shim on Finger Pad, Frequency Plot

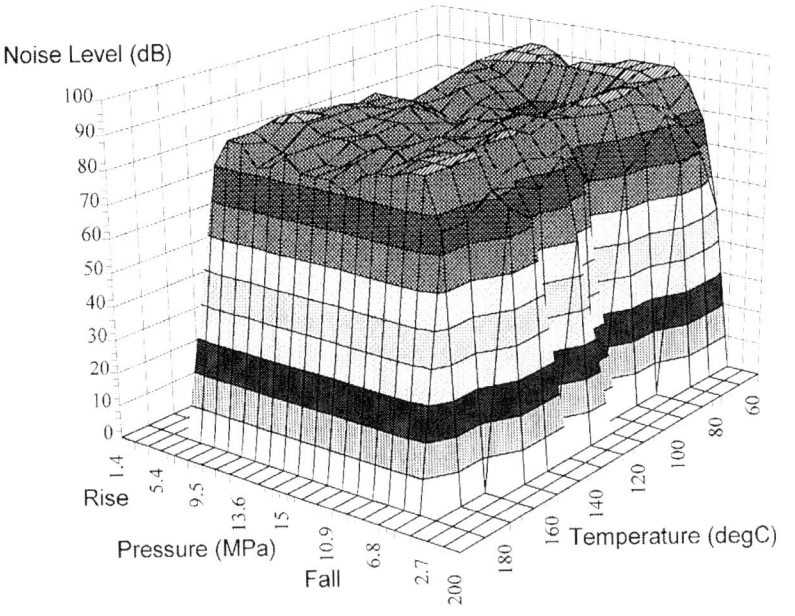

Figure 14a - Type 2 Shim on Piston Pad, Noise Signature

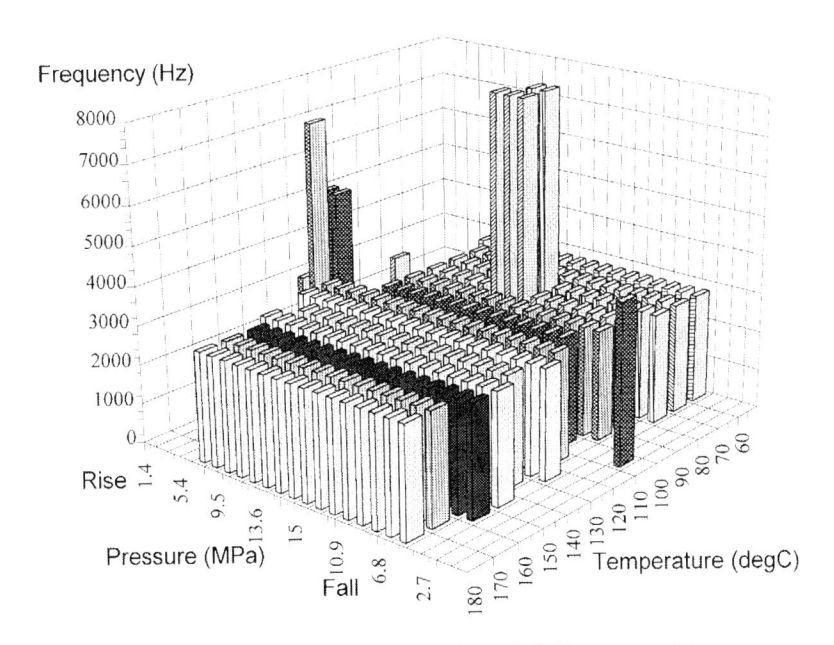

Figure 14b - Type 2 Shim on Piston Pad, Frequency Plot

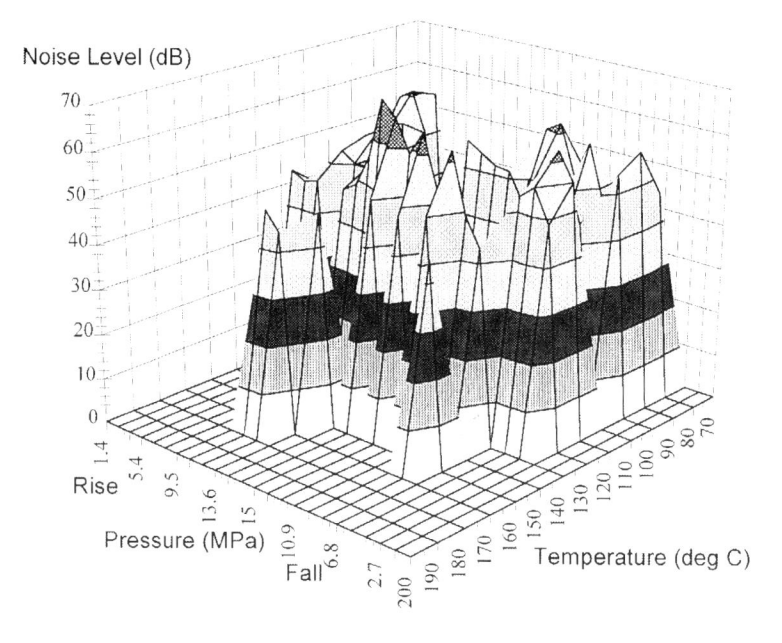

Figure 15a - Type 2 Shim on Both Pads, Noise Signature

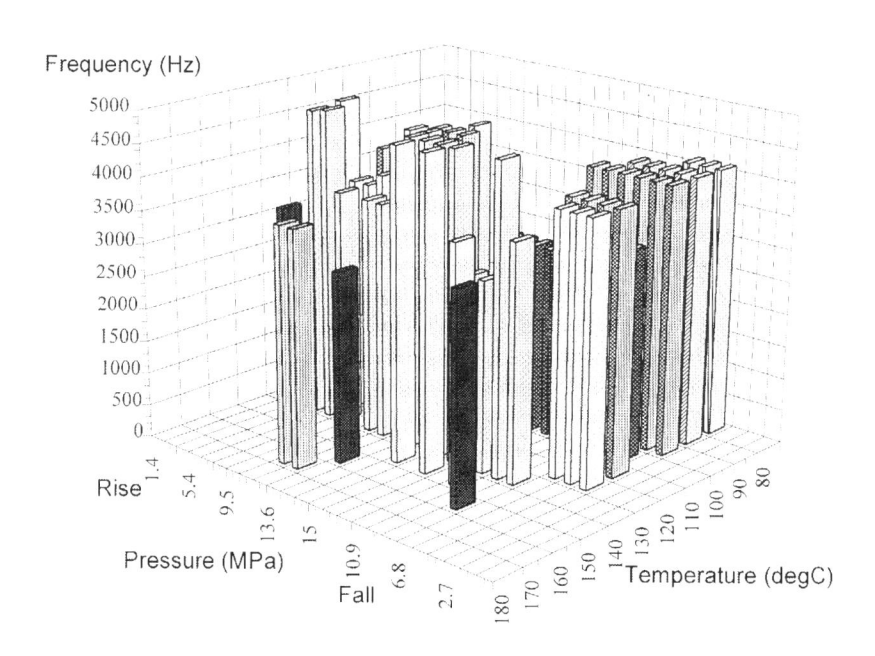

Figure 15b - Type 2 Shim on Both Pads, Frequency Plot

3.3 Equivalent Noise Dose Rating:

The comparative "equivalent noise dose rating" values for each condition was calculated and listed as shown in Table 1. This may also be shown graphically as in Figure 16 which gives a visual representation of the consequences of each series of tests.

Table 1 - Equivalent Noise Dose Rating for the Same Pads Operating with Differing Shim Arrangements.

	Equivalent Noise Dose Rating		
	Logarithmic	Log/Linear	Linear
	Standard Pads		
	79.82	45.18	41.72
Shim Position	With Type 1 Shim		
Finger	63.25	11.83	9.96
Piston	71.50	38.38	35.97
Both	73.79	17.04	14.12
	With Type 2 Shim		
Finger	66.44	16.58	13.05
Piston	85.59	61.42	59.17
Both	50.25	12.39	11.23

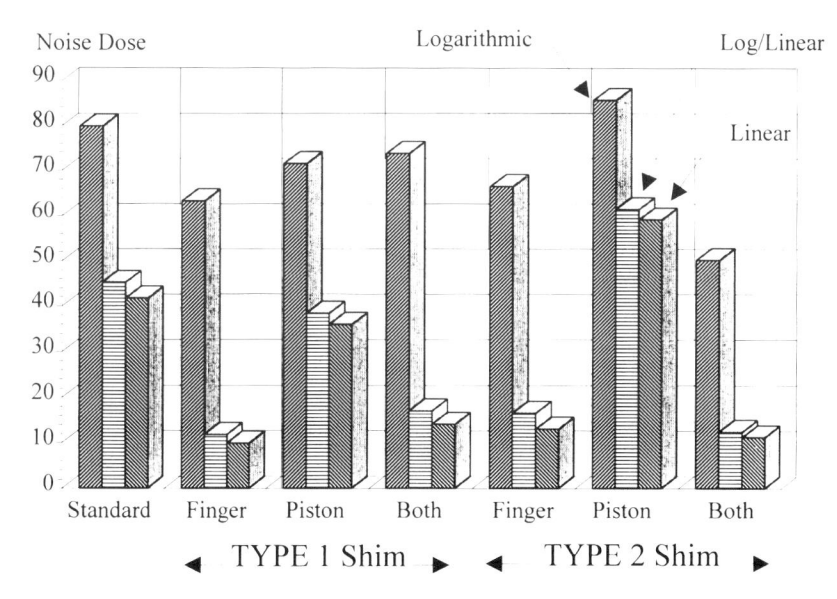

Figure 16 - Comparison of Equivalent Noise Dose Rating Values for each Condition

4.0 DISCUSSION OF RESULTS

4.1 Equivalent Noise Dose Rating Figures.

The graphical representations of the noise level equivalent values are clearly useful in providing the investigator with a visual impression of the problem. To make a reasoned judgement on which is the best improvement on the situation as recorded for the standard pad in Figure 9a is somewhat more difficult than it first appears when referring to Figures 10a, 12a, 13a and 15a. What may be said at the outset is that applying a noise fix shim on the finger pad does make a significant improvement in all applications.

Applying either type of noise fix shim to the piston pad alone has little positive effect and in the case of the Type 2 shim it appears to have an observable detrimental effect.

The application of noise fix shims to both pads does show improvement but it is difficult from the figures to provide clear evidence of which is the more economical application and which is the more effective. The "equivalent noise dose rating" allows this degree of judgement to be applied and this is discussed below.

4.2 Frequency Plot Figures.

Since this paper is concerned with the effects of applying noise fix shims to pads it is worth noting the frequency changes which take place when shims are applied.

Standard Pad - The standard pad has a predominant frequency of 2900 Hz which is most apparent at high pressures and over the full temperature range. At lower pressures frequencies of 8800, 5900 and 1750 Hz are detected in that order as the pressure reduces. There are other minor occurrences of frequencies around 7800 and 4400 Hz, again at the lower pressures.

Finger Shim - In both instances of applying a shim a horseshoe profile is apparent with noise over the full temperature range at medium pressures and only at the higher pressures at lower temperatures. The Type 1 shim promotes 4400 Hz at the toe of the shoe and 2900 Hz in the arms. The Type 2 shim is somewhat more complex with 6650 and 3200 Hz at the toe and 10500 Hz in the arms.

Piston Shim - With the exception of a few instances the predominant frequency is about 2900 Hz over much of the test area. Possibly the piston pad is responsible for promoting the higher frequencies when "coupled" to the piston, this giving a "stiffer" piston pad.

Shims on Both Pads - With the Type 1 shim the horseshoe profile is re-established but there is a significant frequency shift towards 2750 Hz with a few occurrences of 8750 Hz central and 4400 Hz in the rising pressure arm. The Type 2 shim differs with the horseshoe reversed (ironically almost a μ shape). Now the mid-temperatures, high pressures generate a frequency of 4700 Hz whereas the arms, (low pressures, mid to low temperatures) promote frequencies of around 4100 Hz.

This variation in frequency content and shift deserves further investigation and possibly indicates that some steps should be taken to account for frequency level in the "noise dose" calculations.

4.3 Equivalent Noise Dose Rating.

It is clear from the graphs that without a measurable value it is difficult to make a valid judgement on the progress being made on brake noise eradication. Table 1 and Figure 16 provide three "equivalent noise dose rating" values to allow that judgement to be made. The "Logarithmic Average" is scientifically correct if frequency affects are ignored. Both the "Log/Linear and Linear Averages" are not scientifically correct but reference to Figure 16 would indicate that their trend is very similar and almost identical. Comparison with the Logarithmic Average would indicate that again the trend is similar other than the Logarithmic Average is less sensitive to change. The Log/Linear and Linear Averages appear to react more to changes and as such could prove advantageous to the experimenter.

Considering the objectives of the case study investigation it was possible to suggest that the most economical solution to reduce noise on this brake would be to use a Type 1 shim on the finger pad only. The most overall effective arrangement to reduce noise would be to use the Type 2 shim on both pads, unfortunately this being also the most expensive solution.

5.0 ANALYSIS OF RESULTS

5.1 Pad Modes of Vibration:

It has to be questioned why such a noise reduction may be achieved by applying a shim to the finger pad whereas the same shim and pad applied to the piston side shows little to no improvement and may indeed increase the noise emissions.

The reason for the significant improvement in noise reduction when a shim is applied to the finger pad is due to the mode of vibration of the pads. It has been shown using holographic interferometry (2, 3) that the finger shoe is excited in a bending or torsional mode (or a combination of the two) whereas the piston shoe moves bodily, normal to the disc surface. It is felt that the calliper piston provides substantial support to the pad so preventing flexure. Support for the finger pad is somewhat different with the calliper fingers providing pad end support at two positions. This allows the central section to remain relatively unsupported thus promoting a bending mode. This bending mode on the finger pad side allows the noise fix shim to operate effectively so maximising its damping action - system instability being restricted and noise subsequently reduced in such cases. The whole body movement of the piston pad does not allow the shim to operate effectively and as a consequence system instability persists. If this is accepted it may be concluded that the application of noise fix shims to single opposed piston type callipers may not prove very effective because of the absence of pad bending. This may demand a more subtle approach to pad backplate design such that pad bending is positively encouraged with single piston opposed callipers. This may be extrapolated to four piston opposed callipers such that bending between the pistons is encouraged and full face shims employed. Other tests have shown that full face shims on four piston opposed callipers are more effective than half face shims under each piston.

A possible reason why noise may in fact increase when a noise fix shim is applied to the piston pad is that the piston is now "decoupled" from the pad and system instability increases. The thicker multi-layered Type 2 shim has a greater "decoupling" affect and so system instability is greater, leading to a greater propensity to generate noise. To extend this explanation to the limit it may be concluded that an additional approach to that proposed above would be to increase the stiffness of pads and not to use shims. This would encourage higher frequencies to be promoted and may indeed extend them outside the audible range.

5.2 Shim Delamination:

The shims used this investigation were fixed to the pads using cold pressure adhesive. In a separate investigation where a thermal bond adhesive was used it was noted that some of the shims delaminated during the tests. Indeed the effectiveness of the bond deteriorated with increasing temperature such that the adhesive was seen to extrude from the edges of the shim. In such instances the brake behaved as though the shims were not present and general instability persisted. An unstable brake generating a high frequency noise of around 11,000 Hz with a 9 diameter mode of vibration is shown in Figure 17. The delaminated shim shows a "ribbed" form of "blistering", a closer image being shown in Figure 21.

Figure 17 - Holographic Reconstruction Showing High Frequency (circa 11,000 Hz), 9 diameter mode of vibration. Shim Delamination Apparent (refer also to Figure 21).

This delamination shows itself on the holographic reconstruction where the shim exhibits independent excitation on the pad face. Typical observations of this independent excitation are shown in the following Figures 18 to 21.

Figure 18 - "Blistering" of Shim at the Pad Inner Edge.

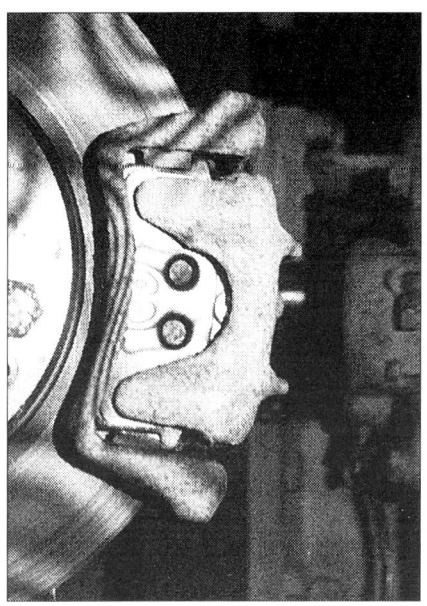

Figure 19 - "Blistering" of Shim at the Pad Inner Edge (higher amplitude).

Figure 20 - Central "Blistering" of Shim.

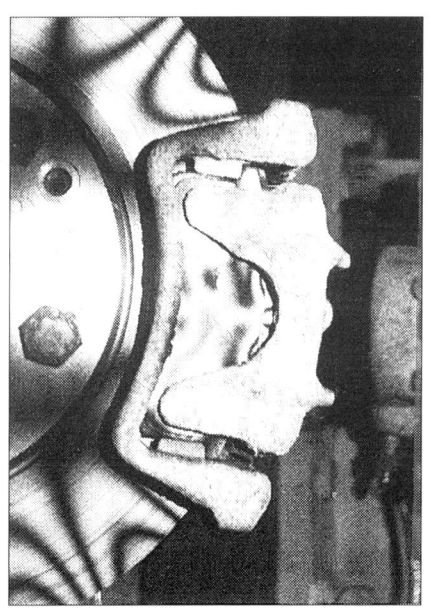

Figure 21 - Central "Ribbed Blistering".

The figures demonstrate that care needs to observed in the correct selection of bonding medium. It a cold state the shim may in fact indicate a firm attachment to the pads and behave properly as a "noise fix". In the operating state, at elevated temperatures, the adhesive may soften allowing the shim to detach itself from the pad and so prove ineffective under such conditions. On cooling the shim would re-bond itself to the pad leading to the investigator assuming that the shim is behaving correctly. Such delamination will result in the shims proving ineffective as a damping agent and may indeed add to the complexities of system instability or even a misunderstanding of the problem. A ready conclusion would be that the shims were effective at lower temperatures but the damping medium was ineffective at elevated temperatures. The chemist would then undertake research to increase the effective damping at elevated temperatures whereas the real problem is poor adhesion at elevated temperatures.

6.0 CONCLUSIONS:

- Noise fix shims do reduce the propensity of a brake to generate noise through their damping effect on pad excitation and thus system instability.
- The application of shims influence the frequencies which a brake is able to excite. A brake without shims is able to be excited over a fair range of frequencies whereas a shim applied to the finger pad alone allows higher frequencies to persist and a shim applied to the piston pad alone encourages the lower frequencies to prevail.
- With a sliding fist type calliper it has been shown that the application of a noise fix shim to the finger pad alone is a very economical first step to reducing brake noise.
- The application of a shim needs to be carefully considered. It has been shown with both types of shim that the application of a shim to the piston pad alone may indeed have a detrimental effect on brake noise reduction.
- Shims applied to both pads proved most effective in this investigation but their application tended to promote frequencies to which the human ear is most sensitive, 2000 to 4000 Hz.
- If it is accepted that the piston provides "stiffening" to the piston pad then it may be anticipated that noise fix shims may not prove to be an effective solution to brake noise on single opposed piston callipers. With such brake designs some thought to pad design would be necessary to allow increased flexure of the pad backplate.
- The possibility of delamination of the shims under operating conditions needs further consideration.
- A convenient yet scientific approach needs to be agreed and adopted so that some standardisation is used to classify brake noise. Without such an approach comparisons of differing situations will be purely subjective.

7.0 ACKNOWLEDGEMENTS:

The authors and Huddersfield University would like to thank BBA Friction for their continued support of our work in the area of brake noise research and investigations.

8.0 REFERENCES:

1. Fieldhouse JD and Rennison M "Proposals for the Classification of Brake Noise - An Equivalent Noise Dose Rating" Proceedings of the 15th Annual SAE Brake Colloquium, pp 87 - 94, October 1997. SAE Paper Number 973027.
2. Fieldhouse JD and Newcomb, TP "Holographic and Analytical Techniques Applied to Drum Brake Squeal" - EAEC 5th International Congress " The Automotive Industry meets the Challenges of the year 2000". 21st - 23rd June 1995, Strasbourg.
3. Fieldhouse JD and Newcomb TP, "The development of a visual recording process to aid the understanding of disc brake noise" 4th International EAEC Conference - Vehicle and Traffic Systems Technology, Volume 1 pp 483-503, 16-18 June 1993, Strasbourg.

Brake pull – a vehicle-based factorial experiment

R BARTHOLOMEW
Jaguar Cars Limited, Coventry, UK

SYNOPSIS

The paper will detail the use of a factorial experiment to investigate the possible effects and interactions of six factors on vehicle brake pull. Four of the chosen factors are brake related - *front brake pad composition, front brake disc finish, front / rear brake balance* and *ABS*. The remaining two being chassis related - *lower front wishbone bush stiffness* and *steering rack bush stiffness*.

Confirmation was obtained on three factors that the current specification was optimal and also that vehicle performance could be improved by the adoption of alternative specifications, or combinations thereof, used in the experiment.

1. INTRODUCTION

The need for an experiment was customer driven. Product engineering, through customer satisfaction tracking, had received a number of customer comments from the German market regarding brake pull on the "Sports Coupe", (XJ-S) .

Investigation found that it was primarily the German market that was making these comments, and that was attributed to their very aggressive driving style. The luxury car market provides high performance vehicles that the Germans use to their full potential. Local dealers reported that owners regularly drove at speeds in excess of 200 Kph and also left their braking relatively late, resulting in high levels of vehicle deceleration being achieved.

Initial vehicle tests of our own did not however show that there was a problem. Further discussions did reveal what was perhaps the true nature of the comments. The majority of the reports were being made by drivers new to our marque. Typically they had previously owned and driven the larger Mercedes and BMW vehicles. Appraisals of the competitor vehicles made it clear what these customers were telling us. Our vehicle had become inferior in this area of dynamics compared to our competitors, rather than having a specific fault. Clearly if we were to continue to take sales from our competitors and hold on to them, improvements in this area would have to be made.

The choice of a factorial experiment is significant as previously only traditional "change one thing at a time" methodology had been used. Producing a luxury, high performance vehicle demands that the highest levels of performance are achieved, and with the complex interactions of system elements contributing to the overall result, the optimisation process becomes extremely difficult. The factorial methodology was seen as a way of delivering the necessary optimisation.

2. OBJECT

2.1 To select a number of braking system/chassis components and test them for effect on high speed, high deceleration braking vehicle stability, using a factorial experiment.

2.2 To select an experimental plan that would estimate single factor significance and the significance of any two-way interaction of factors.

3. COMPONENT SELECTION

Component selection was made after discussion with design and development engineers from braking and front suspension departments within Chassis Engineering. It was decided that only changeable components should be selected. Major redesigns affecting suspension geometry were ruled out for this exercise.

3.1 The selected components / systems

The following list shows the component or system selected for investigation. The listing identifier (A to F) was also used as the identifier in the experiment.

A. *Front suspension lower wishbone bush.*
B. *Steering rack bushes.*
C. *Front / Rear brake balance.*
D. *Anti-lock Braking system (ABS).*
E. *Front brake disc surface finish.*
F. *Front brake pad material.*

3.2 Selection rationale

The *front suspension lower wishbone bush* (factor A) and the *steering rack bush* (factor B) were selected because they impart compliance into the front suspension assembly. It was argued that changes to the vehicle steering geometry would take place under the forces of braking.

The *front to rear braking balance* (factor C) was chosen as this is fundamental to a vehicle's stability under braking. The experiment therefore included braking the vehicle not only with all four wheel brakes as normal, but also braking on only the front wheel brakes.

The *ABS* (factor D) was chosen as it divides the hydraulic brake lines into three circuits. This is required for individual control of each front brake and select low control of the rear brakes during operation. Although braking during the experiment would not intentionally invoke operation, this factor was included to eliminate any unknown operation.

The *front brake disc surface finish* (factor E). The standard finish is one of fine "cross hatching" produced by Axial grinding. This effectively means the brake pads on one side of the vehicle are applied against the direction of the surface finish and the other side with it. Radial grinding was used producing discs finished in clockwise and anti-clockwise directions, so when fitted on their respective side of the vehicle the pads are braking with the direction of the surface finish.

Finally, although the first to be suggested, the *front brake pad material* (factor F) would be included. With an almost limitless number of lining materials to chose from who is to say that the best material for this application had been selected. A brake pad material with a stable coefficient of friction with temperature rise (regardless of refinement characteristics) was included.

4. THE FACTORIAL EXPERIMENT

The type of plan chosen reflected the objective of the experiment. A two level L32 orthogonal array was chosen. Each factor was tested at two levels to determine its significance and 32 tests conducted to ensure results to estimate any two-way interaction of factors were obtained.

4.1 The factor levels

Testing each factor at two levels would provide results on it's significance, but not its optimum setting. The two levels for each factor are identified with either a - (negative) or + (positive) sign and are shown below:

Factor A - *Front suspension lower wishbone bush*
- (negative) Standard bush.
+ (positive) Solid bronze bush.

Factor B - *Steering rack bushes*
- (negative)Standard bush.
+ (positive) Solid aluminium bush.

Factor C - *Front / rear brake balance*
- (negative) Front brakes only.
+ (positive) All brakes.

Factor D - *Anti-lock braking system*
- (negative) ABS off.
+ (positive) ABS on.

Factor E - *Front brake disc surface finish*
- (negative) Normally ground brake disc.
+ (positive) Directionally ground brake disc.

Factor F - *Front brake pad material*
- (negative) Standard front brake pad.
+ (positive) Temperature stable front brake pad.

4.2 The L32 orthogonal array

To achieve the second objective with six factors at two levels, a L32 orthogonal array was needed. This is a resolution V (Five) plan. This has main effects clear of two-way interactions and no confounding between two-way interactions.

This is still very economic when compared to the traditional test. The number of tests being reduced from 64 to 32.

4.3 Measured response

Two measured responses were recorded for each run of the experiment. One objective measurement, the number of carriage way lanes moved (deviation) from the straight ahead position and the other subjective, a rating of one to ten assessing the feel of any movement. One being very poor and ten excellent. The subjective rating was considered necessary as this would directly reflect the feeling the customer would experience. It is not necessarily more significant than the other measure, but it is an essential part of the evaluation.

The above responses were measured with no steering control from the driver and the subjective rating being allocated as detailed by the in house Jaguar Rating System, or JRS.

4.4 Test procedure

The test procedure needed to represent the customer criticisms, allow recording of the measured responses, be safe and relatively quick to perform. The procedure was as follows:
The vehicle would be braked to rest from 200 Kph.
The rate of deceleration would be 0.8g for an "all brake" condition and 0.6g for a "front brake only" condition.
To eliminate power train influences, the vehicle would be braked in neutral.
An initial front brake disc temperature of below 100°C would be required before each run.

Track surface condition must be dry and wind conditions calm.
From a virgin brake condition, four runs to be completed.
An abbreviated brake bedding procedure completed after virgin brake test runs.
10 runs completed after brake bedding.
The above would be completed with driver and observer.
The driver and observer would rate any brake pull independently.

4.5 Other considerations

To reduce test time and the effects of component variability the following were implemented.
Two vehicles of identical specification were used to speed up the exercise.

The test site, Bruntingthorpe Proving Ground, was chosen for it's 3.2 kilometre long by 60 metre wide straight. It had a slight slope running across it's test surface and to eliminate the effect of this, the test runs were conducted in both direction's, alternating between runs.

The test run order was changed to minimise the number of reworks to the suspension and steering systems.

The consumable components needed for the experiment were procured from a single batch. The tyres were included in this, being replaced every test to eliminate any effect of their wearing.

5. RESULTS

Average measured responses and average signal / noise ratio values, for each measured response, were calculated for each factor at both levels. The JRS score measured response, was calculated with the scores of the driver and observer added together for increased resolution. Analysis was by means of graphical interpretation. Effects plots were used to contrast the average response of both levels of each factor. Factors worth noting showing a large difference between their positive and negative signs. Normal and Half-Normal plots were used to interpret the significance of any factors, indicated by their position away from the straight line. All plots made reference to in the following "Discussion of results", are presented in the appendix.

5.1 Discussion of results

Analysis of the results for the virgin brake condition will be discussed first, followed by discussion of the bedded brake condition and finally as a whole looking for any commonality.

5.1.1 Virgin brake results

Discovering any significant factor in the virgin brake condition proved difficult. Study of the JRS rating Effects plot (figure 1), would appear to indicate that the interactions of factors BE, *steering rack bush* with *front brake disc surface finish*, and CE, *front brakes only or all brakes* with *front brake disc surface finish*, have a substantial effect. However, only the BE interaction appears as significant. This is shown on the Half-Normal scores plot (figure 2). All factors with the exception of BE are lost to noise, common cause variation, lying on or close to the straight line. Further evidence to support BE's effect is found on the Signal to Noise ratio - Lanes moved Effects plot (figure 3). Analysis of the sign (positive or negative) finds both Effects plots showing BE is best set to positive. When interpreting the Effects plots, the higher the JRS rating the better and with Taguchi's Signal to Noise ratio numbers the higher value is always the better. Looking at the signs of the individual factors, B and E, finds both are best at the positive setting confirming the setting of the BE interaction. So it would appear that already two components under investigation are best set to a condition other than their current production setting. The *steering rack bush* should be solid and the *front brake disc* produced with a directionally ground surface finish.

Surprisingly, one factor in particular can be seen to be insignificant in the virgin brake condition. Factor F, the *front brake pad*, shows no contrast on both Effects plots and was completely lost noise on all the Normal scores plots. All other factors are also lost to noise showing no significance in the virgin brake condition.

5.1.2 Bedded brake results

Analysis of the bedded brake results finds much clearer contrasts on the Effects plots and significance on the Normal plots. Six factors are clearly visible from the rest on the Signal to Noise ratio - Lanes moved Effects plot (figure 4). Factors C, *front brakes only or all brakes*, CE, the interaction of C with the *front disc brake surface finish*, F, the *front brake pad* and also three unknown factors allocated to columns 4, 7 and 10. Referring to the Half-Normal plot (figure 5) finds all six factors to be significant, they all lie well away from the straight line and the remaining factors lost to noise. Confirmation is found for all except factor 7 on the Normal scores plot (figure 6). Further evidence is found on the Signal to Noise ratio JRS rating Effects plot (figure 7), the JRS rating Effects plot (figures 8) and the JRS rating Half-Normal scores plot (figure 9) for effect and significance of factors F, 4, 7 and 10.

Factor F is now quite clearly the single most significant factor, showing a large contrast on both Effects plots and lying away from the common cause variation on all the Normal scores plots. Looking at the setting for the *front brake pad* finds that it is best set to positive, a composition with a stable coefficient of friction with temperature rise. Now this is the third factor that has been identified to be best at a setting other than the current production one.

Analysis of factor C, *front brakes only or all brakes*, finds that it is best set to it's positive condition, all brakes. This follows nicely to its interaction with factor E, the *front brake disc surface finish*. It is clear that the interaction CE is also best set at it's positive condition, a front brake disc produced with a directionally ground surface finish, and so has the requirement of C to also have a positive setting.

5.1.3 Overview

Looking for any commonality between the virgin brake and the bedded brake conditions finds it easier to pick out the factors that had no effect on brake pull during this experiment. The *front suspension lower wishbone bush*, factor A, tested at standard setting or a solid no compliance setting, showed no or very little effect and was always lost to common cause variation. Likewise factor D, the *Anti-lock Brake System* tested "on" as standard or "off" by removal of it's electronic controller, sometimes showing a slight effect but not any significance. However, this shows that these two factors do not influence brake pull when braking in a straight line below the point of ABS activation and so require no further investigation with regard to this issue.

The significance of the BE interaction does not translate from the virgin brake condition to the bedded condition. However, as discussed earlier, to meet the virgin brake condition factor

E needs to be set positive. This at least does not compromise the setting of factor E for the bedded brake condition of the CE interaction, which also requires factor E to be set positive. This being the case, further investigation into the factors B and E can be made without the need to redefine factor C. Further tests can be completed with all brakes connected, as standard. This experiment has only tested these factors at two levels and while it is conceivable to change to a production method of grinding that gives a directional surface finish quickly, as the finish design specification can still be easily achieved, a move to a solid steering rack bush is certainly not yet feasible. These have been implemented to give an acceptable level of vibration isolation to the steering gear. To achieve the implementation of a bush with less lateral compliance a separate design of experiments could be completed on bushes with differing rates of compliance whilst still achieving their isolation targets.

So far no discussion has been made about the unknown factors 4, 7 and 10. Certainly they have shown themselves to have a significant effect. These showings are the result of genuine calculation because the process demands that all columns have their effects calculated. The problem with these factors is that they are the result of more complex interactions. Each of the factors allocated a simple number is at least a three way interaction (e.g. ABC) and will be confounded by other possible combinations (e.g. DEF). It is therefore not possible to identify with any certainty their true make up. The original objective did not set out to look for any combinations more complex than two-way interactions and could not have foreseen this.

Three factors have been identified that are worthy of further investigation. The *steering rack bush* (factor B), the *front brake disc surface finish* (factor E) and the *front brake pad* (factor F). Before any detailed investigation is made into these component areas, a confirmation experiment should be conducted to verify these results. Three factors at two levels could be conducted very quickly, adhering to a L8 orthogonal array. This would again provide estimation of single factor significance and the significance of any two-way interaction of factors. While this will not investigate the more complex interactions identified above, it does provide a spare column that the two test vehicle's could be assigned to and so provide an estimate of their significance. The remaining vehicle settings should be as standard production as these have been shown to either have no effect or a beneficial effect during this experiment. I.e. The vehicle fitted with standard *front suspension lower wishbone bush* (factor A), have *all wheel brakes connected* (factor C) and have the *anti-lock braking system* (factor D) left operational.

6. CONCLUSIONS

6.1 The *front suspension lower wishbone bush* (factor A), had no effect.

6.2 The *steering rack bush* (factor B), had an effect through its interaction with the *front brake disc surface finish* (factor E) in the virgin brake condition.
6.2.1 The *steering rack bush* is best set to a condition with no compliance.

6.3 The *front to rear brake balance* (factor C), had an effect in the bedded brake condition.
6.3.1 The *brake balance* is best set to an all wheel brake condition.

6.4 The *anti-lock brake system* (factor D), had no effect.

6.5　The *front brake disc surface finish* (factor E), had an effect through its interaction with the *steering rack bush* (factor B) in the virgin brake condition and through its interaction with the *front to rear brake balance* (factor C) in the bedded brake condition.

6.5.1　The *front brake disc finish* is best set to a directionally ground condition.

6.6　The *front brake pad* (factor F), had an effect in the bedded brake condition.

6.6.1　The *front brake pad* is best with a stable coefficient of friction with temperature rise characteristic.

6.7　Factorial experiments can be successfully applied to motor vehicles and need not be restricted to manufacturing and workshop environments.

7. RECOMMENDATIONS

7.1　The *steering rack bush* (factor B), the *front brake disc surface finish* (factor E) and the *front brake pad* (factor F) should be subjected to a confirmation experiment.

7.2　The factors should be tested at the same two levels.

7.3　The experiment should be a full factorial L8 orthogonal array.

8. ACKNOWLEDGEMENTS

The author would like to thank Allied-Signal Automotive for their support with supplying a suitable front brake lining for inclusion in the experiment.

9. BIBLIOGRAPHY

9.1　Engineering Quality & Experimental Designs - D M Grove & T P Davis.

9.2　Engineering Experiments - C Lipson & N J Sheth

9.3　FORD User Guide to 2 Level Orthogonal Arrays - N Buswell & THE Davis.

9.4　Design & Analysis of Experiments (2nd Edition) - D C Montgomery.

APPENDIX

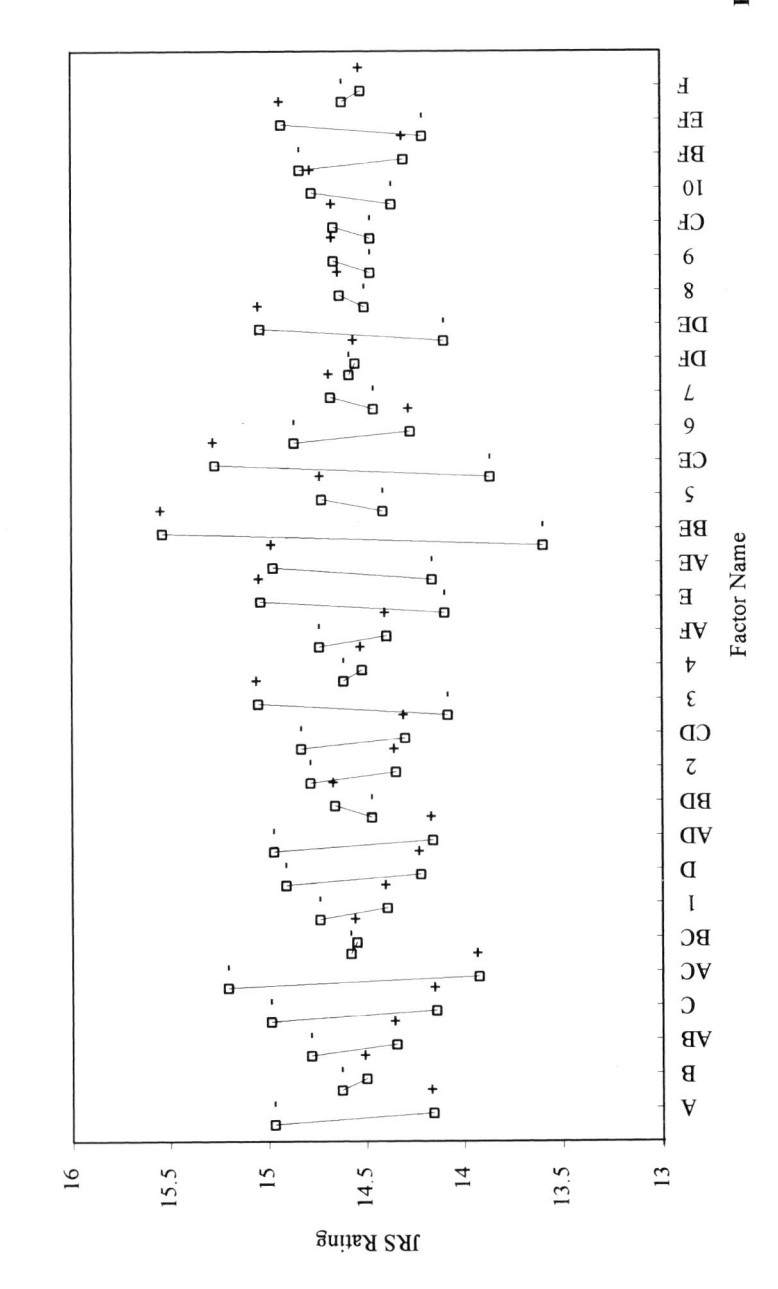

Virgin Brakes
JRS Rating

Figure 1

**Virgin Brakes
JRS Rating**

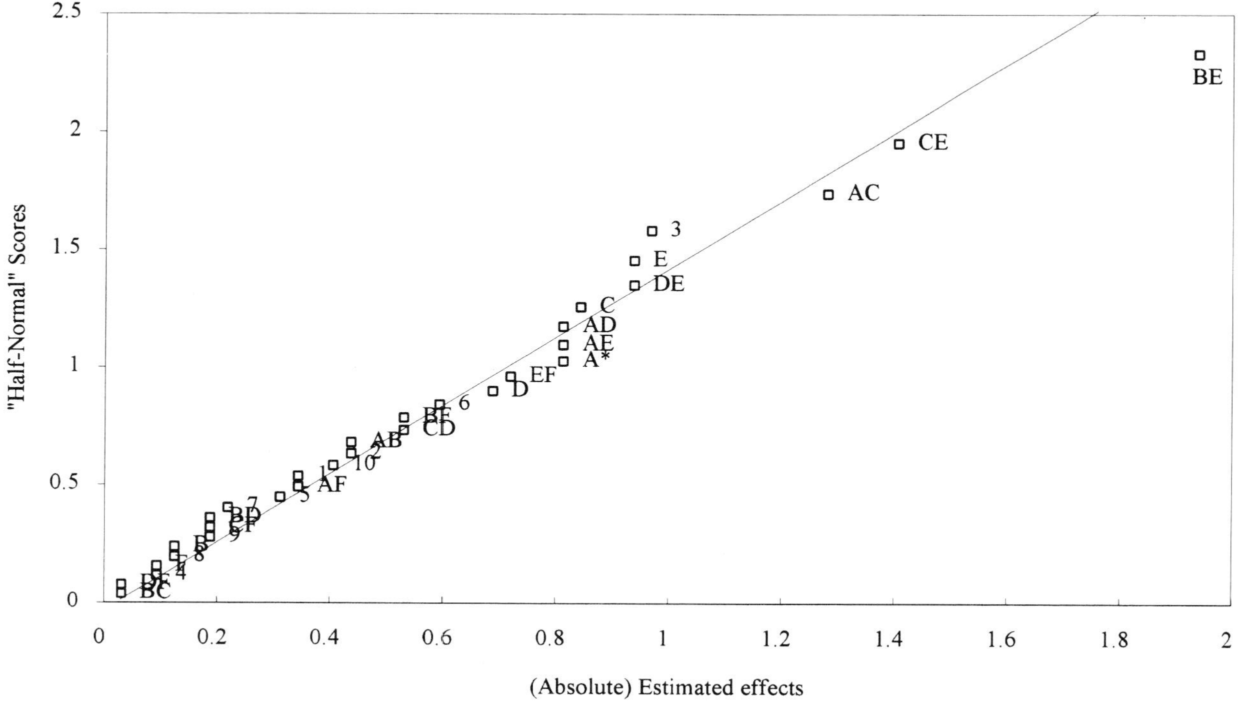

Figure 2

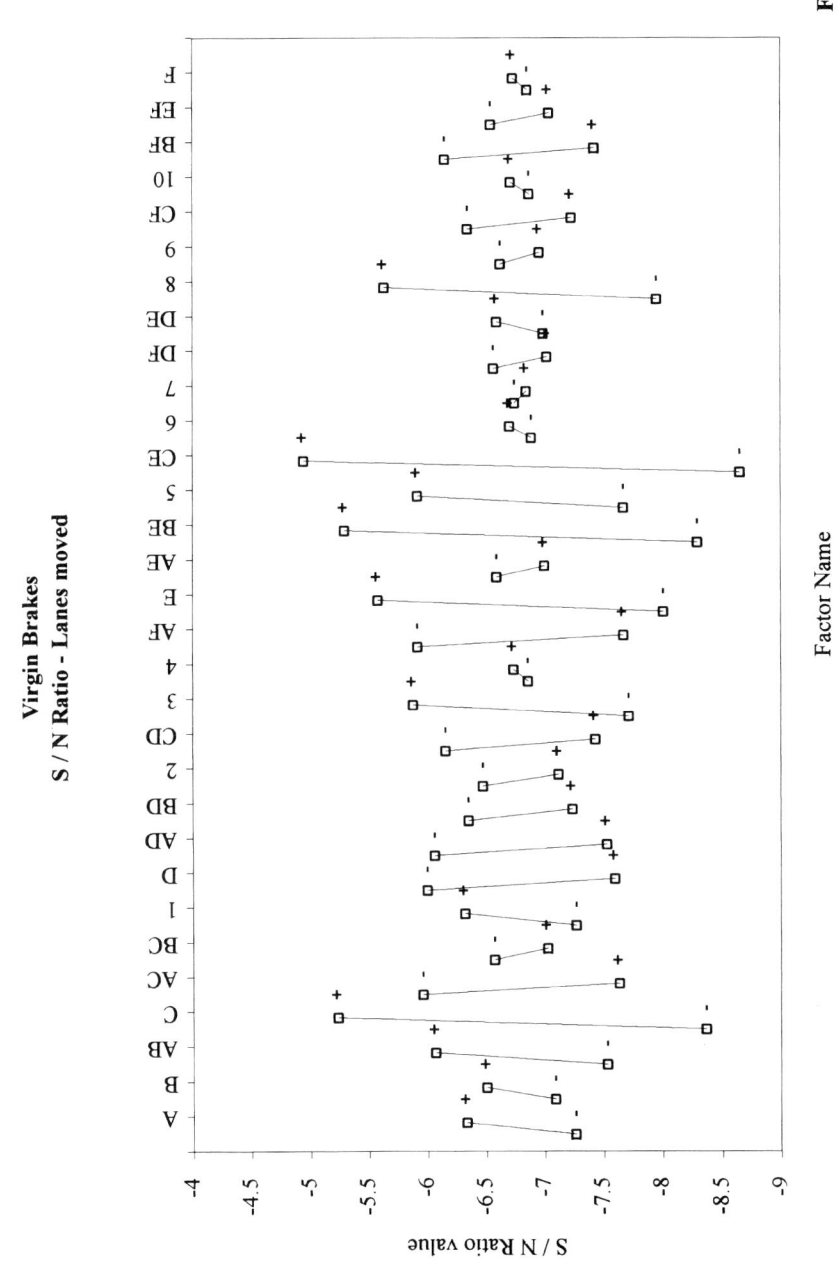

Figure 3

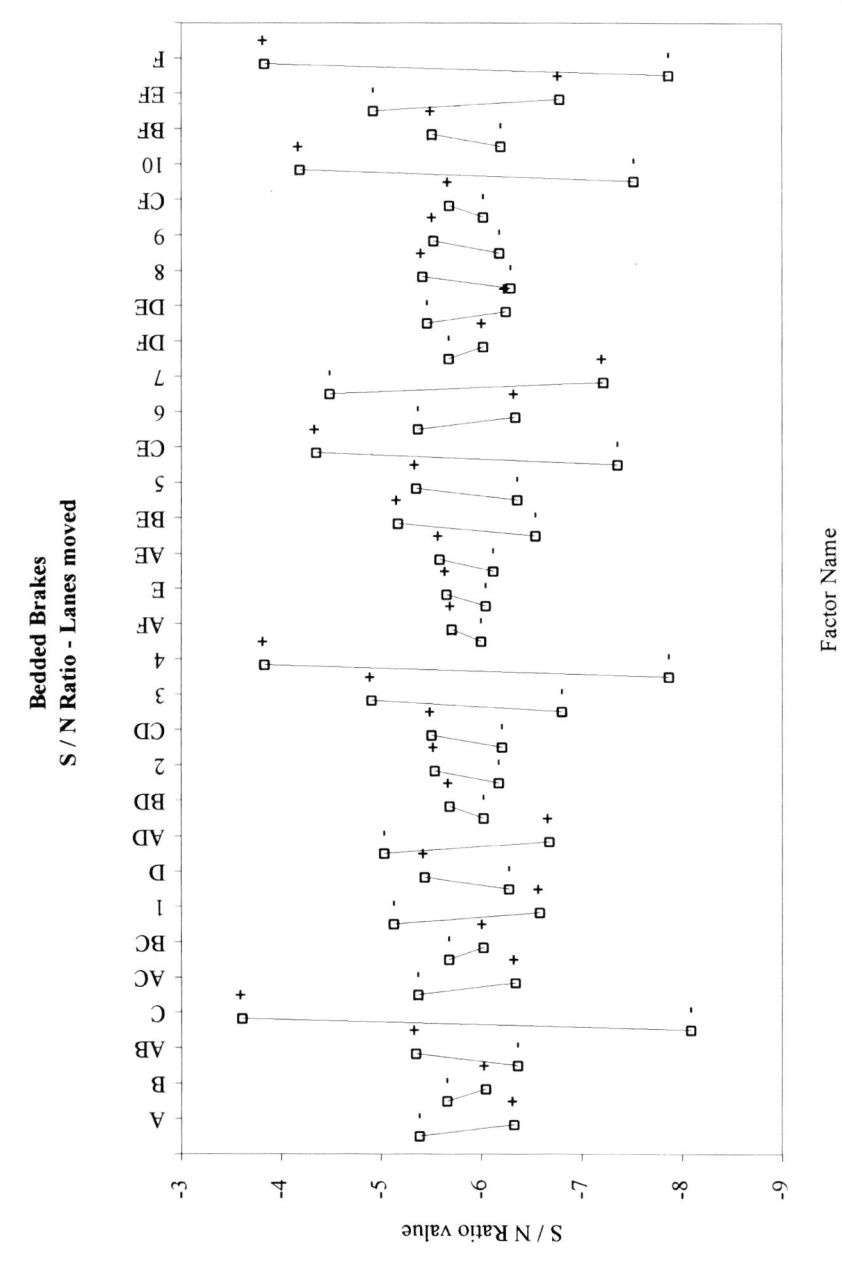

Bedded Brakes
S / N Ratio - Lanes moved

Factor Name

Figure 4

Bedded Brakes
S / N Ratio - Lanes moved

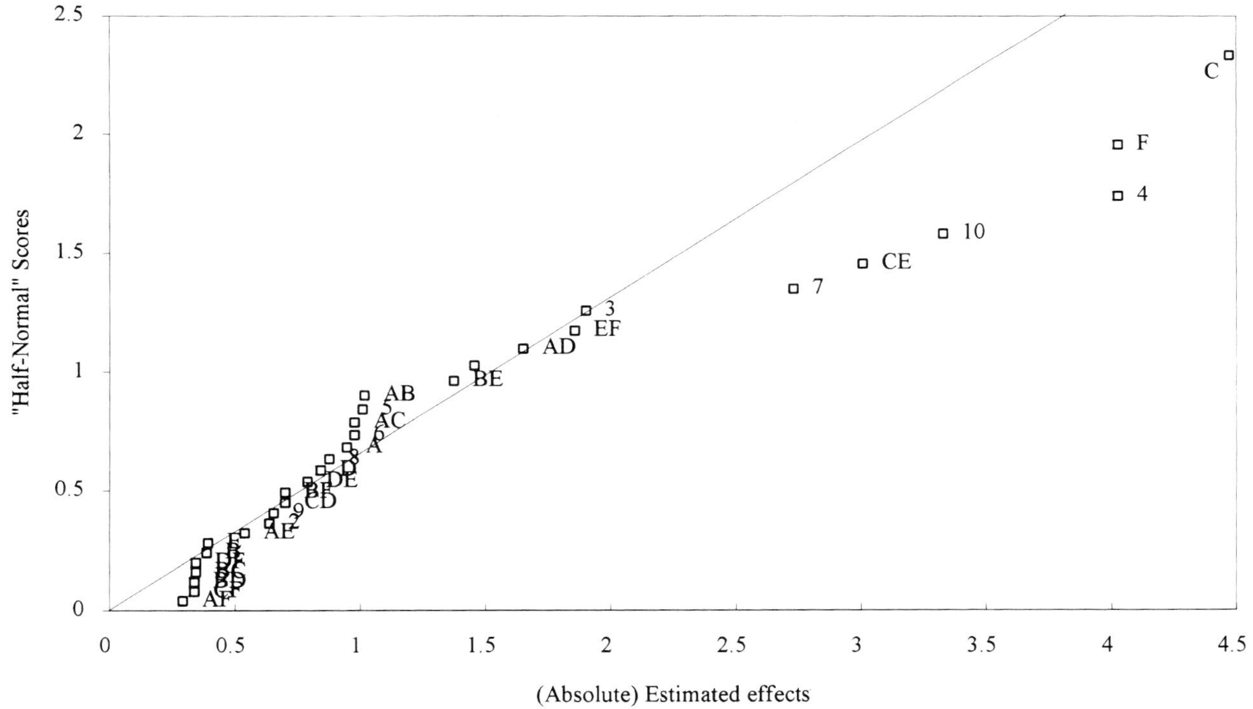

Figure 5

Bedded Brakes
S / N Ratio - Lanes moved

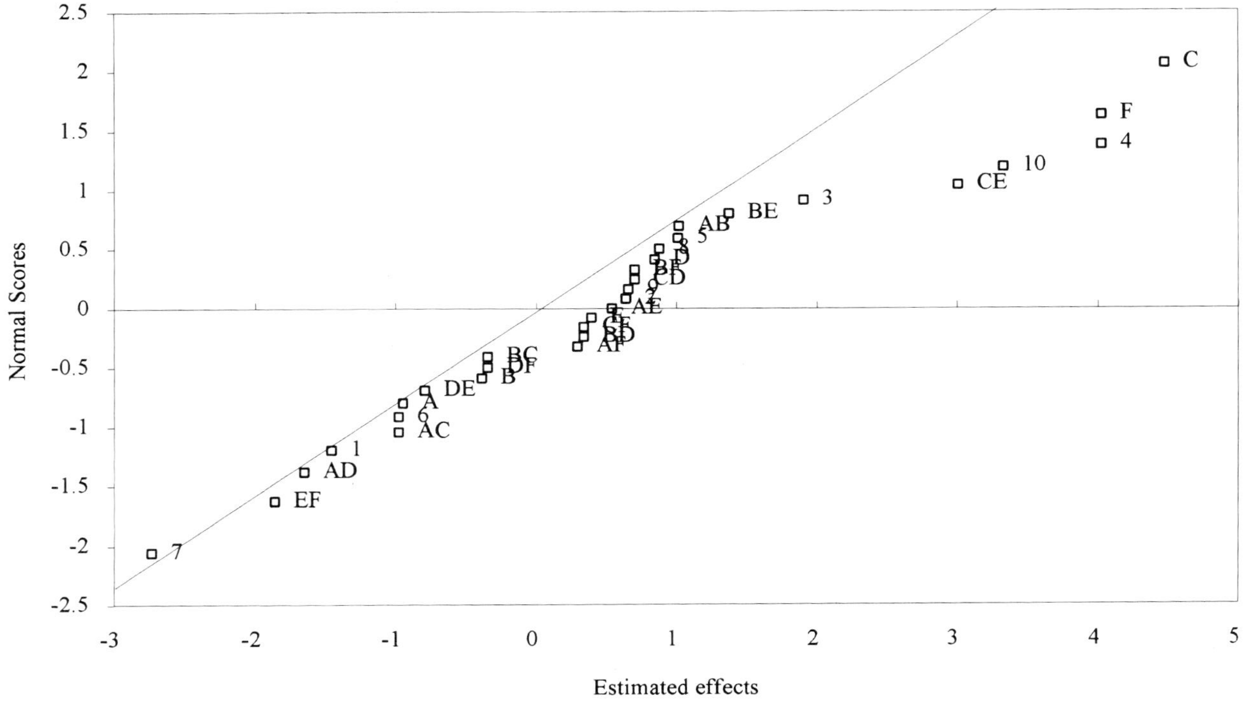

Figure 6

Bedded Brakes
S / N Ratio - JRS Rating

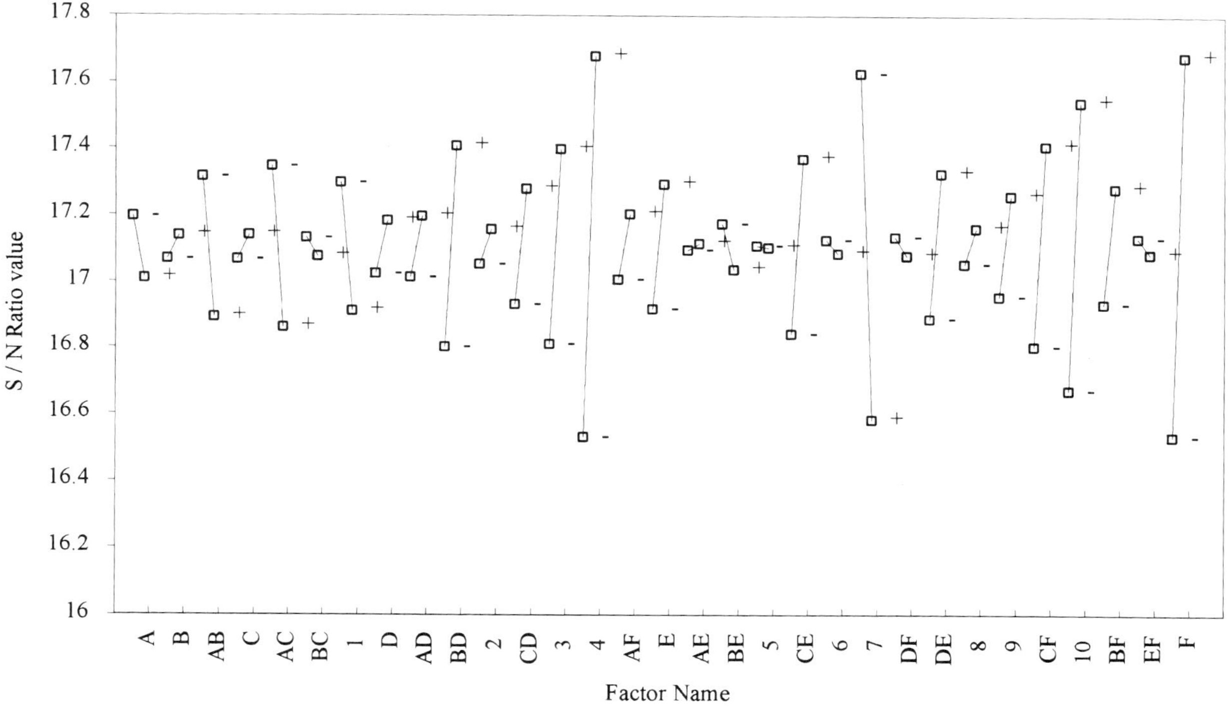

Figure 7

**Bedded Brakes
JRS Rating**

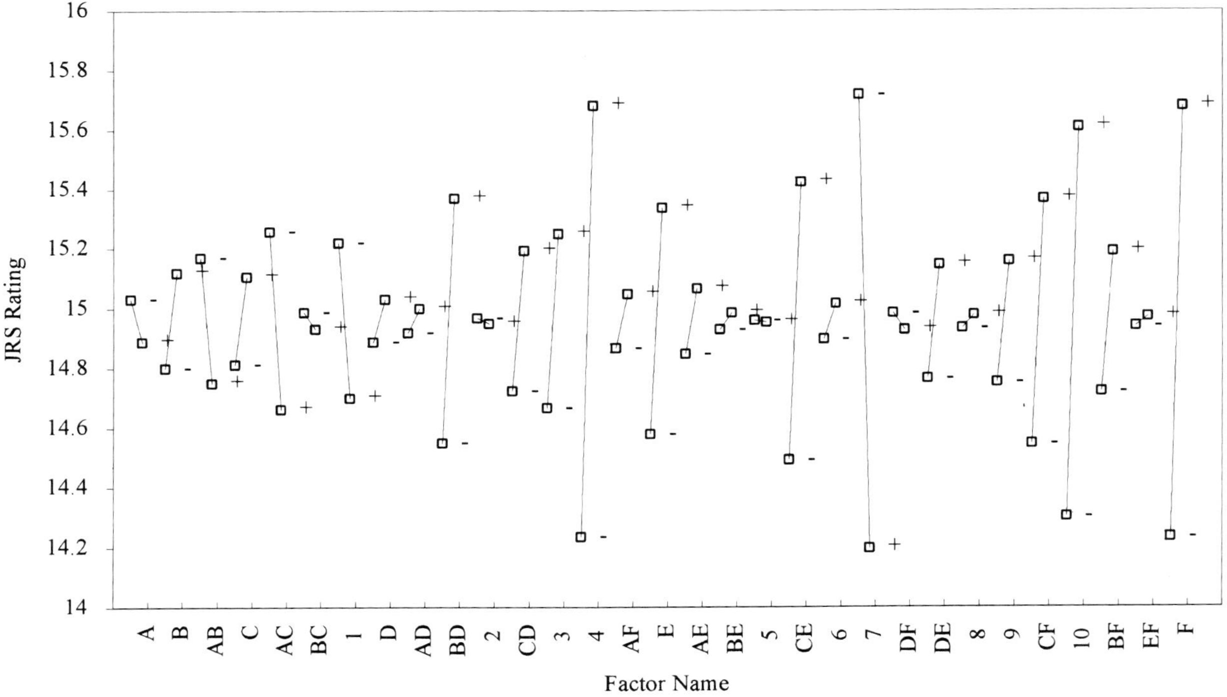

Figure 8

Bedded Brakes
JRS Rating

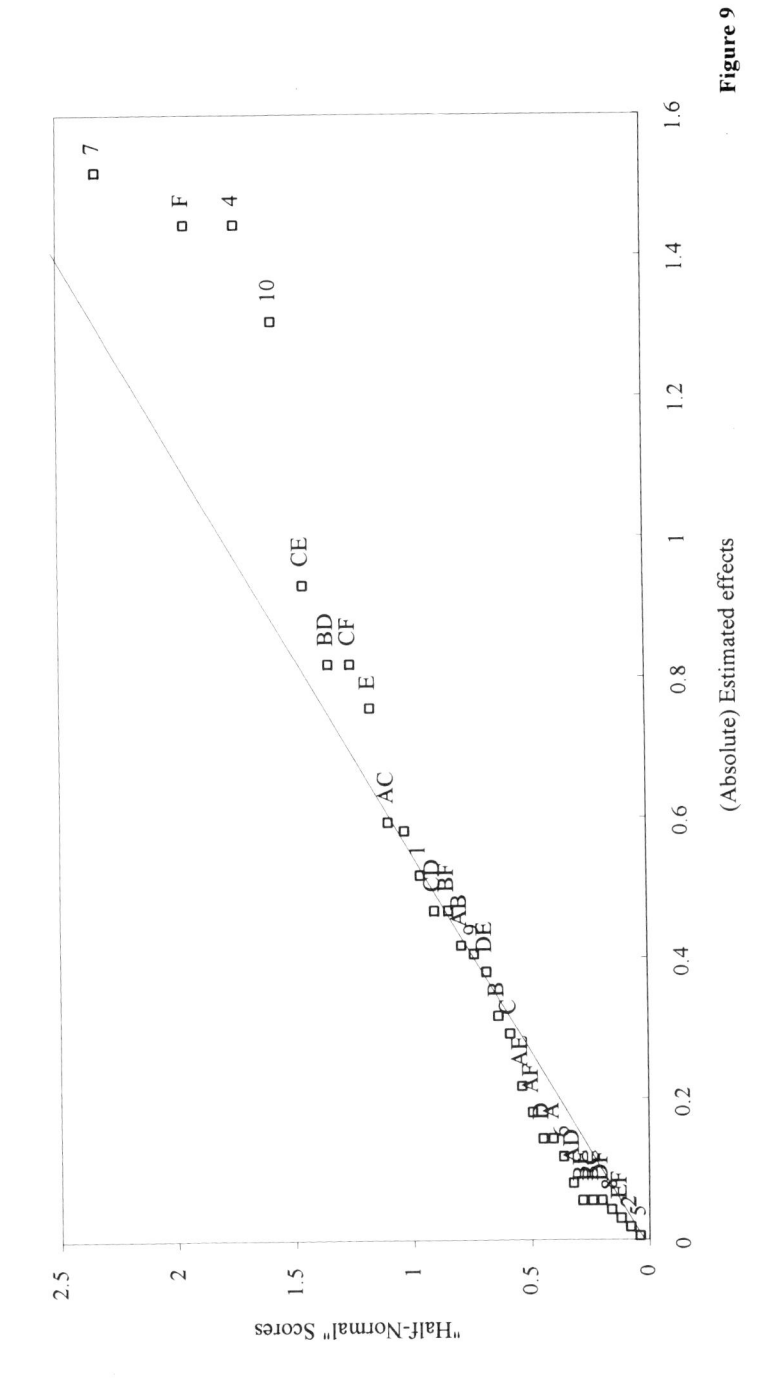

Figure 9

Rear disc brake moan – experimental investigation and ADAMS simulation

D RIESLAND and **J JANEVIC**
Mechanical Dynamics Inc., Ann Arbor, Michigan, USA
J MALOSH and **W STRINGHAM**
Bosch Braking Systems Inc., South Bend, Indiana, USA

SYNOPSIS

Low frequency brake moan is rarely associated with functional problems, yet brake moan and other brake noises are a significant customer satisfaction issue presenting significant expense to brake and vehicle manufacturers. Bosch Braking Systems and Mechanical Dynamics, Inc. (MDI) have developed and experimentally validated a fully non-linear dynamic model of a rear disc brake system which replicated low-frequency brake moan, using the ADAMS® mechanical system simulation software. The model correlated to vehicle and laboratory measurements of moan frequency and amplitude, and demonstrated signature features such as stick-slip pulsing and moan modulation. Parametric variations of the model showed sensitivities mimicking those observed experimentally.

INTRODUCTION

Brake moan is an audible noise in the 180 to 200 Hz range which does not normally indicate a brake system malfunction, yet brake noises are a significant cause of warranty claims possibly greater than those of any other single automobile component. The reduction or elimination of brake moan is the goal of every automotive manufacture and brake system supplier.

Bosch and MDI have experimentally validated a novel approach to simulation of low-frequency brake moan, using ADAMS mechanical system simulation software. This approach entailed five steps:

- *Basic Modeling*—Existing CAD models of a Bosch rear disc brake system prone to exhibit brake moan were imported directly into ADAMS and parameterized. Components included the caliper, inner/outer pad, rotor, axle tube, axle shaft, and axle bearing.
- *Model Enhancement*—Compliant elements, and non-linear contact and friction algorithms were added to the basic model.
- *Simulation*—A fully non-linear dynamic simulation of brake system behavior was performed for a prescribed set of operating conditions.
- *Correlation/Validation*—Simulation results (frequency response, time histories, and model parameters) were validated using data from vehicle and laboratory experimental testing.
- *Design Studies/Optimization*—The validated model was examined over a prescribed range of operating conditions to determine the effects of parameter variations, predict system-level performance and find parameter values which give desired performance changes.

Brake moan frequency and amplitude calculated by the ADAMS brake system model correlated with real world measurements taken from brakes installed on board vehicles as well those obtained under laboratory conditions. Other time history and spectral signature features such as stick-slip pulsing, and moan modulation were present in both the model and in the experimental data. System modal analysis showed a high degree of correlation between experimentally measured modes and those predicted by simulation. As a final validation step, trends observed in parametric variations of the model were found to compare favorably to trends observed during experiment.

The methodologies developed and software used were found to have several significant advantages over traditional low frequency brake moan analysis. Unlike finite element (FE) models, the ADAMS model is neither large nor cumbersome. The vibratory response predicted by the ADAMS model is self-exciting, and does not require input excitation at output frequencies. Finally, ADAMS provided non-linear coupling between structural components; which facilitated a system modeling approach rather than a component-focused approach. This proved important, as sustained brake moan was not observed in any simulation until the axle was added to the brake assembly.

BRAKE MOAN OVERVIEW

The operational definition of brake moan at Bosch Braking Systems is an audible noise that occurs at low speed and light brake apply conditions. Typically, brake moan consists of a single stationary frequency between 180 to 200 Hz, although other frequencies up to 1000 Hz are regularly measured. These frequencies usually correspond to one or more of the rigid or flexible body modes of the brake, axle, and suspension system. Modulation of the moan signature by some of these resonances can also occur. Moan can occur on one or both brakes, in the front or rear.

The operational conditions that produce brake moan on a vehicle vary. For example, brake moan will occur on brand new vehicles as well as vehicles with several years or thousands of kilometers of service. The Bosch brake system under investigation in this study was noted to produce moan with new and worn lining materials. Typical vehicle speeds at which the rear brakes moaned were less than 10 km/h, in both forward and reverse. Brake pressures were less than 700kPa, and lining temperatures were less than 25° C.

Brake moan was also achieved in the same brake system using a laboratory bench configuration in which a single brake and half axle system was isolated. The axle tube and shaft were cantilevered to the bench, and the tire and wheel were eliminated. A systematic reduction in complexity was undertaken to establish the simplest possible configuration in which moan would occur. Use of a hand-crank to impart low-speed rotational motion to the rotor and simple clamps to apply light pressure to the brake pads proved sufficient to repeatedly generate brake moan of the same frequencies, amplitudes, and signature features as observed on the vehicle.

Brake moan was measured experimentally both on the vehicle and in the laboratory using accelerometer transducers located on the outer brake pad, and measuring vibration in the tangential direction. A time history of a moan event of 500 msec duration measured on a vehicle is shown in Figure 1, below:

Outer Pad Vibration -- Vehicle

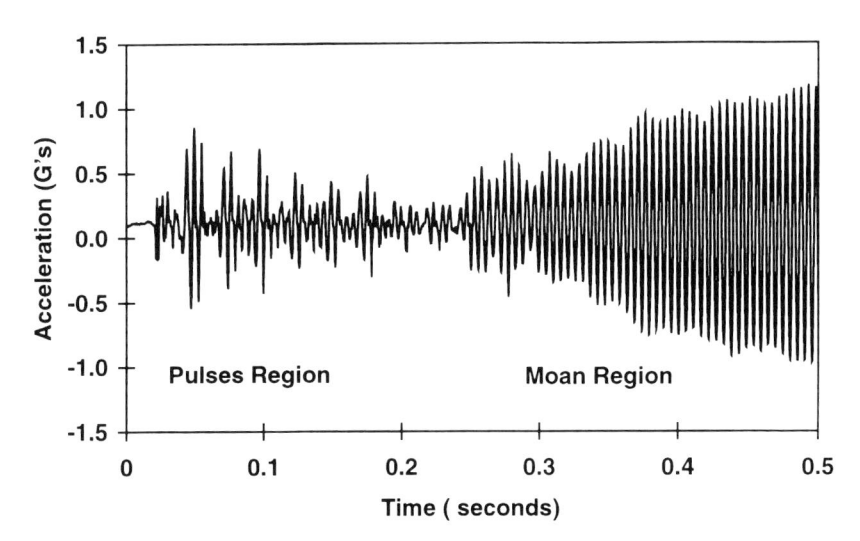

Figure 1: Brake moan time signature

As noted in Figure 1, the moan event is characterized by a series of stick-slip pulses (0.00 to 0.20 sec) which initiate the sustained moan vibration (0.25 to 0.50 sec). A spectral analysis of the same moan event above revealed that the vibration occurs at approximately 187 Hz, and modulation of this vibration approximately every 50 Hz. This is illustrated in Figure 2.

Outer Pad Vibration -- Vehicle

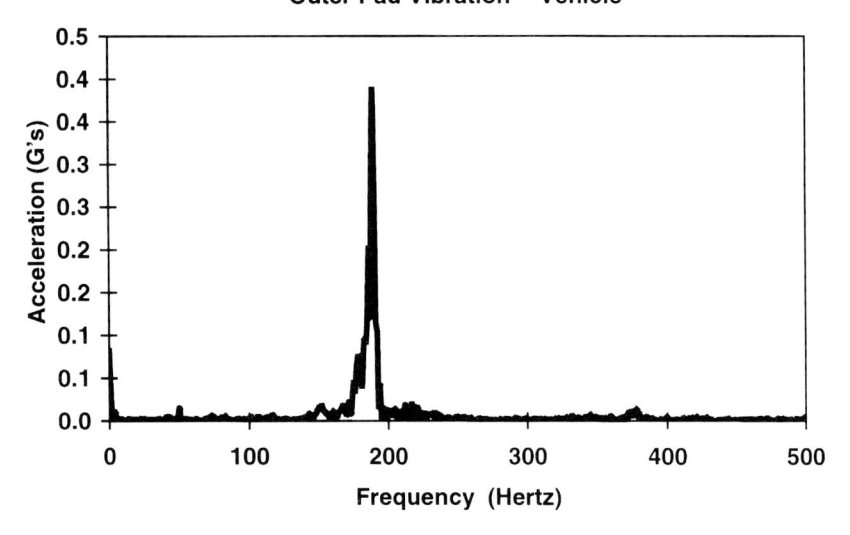

Figure 2: Brake moan spectrum

REAR DISC BRAKE MODELING

Successful simulation of low frequency brake moan has eluded many previous attempts using many different types of software. The ADAMS multibody system simulation (MSS) software is an appropriate analysis tool for low frequency issues in brake systems because of the non-linear nature of the mechanisms, the rigid body motion which most components exhibit, and the ease with which the software produces fully parameterized models and performs design studies.

About ADAMS

The ADAMS (Automatic Dynamic Analysis of Mechanical Systems) simulation code is an important computational tool for virtual prototyping. The laws of Newtonian mechanics are the basis for ADAMS, which automatically formulates the Euler-Lagrange equations of motion for a mechanical system and solves these three-dimensional non-linear equations using sophisticated numerical methods.(1) ADAMS requires the following to specify the mechanical model for a simulation:
- The mass and inertia of the rigid bodies or parts
- The definition of the system geometry
- The connectivity for the system (the mechanisms for connecting the parts) defined in terms of mechanical joints, higher-pair contacts, other constraints and elastic elements
- A description of the external forces and excitations acting on the system

ADAMS can perform static equilibrium, kinematic, linear, and dynamic analysis. During a full dynamic analysis, ADAMS calculates time-dependent translational and angular displacements, velocities, accelerations, and applied forces, which can then be used to verify system performance characteristics or system response to a set of operating conditions.

Model Construction Process

Creating a simulation model has two particularly challenging aspects: determining the appropriate level of fidelity of the model, and then obtaining the appropriate input data. At each step in the modeling process, model complexity was incrementally increased and the model was verified. An initial kinematic model was built to verify gross motion characteristics. Dynamic effects such as compliant elements, friction forces, and pad-to-rotor contact were added to the kinematic model in a stepwise fashion. Extreme care and attention to detail was taken during the construction of the model.

Through the various stages of model development, data input for the ADAMS model of the Bosch rear disc brake system required data from a variety of sources. The final ADAMS model is depicted in Figure 3, below:

Figure 3: Final ADAMS model of rear disc brake system

The geometric definition of system components was imported directly into ADAMS from a commercial solid modeling software package. Within the solid modeler, all geometric entities comprising any component that could exhibit motion relative to any other component were lumped into a single rigid part prior to export. This enabled precise definition of the mass tensor for each part.

Once in ADAMS, rigid parts were connected using standard mechanical joints to create a kinematic model. Simple kinematic analyses with prescribed motion of the rotor were performed to verify the correct topology, assembly, and gross motion characteristics of the model. A kinematic analysis is used for zero degree of freedom systems in which all motions of the system are algebraically determined and not influenced by external forces. The kinematic model was not sufficient to replicate the brake moan signature.

The mechanical joints in the model were then replaced by dynamic elements requiring precise definition of component interactions, thereby releasing system degrees of freedom. For example, the interaction between each brake pad and the rotor was modeled with a series of non-linear contact force elements which accounted for the normal stiffness and damping properties of the interface, as well as friction in the interbody plane. The friction force was a complex function of pad to rotor relative velocity and other parameters. Compliant elements, such as caliper pin bushings, were added to the model at this time. Preliminary design studies (discussed in the following section) indicated that the axle tube and shaft needed to be modeled as flexible components in order to replicate the sustained brake moan. All other parts in the system were modeled as rigid bodies, an assumption that was substantiated by laser vibrometer studies.

The model configuration was selected to match up with that used in laboratory experimentation, which exhibited brake moan while isolated from the vehicle as well as other complicating factors such as the hydraulic system and the wheel and tire.

In most cases, data required for input to the model was available from vendor specifications, such as compressibility of the brake pads. In other cases, additional tests were required to define the required input data, such as obtaining the correct stiffness for the caliper pin bushings and axle bearing. The table below summarizes the model configuration used in this study.

Table 1: Model configuration summary

MODEL FEATURE	DESCRIPTION
Components	caliper, inner/outer pad, rotor, axle tube, axle shaft, axle bearing
Axle	½ axle –tube, shaft and bearing
Coefficient of friction	static (0.48), dynamic (0.32)
Pressure apply	ramp: 0 – 172.4 kPa in 0.200 sec
Rotor rotational speed	5 – 10 deg/sec (vehicle 0.3 km/h)
Assumptions	all rigid parts except axle tube and shaft
	cantilevered axle tube and shaft
	no tire/wheel
	no hydraulics
	thermal effects neglected (ambient temperature)

The final ADAMS model of the rear disc brake system had 92 degrees of freedom, and required only about 30 minutes of analysis time for 1 second of simulation time on a 200 MHz Pentium Pro workstation, even when the complex normal force and friction representation was included. This model was able to produce a brake moan vibration signature that closely matched experimental measurements. A representative time history and spectrum can be seen in Figures 4 and 5, below.

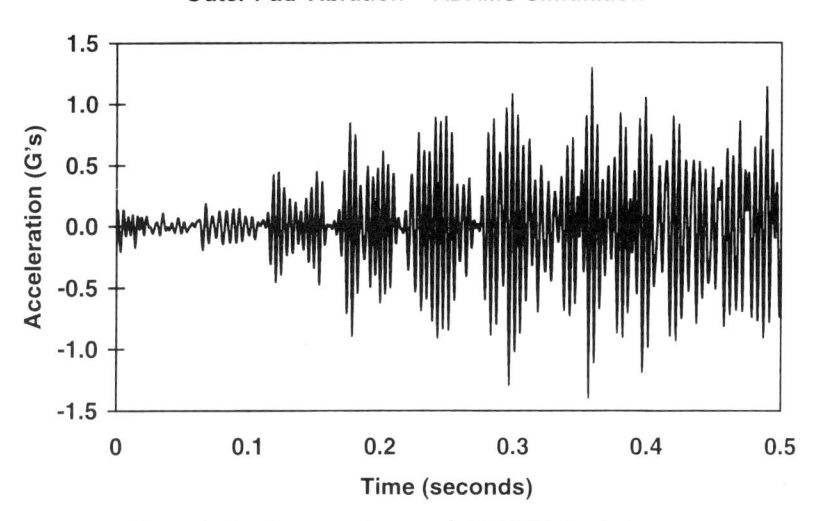

Figure 4: Brake moan signature, ADAMS simulation

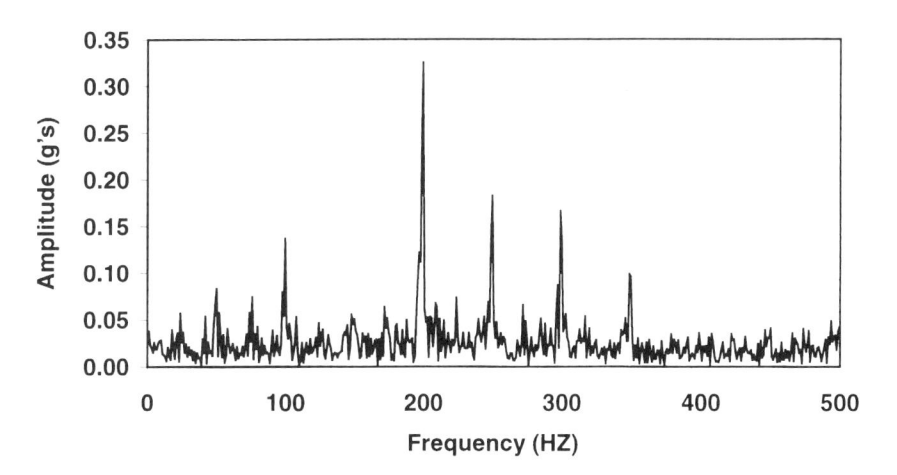

Figure 5: Brake moan spectrum, ADAMS simulation

Note that that the key features discussed in the "Brake Moan Overview" section, such as stick-slip pulses, a sustained moan vibration, primary frequency near 200 Hz, and modulation of the primary frequency at 48 Hz, are all present in the simulation results.

Model Configuration Study

A computer model provides the ability to measure data difficult to obtain experimentally, or to construct topologies not easily physically realizable yet greatly enlightening. The ADAMS model of the Bosch rear disc brake system was used to investigate the role played by the axle in the moan vibration. Four model configurations were examined: the full model, the model without the axle tube, without axle shaft, and without either the axle tube or shaft. The results of this study are shown below in Table 2.

Table 2: Model configuration study results

	Full Model	Without Axle Tube	Without Axle Shaft	Without Axle Tube and Shaft
Pulses	Yes	Yes	Yes	No
Moan signature	199-250 Hz	No	No	No
Moan modulation	48 Hz	No	No	No

The stick-slip pulses which precede and initiate sustained moan were present in all model configurations, except the model without the axle tube and shaft (i.e. the brake assembly only). However, the full model was the only configuration in which the stick-slip pulses actually excited the characteristic moan signature: a sustained vibration near 200 Hz and 48 Hz modulation of the primary frequency.

EXPERIMENTAL VERIFICATION

A comprehensive comparison of the simulation data to experimental data was performed to verify model behavior. The comparison centered on three areas: comparison of essential moan vibration signature features in the time history and spectrum, system mode shape(s) and frequencies, and parametric design study results. It is important to note that while the laboratory and in-vehicle experimental testing could indeed repeatably produce brake moan, the test-to-test variation was quite high. Consequently, all comparison of time history and spectral data is qualitative. The modal analysis comparison was quantitative.

Vibration Signature Comparison

The key features in the tangential vibration of the outer brake pad which characterize brake moan are discussed in detail in the "Brake Moan Overview" section. A summary of the comparison of typical simulation results to both vehicle and laboratory test data is presented in the table below:

Table 3: Vibration signature comparison

Feature	ADAMS Model	Vehicle Test	Laboratory Test
Stick-slip pulses present	Yes	Yes	Yes
Moan frequency (Hz)	199-250	190	195
Moan modulation (Hz)	48	50	50

Figures 1-2 and 4-5 illustrate typical simulation and laboratory time history and spectral plots. In addition to the vibration signature components noted in Table 3, vibration amplitudes calculated by the ADAMS model fall well within the confidence range of the experimental vibration amplitude measurements. The moan modulation roughly corresponds to the natural frequency of the axle assembly first bending mode shape.

Modal Analysis Comparison

Experimental modal analysis was performed on the rear disc brake system of interest, and compared to the results from the ADAMS model. To perform the eigen-analysis in ADAMS, the non-linear equations of motion were first linearized about the static equilibrium configuration of the model. Mode shapes calculated by ADAMS were similar to the mode shapes determined experimentally. These results are summarized in Table 4 below.

Table 4: Modal analysis comparison of ADAMS model to experimental data

Mode	Natural frequency, ADAMS model (Hz)	Natural frequency, experimental (Hz)
Axle shaft torsion	19	25
Axle shaft and tube first bending	54	55
Shaft second bending	185	187*
Axle tube torsion	199	205
Axle shaft third bending	275	unknown

* frequency measured, axle shaft shape undetermined

Design Studies

At each stage of model development, essential model structures and data, such as brake pad contact stiffness, were defined parametrically to facilitate later design studies. Once a model is defined parametrically, ADAMS is able to perform a series of simulations that sweep values through a defined range, and automatically determine parameter sensitivity. Five design studies were performed for model verification: an axle bearing stiffness, axle bearing clearance, axle tube stiffness, axle shaft stiffness, and a pad contact stiffness study. In each case, these parameters were varied over a range corresponding to physically realistic variations in componentry. The parameters were selected because design study experiments on the brake system hardware had shown them relevant to the moan characteristics. Table 5 below summarizes the design study comparisons.

Table 5: Design study correlation

Brake System Parameter	Moan Characteristic	Did Parameter affect Moan Characteristic?	Does this correspond to Experimental Testing?
Axle bearing stiffness	Frequency	No	Correlates
	Amplitude	Yes	Not Tested*
Axle bearing clearance	Frequency	Yes	Correlates
	Amplitude	Yes	Not Repeatable
Axle tube stiffness	Frequency	No	Correlates
	Amplitude	No	Correlates
Axle shaft stiffness	Frequency	No	Not Tested*
	Amplitude	Yes	Not Tested*
Pad contact stiffness	Frequency	No	Correlates
	Amplitude	Yes	Correlates

*While these quantities were not specifically tested, trends were observed to be consistent with model behavior

DISCUSSION

Careful examination of the simulation results suggests that the ADAMS model of the Bosch rear disc brake system accurately portrays the system's real-world brake moan behavior. The model correlated closely with both vehicle and laboratory experiments, and exhibited characteristic vibration frequency and amplitude, as well as critical moan signature features such as stick-slip pulsation and modulation.

Differences in the moan signature between test and analysis can be attributed to rotor rotational speed and/or brake pressure variations. While ADAMS provides constant operating parameters in the modeling environment, actual tests do not. Actual vehicle moan is sometimes difficult to achieve due to the narrow range of vehicle speed and braking pressure required to sustain moan. Oxidation also factors into whether or not moan will occur. Frictional coefficients reflecting oxidized contact surfaces were used in the ADAMS model.

This study emphasized the importance of modeling from a systems rather than a component focus, as the moan signature did not occur with the brake assembly only, but required the presence of axle tube and shaft. The axle tube properties, as well as the brake pad contact stiffness were found to significantly influence moan. These may warrant further investigation as potential candidates for design changes that may alleviate or completely mitigate brake moan in this rear disc brake system.

This approach to the investigation of rear disc brake moan was novel, yet the resulting simulation model did not need to be as complex as originally expected in order to achieve useful results. While most of the mechanical system data required for populating the model was readily available, in some cases, special tests were devised to measure component mechanical characteristics such as stiffness and damping which were not otherwise available.

A validated brake system model provides the capability to predict moan characteristics while designing new products. This "virtual prototyping" enables rapid and inexpensive evaluation of many thousands of candidate designs, and facilitates selection of an optimal design much better than relying on traditional hardware build-and-test methodologies.

REFERENCES

(1) An Overview of How to Use ADAMS/Solver version 8.0. Mechanical Dynamics, Inc., Ann Arbor, Michigan, USA. 1994.

Brake vibration and disc thickness variation (DTV)

K VIKULOV, B TRON, and **P BUONFICO**
ITT Automotive Italy SpA, Barge, Italy

SYNOPSIS

The problem of low frequency vibrations during braking (judder) originates from disk thickness variation (DTV). DTV is created either by non uniform wear of the disk surfaces ("cold" judder) or by deposit of friction material on the disk surface ("hot" judder). Such parameters like disk composition, calliper design and friction material formulation and properties show significant impact on DTV generation.

1. INTRODUCTION

ITT Automotive Italy S.p.A. is a company producing friction material for passenger cars and small commercial vehicles. It is successfully operating on the international market supplying brake pads to FIAT, Ford, Volkswagen, BMW and other European car producers. The history of the company starts more than 50 years ago. Born as Galfer in 1950, it was first a small factory located in Turin, supplying asbestos-based friction material mainly for FIAT cars. In 1975 Galfer has been purchased by ITT Corporation. At that time first asbestos-free materials have been developed and introduced in production. The operative company ITT Automotive was founded in 1991, and Galfer became a part of it as ITT Automotive Italy.

Today ITT Automotive Italy has two production units in Italy situated in Barge and in Termoli. The company has more than 500 employees, about 70 working in its R&D department. The successful growth of the company over the last decade contributed to large investments made in R&D dedicated to the development of friction material for new applications. There have been important changes introduced in the car brake system during last 10 years: asbestos-free formulations, increasing power of engines coupled with decreasing weight of the car, light floating callipers, ventilated disks, ABS etc. All these changes were followed by more and more severe requirements of the car manufacturers for the life-time, safety and comfort of braking system. The competition between friction material producers is mainly concentrated in the formulation and technology of new materials, satisfying most severe requests of the market. Basic research becomes essential in understanding of phenomena of wear, friction stability, noise and vibration generation and in finding out new criteria and testing procedures for material selection.

In this paper the origin of brake vibration called "judder" will be discussed together with the description of the testing procedures applied in ITT Automotive Italy for selection of new materials.

2. THE ORIGIN OF LOW-FREQUENCY VIBRATIONS (JUDDER)

Judder may be defined as vibrations with frequencies which are proportional to the turnover frequency of the wheel and consequently proportional to the speed of the car during braking. In most cases these vibrations have frequencies between 5 and 50 Hz. The lowest frequency is equal to the turnover frequency of the wheel, and others being its harmonics. Typical set is shown in Fig.1. The steering wheel and the brake pedal are those elements of the car where the driver feels the effect. In some cases one can hear the so-called acoustic judder or sound produced by vibrations of the body of the car. In all the cases the origin of judder has to be found in the interaction between the surface of the brake disk and the friction material.

When perception judder vibrations occurs through the brake pedal, their origin is due to the variation of the pressure of brake liquid produced by the vibrations of the piston of the calliper in contact with the brake pad. Consequently, this type of vibrations may be described as periodic displacement of the brake pad in the direction normal to the disk surface. The amplitude of the brake pressure variation (BPV) is the measure of the effect. Two principal mechanisms of BPV may be considered. First is due to the mechanical defect known as "run-out" of the brake disk which can be considerably but not totally reduced by a proper check of the vertical alignment of the brake disk during assembling. The second mechanism is due to the non uniform wear of the disk surface discussed below.

The vibrations of the steering wheel during braking are associated with the displacement of the calliper in the direction tangential to the disk surface. These vibrations are due to the brake torque variation (BTV) over the working surface of the disk. The BTV and BPV phenomena are coupled because the friction force is proportional to the normal pressure. The relation between the amplitudes of BPV and BTV depends on the parameters of the car. In some specific cases strong BTV can be observed without significant BPV because friction coefficient is not uniform over the disk working surface and thus BTV is sensitive to the composition and structure of friction interface.

It often happens that strong judder is observed after the brake disk has been heated up above 500°C (fading). Such a treatment leads to the formation of a deposit of friction material on the disk surface or so-called "hot spots". These deposits modify friction coefficient and thus cause BTV. At the same time, BPV can be observed when the thickness of such deposit becomes high (above 10 micron). This specific vibration is called "hot" judder, indicating that it is produced after a heat treatment of the disk surface in contact with friction material. It may be improved by an assessment of the friction interface at lower temperatures when deposits are removed from the surface.

In most of the applications judder is progressively increasing with the wear of the disk. It happens that two surfaces of the disk become not perfectly planar and parallel due to non uniform wear. This is expressed as disk thickness variation (DTV) over the working surface. The empirical rule shows that disks with DTV above 20 micron exhibit noticeable vibrations (BPV and BTV).

3. DTV GENERATION

The tolerance limit for the initial DTV of a new disk is usually below 5 micron. The ideal brake system should operate in a way that DTV is not increasing. In practice, DTV generated at the end of the life-time of a disk should not exceed 20 micron. Considering that the average life-time of a brake pad is approximately one half of that of a disk, above limit is even lower.

The demand for friction material that does not generate DTV is continuously increasing on the market. ITT Automotive Italy has started research in that field with development of experimental procedures for detection and monitoring of DTV generation on a dynamometer. The task was to develop discriminative testing procedures for selection of friction material that does not create DTV.

The first fundamental measurement is disk profile geometry in radial and circular directions. The design of the apparatus is shown in Fig.2. It consists of two micrometers (upper and lower) places on a hand, allowed to move in horizontal plane, in a way that two micrometers touch the two surfaces of the disk, the measurement giving the radial profile. The disk in turn can rotate around its axis from 0 to 360° with two micrometers fixes at certain radius. This gives the circular profile. After the disk is placed in a horizontal plane and reference point is selected, the apparatus automatically obtains data of the disk surface planarity, parallelism (DTV) and radial profile.

Selection of friction material is based on the maximum value of DTV generated during a standard test on a dynamometer. The testing procedure consists of up to 20 identical cycles, each being a sequence of 200 braking at different initial temperatures, speeds and pressures. All the test program is controlled by a PC, collecting experimental data with frequencies up to 5000 Hz. Thus, BTV and BPV measurements are allowed with high time resolution. Typical data for a material generating DTV are presented in Fig.3. Each window shows brake pressure (Bar), brake torque (dN·m) and speed (km/h) registered for one single braking. It can be seen that the intensity of BPV and BTV is growing with the number of braking progressively. The material used in this test has developed maximum DTV of 54 micron after 2800 braking. Figure 4 shows the results obtained with a material that does not create DTV. In this case DTV as low as 7 micron was measured after the test.

4. DISCUSSION

The physical process that contributes to DTV generation is poorly understood. The following problems should be considered: disk composition, performance of the caliper and properties of the friction material. It has been noticed that wear resistance of grey cast iron strongly differs with its composition and structure (1). Our experimental evidence shows that applications, where the disk material has higher values of HB hardness show less DTV generation. It is difficult to investigate the influence of the calliper design because this is in most cases a fixed parameter for a friction material manufacturer as we are. However, it should be noticed that fixed (not floating) calliper shows better performance. Composition and physical properties of friction material are of primary importance for the problem, being the essential part of the "know-how" principle.

5. LITERATURE

1. Y. C. Liu, J. M. Schissler, T. G. Mathia. The influence of surface oxidation on the wear resistance of cast iron. *Tribol. Intern.*, 28 (1995) 433-438.

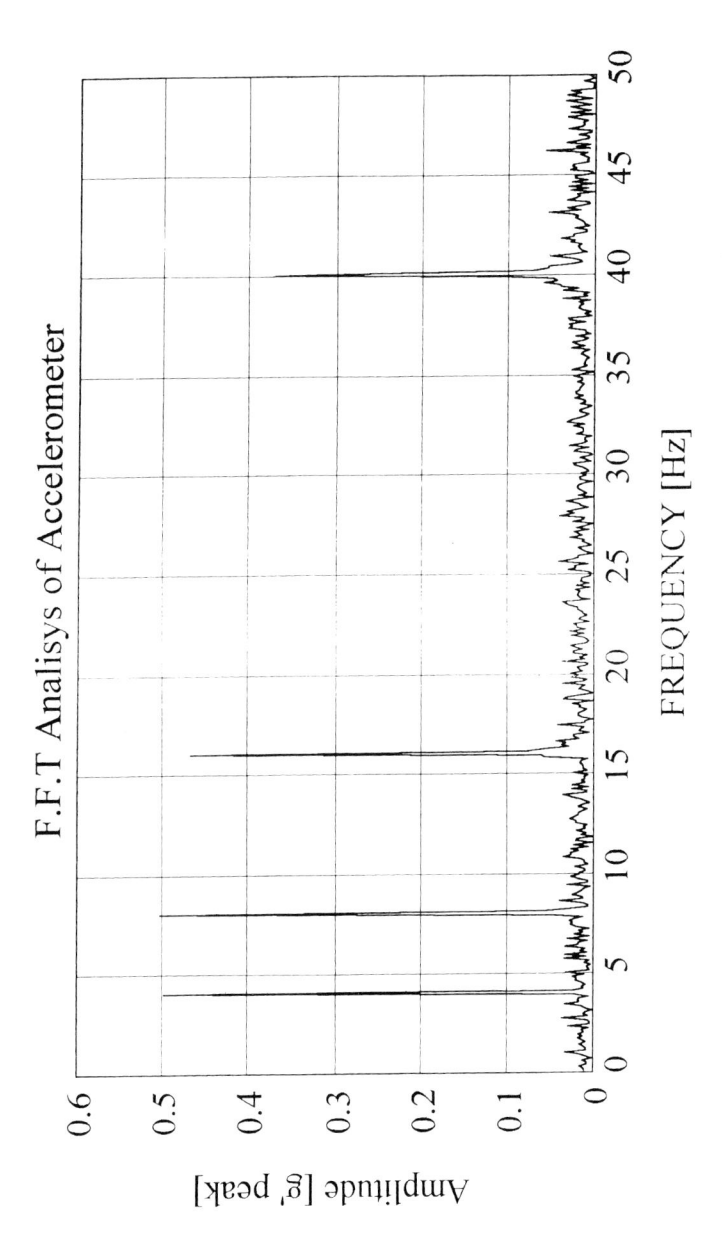

Fig. 1 FFT Spectrum of Judder Vibration

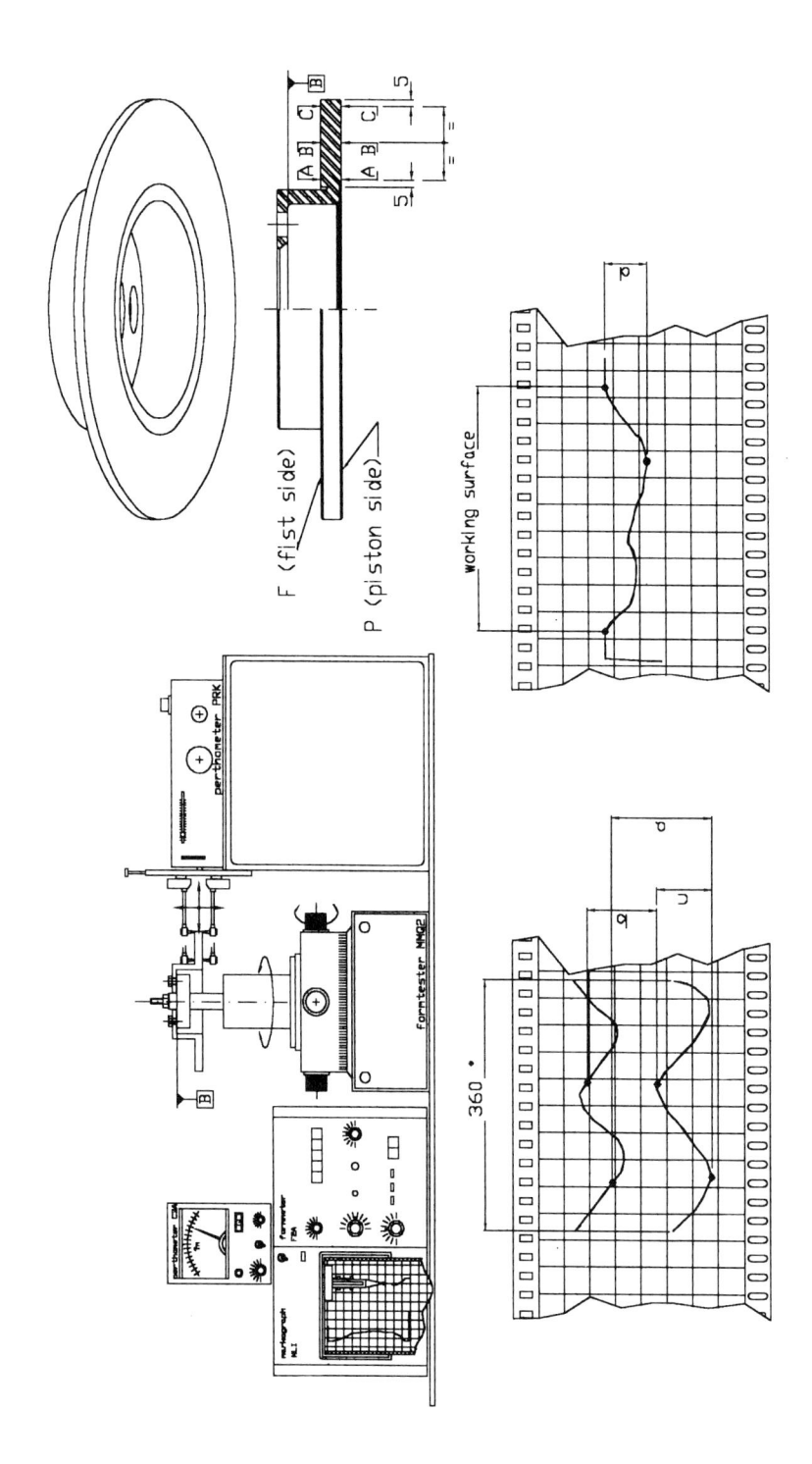

Fig. 2 Disc profile measurement. Apparatus design and operating principle.

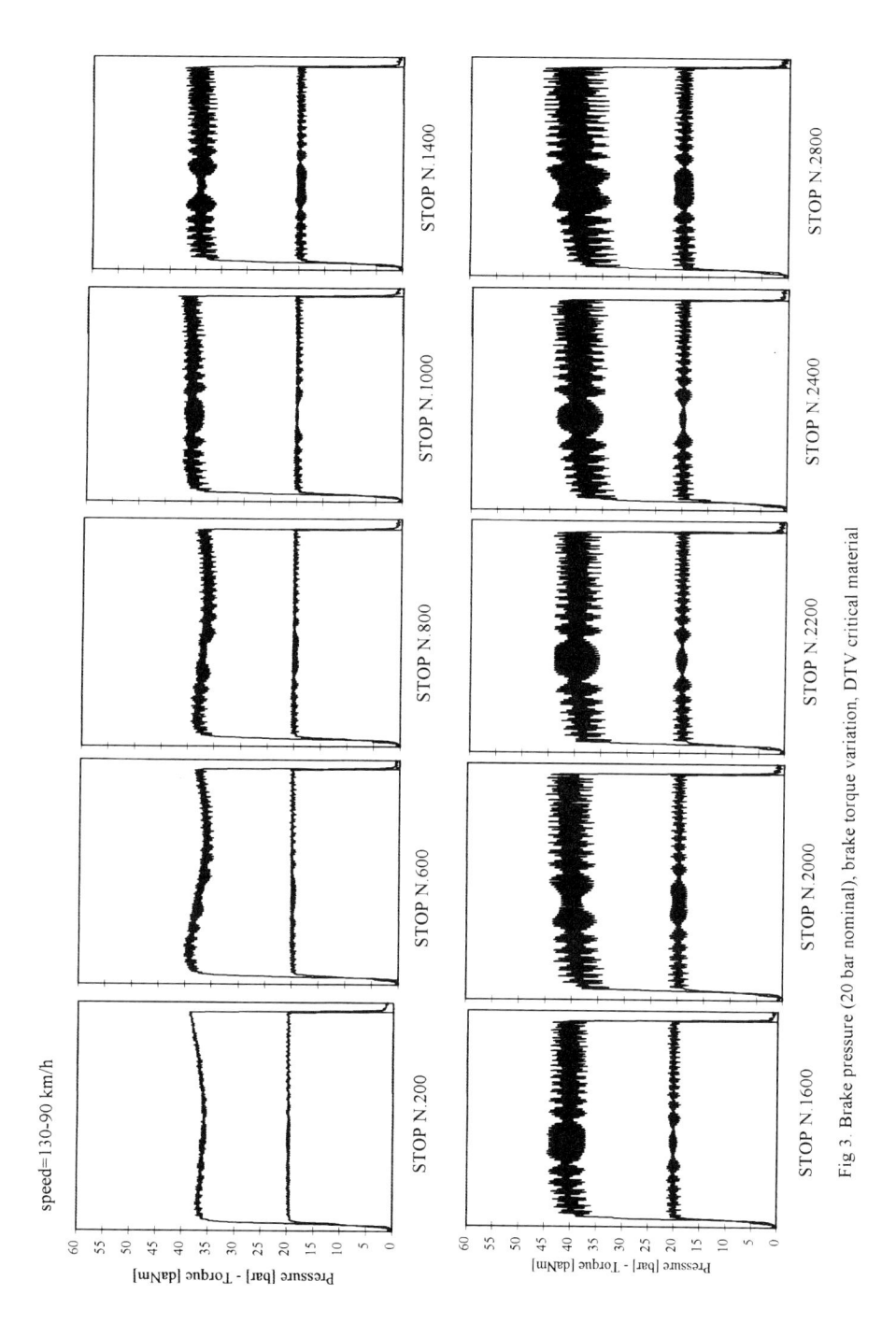

Fig 3 Brake pressure (20 bar nominal), brake torque variation, DTV critical material

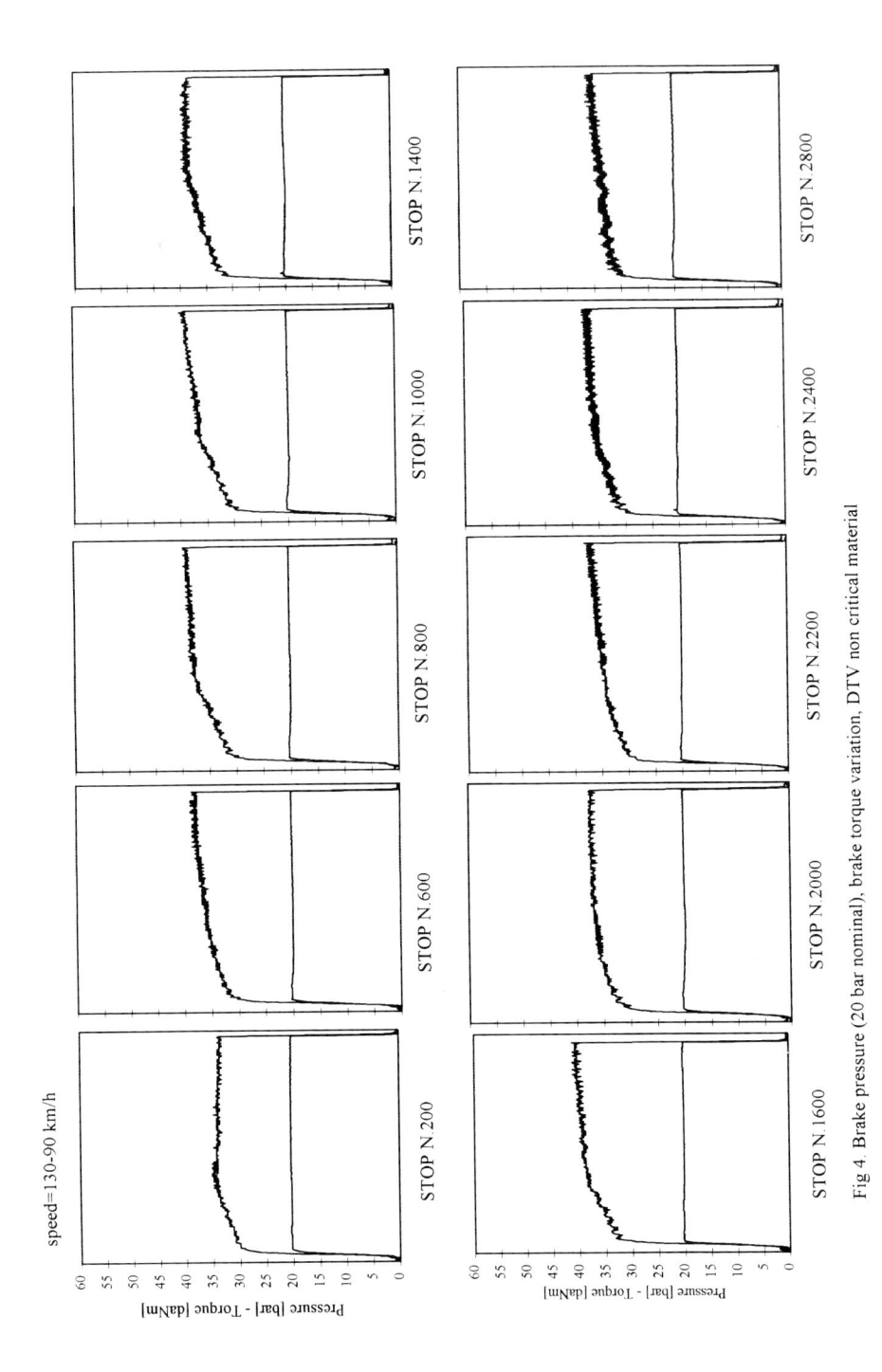

Fig 4. Brake pressure (20 bar nominal), brake torque variation, DTV non critical material

Testing and Miscellaneous Topics

Prediction of whole vehicle braking performance using dual dynamometer differential effectiveness analysis (D^{3}EA^{TM})

C W GREENING Jr
Greening Test Laboratories, Detroit, Michigan, USA

1 ABSTRACT

A new certification procedure has been developed for passenger car and light truck brake friction materials. This technology provides utility for both original equipment and replacement friction product providers. The program utilizes a specially designed dual-ended inertia brake dynamometer to evaluate brake performance, and a computer software program and analysis technology to establish vehicle specific acceptance criteria.

2 BACKGROUND

For years, friction material manufacturers around the world have explored methods to predict the performance of a friction material in a specific vehicle braking system. While varying degrees of success have been achieved, there has not been a reliable process nor performance criterion by which to objectively gauge the test result. Because of increasing economic pressures and governmental regulations on vehicle manufacturers, it is now more important than ever for friction producers to understand the functional demands of current production vehicle brake systems and the resultant requirements of brake friction materials.

Assessing the performance of brake friction materials in vehicles is a complex process that includes vehicle platform specific features and interactions with brake system characteristics. Important attributes that need be considered in the overall evaluation of brake friction material suitability include stopping distance, vehicle brake balance, thermal conductivity, system pedal force gain, compressive compliance, failed system performance, and wear. Competing economic and performance criteria and their subsequent compromises are manifested in the resulting friction products.

Each year in the United States, approximately 80 million axle sets of disc brake pads and drum brake linings are installed on in-service motor vehicles to replace worn or damaged parts. Testing performed by the U.S. National Highway Traffic Safety Administration (NHTSA) and others have shown that these replacement parts in many cases deviate substantially in output relative to the original equipment materials originally installed on these vehicles. This deviation in turn can have significant adverse effects on stopping distances, brake balance, thermal performance, interactions with anti-lock braking system algorithms, etc. Given the direct relationship between braking performance and crash avoidance, it is desirable that replacement brake friction materials be capable of restoring a vehicle to an in-compliance condition with the brake standard (FMVSS 105, 121, 122, 135, ECE R13H) to which the vehicle was originally certified. It is virtually impossible, however, to assess the efficacy of brake friction material performance independent of specific vehicle characteristics. A brake friction material may be acceptable in one vehicle model application while being unacceptable in another due to differences in vehicle design characteristics such as weight, static and dynamic weight distributions, foundation brake sizes and hardware designs, system pedal force gain, etc. The inability to meaningfully quantify friction material performance independent of these vehicle characteristics has been a major impediment to the development of industry or regulatory standards for replacement brake friction materials.

As a result, consumers have no objective way to determine which replacements will provide suitable performance on their vehicle. Most consumers therefore select replacement linings on the basis of cost, which does not necessarily correlate with safety performance. Further, most consumers seem to make the assumption that replacement brake friction materials are currently regulated for safety performance and therefore can be reasonably selected on a cost basis. There are no federal standards for performance of replacement brake friction materials in the U.S., but several states require Automotive Manufacturers Equipment Compliance Agency (AMECA) edge code labeling based upon Vehicle Equipment Safety Commission (VESC) Regulation V-3, which in turn is based on Society of Automotive Engineers (SAE) J661 and SAE J866. It has been documented by the SAE and others that the current edge code labeling criteria do not correlate with whole vehicle performance, and thus is not a credible means of selecting replacement friction products. The lack of an existing, independent, credible technical authority within the U.S. friction material industry has hampered the implementation of any objective performance requirements for replacement brake friction materials. The NHTSA has been working for more than 25 years to develop a reliable test and meaningful, vehicle relevant performance criteria for replacement friction products in the U.S. The rule making process in the U.S. is far more complex and difficult than that utilized in Europe, and the self certification style of regulation has further confounded the development of a replacement standard.

3 DUAL DYNAMOMETER DIFFERENTIAL EFFECTIVENESS ANALYSIS

An obvious way to determine whether a replacement friction material is capable of restoring a vehicle to an in-compliance condition is to install that material in the actual vehicle and conduct a certification test to the applicable brake standard. While this approach would provide definitive information as to the suitability of a particular material in a particular vehicle model, it has two shortcomings. First, it is very expensive to conduct whole vehicle certification tests, and therefore impractical for the replacement or aftermarket industry. Secondly, it is not practical for friction material manufacturers to obtain an original equipment level vehicle in the years after that vehicle has gone out of production (when sales of replacement parts are increasing) in order to conduct regulation-based vehicle brake performance testing.

Dual Dynamometer Differential Effectiveness Analysis (D³EA™) is a proprietary laboratory procedure for determining whether or not a replacement brake friction material is capable of restoring a vehicle to an in-compliance condition with the brake standard to which the vehicle was originally certified. D³EA™ accommodates the necessity to evaluate replacement friction materials in the context of the vehicle specific design characteristics, while avoiding the practicability problems associated with whole vehicle testing. This validation process for evaluating replacement brake friction materials consists of two parts; a dual-ended inertia brake dynamometer-based laboratory test, and a Differential Effectiveness Analysis-based computer model for establishing vehicle specific acceptance criteria.

In highly simplified terms, the dynamometer procedure is conducted by mounting brake friction materials and associated disc or drum brake hardware (corresponding to the front and rear brakes of the vehicle) on a dual-ended inertia brake dynamometer equipped with specially designed adaptive test control and advanced data acquisition capability. The brakes are then exercised through a series of stops and snubs. The stop and snub sequence, speeds, apply rates, loads, timing intervals, etc., are determined and carefully controlled to emulate the critical sequences of the FMVSS 105 and FMVSS 135/ECE R13H road test procedures, and are also based upon the vehicle specific design characteristics (weight, static and dynamic weight distributions, foundation brake sizes, tire rolling radii, front and rear brake cooling rates, etc.) for the materials/vehicle combination being evaluated. The test sequence is outlined in Table 1.

Table 1
D³EA™ TEST SEQUENCE

STAGE	SOURCE	PURPOSE
Instrument Check	FMVSS 105	Test set up
Green Ramps	FMVSS 105	Effectiveness
Burnish	FMVSS 135	Conditioning
Burnished Ramps	FMVSS 135	Effectiveness
Fade Heating Cycle	FMVSS 135	Conditioning
Hot Ramps	FMVSS 135	Effectiveness
Cooling	FMVSS 135	Conditioning
Recovery Ramps	FMVSS 135	Effectiveness
Reburnish	FMVSS 105	Conditioning
Post Fade Ramps	FMVSS 105 & 135	Effectiveness

The test sequence was designed specifically to condition the brakes as little as possible at the points of effectiveness evaluation. The greatest amounts of work are done during sections of the test chosen to either cause change or stabilize the friction material environment. Effectiveness determinations ("ramp apply" cycles, described below) are conducted immediately following the conditioning cycles and contribute very little to the overall conditioning of the brakes. Figure 1 shows the energy distribution during a typical D³EA™ test on a passenger car.

Energy Distribution In D³EA™ Test

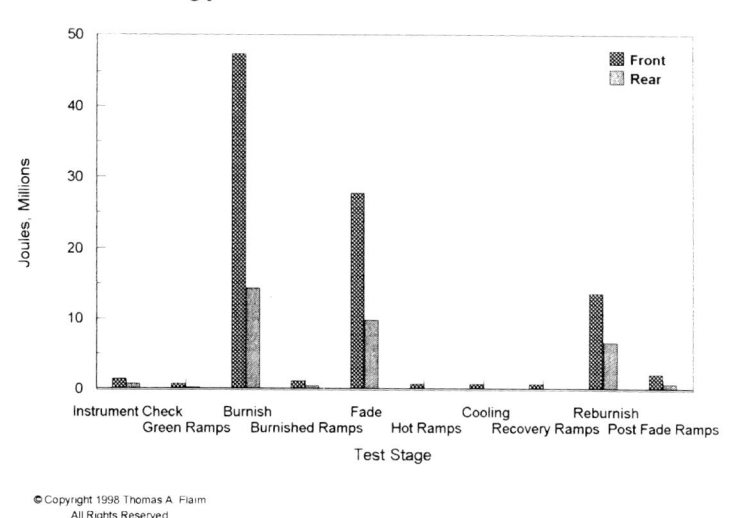

Figure 1

Most of the work is done during the Burnish stage, followed in magnitude by the Fade and Reburnish. The amount of work done by the brakes at the Effectiveness stages (ramp apply cycles) is very small by comparison. It is important not to change the condition of the brakes during the effectiveness measurement in order to accurately determine their performance at that point in time. Over the vehicle platforms studied, typically more than 92% of the work done by either the front or rear brake during a D³EA™ test occurs during the conditioning sections of the test sequence.

4 EFFECTIVENESS METRIC

To measure the effectiveness of the brakes as they progress through the test sequence, "ramp apply" cycles are used. These cycles consist of applying pressure to the brakes at a constant rate until a predetermined maximum output torque (at either brake) is achieved. This maximum torque level is based upon the limit of the tire to road adhesion for the vehicle system under test. At this point, the brakes are released. The output torque of each brake is then plotted versus its respective input pressure, the resulting curve regressed, and the slopes of the resulting best fit straight lines expressed as "regressed specific torque". These key performance data obtained from the dynamometer testing (represented in units of N·m/kPa) is precisely identical to the Objective Brake Factor and vehicle balance requirements measured for burnished brakes in FMVSS 135 and ECE R13H. The vehicle test values for objective brake factors are determined using torque wheels in ramp apply brake applications. Based on the statistical reliability of this whole vehicle effectiveness metric, a similar parameter for effectiveness was adopted in the D³EA™ test. The effectiveness of the brakes at both ends of the dynamometer is evaluated at each phase of the test sequence. This data may then be compared to the effectiveness limits applicable to that particular test phase.

5 WHY USE A DUAL-ENDED DYNAMOMETER?

Historically, the brake industry has relied upon comparative testing to develop replacement brake friction products. Tests in many cases are conducted in small sample test machines that do not closely reproduce the conditions under which the brakes operate while in service. Indeed, even most single-end inertia brake dynamometer test procedures do not fully utilize available technologies in trying to make a reasonable representation of in-service conditions. Credible attempts have been made to develop vehicle relevant test protocols, but have fallen short because of limiting parameters inherent in single-end dynamometers. One such parameter is the test inertia. Calculating the test inertia based upon known vehicle dimensions such as weight on a particular axle and tire rolling radius makes the assumption that the aforementioned variables are constant throughout all braking events and conditions. Clearly, this is not the case. Figure 2 demonstrates the amount of variability that can be present. The graph shows the total test inertia (for one front and one rear brake, representing one-half of a vehicle) to be constant, but the ratio of the amount of inertia (or work) absorbed by the front brake to that absorbed by the rear brake to be highly variable.

Brake Inertia Ratio During D³EA™ Test
Front Material: O.E. / Rear Material: O.E.

Figure 2

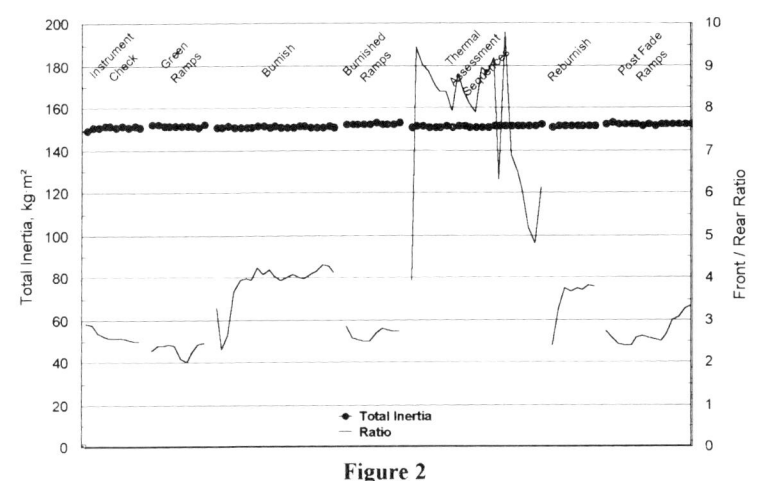

It could be possible to obtain a vehicle where the amount of inertia at the front and rear brakes does not vary as widely as shown in Figure 2. Good correlation between the output of one vehicle brake and a dynamometer test result might be obtainable in such a case. But as soon as the friction material on either axle was changed, the presumption of any "correlation" would be proven false. Figure 3 depicts the same vehicle as Figure 2, but with a different front disc brake friction material. Note that while the total inertia value is the same as shown in Figure 2, the inertia ratios at the individual test stages are very different. Also, the inertia ratio variance (within a test) for the two material combinations is quite different from one material combination to the other.

Brake Inertia Ratio During D³EA™ Test
Front Material: "C" / Rear Material: O.E.

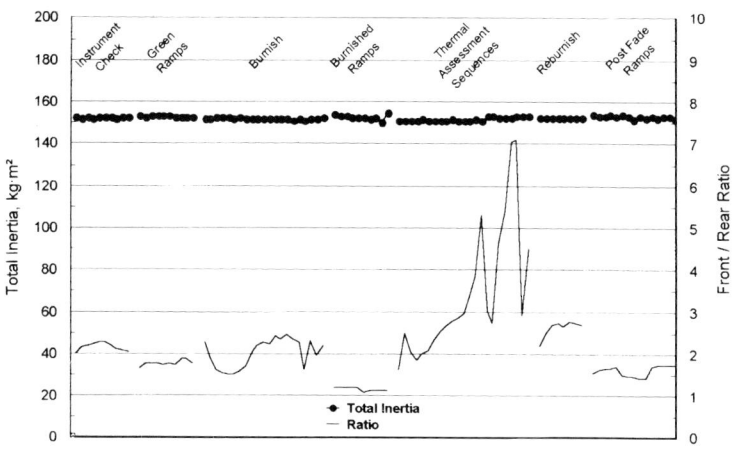

Figure 3

It is for these reasons (amongst others) that the ability to correlate in-service experience with laboratory test results has been lacking. The use of a dual-ended inertia brake dynamometer has proven essential in retaining some of the important aspects of vehicle braking environment and therefore enhancing laboratory to vehicle correlation.

The graphs below help to further illustrate this point. Figures 4 and 5 show the rear fraction of work done by an original equipment rear drum brake friction material when paired with front disc brake materials "A" and "B". Notice how the portion of the rear brake contribution to the overall braking output (during a ramp apply cycle) varies within each singular braking event. The difference in the magnitude of the rear fraction at the two test stages shown (Burnished Effectiveness and Hot Ramps) is very evident, with variation exceeding 50%. Additionally, note the effect the change from front disc brake pad "A" to pad "B" has on the rear brake output contribution fraction.

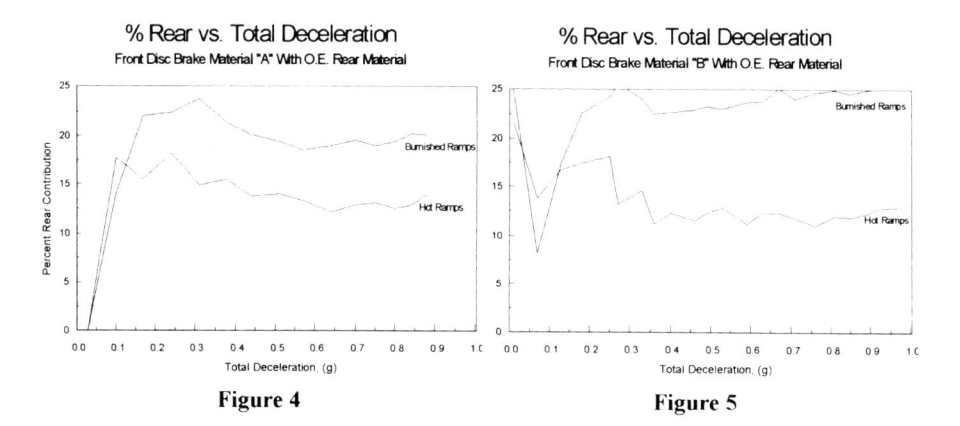

Figure 4

Figure 5

The ability for the brakes to interact and instantaneously and continuously redistribute their work fractions during a brake engagement is an extremely important element in establishing the levels of effectiveness, thermal stability, and durability of the friction materials during subsequent brake applications. Figure 6 demonstrates the "work sharing" that occurs in vehicle brake systems. Data for the same disc brake materials "A" and "B" is shown during a Fade sequence. Observe not only how the amounts of rear brake work contributed to the total braking effort vary over the sequence, but also the effect the different disc brake pad materials have on the same rear brake. The accurate emulation of the working conditions for each brake and preservation of their mutually dependent interactions require both front and rear brakes be tested simultaneously, justifying the need for a dual-ended dynamometer.

% Rear Work During D³EA™ Fade Sequences
"A" vs. "B" Front Disc Brake Materials (Both With O.E. Rear Material)

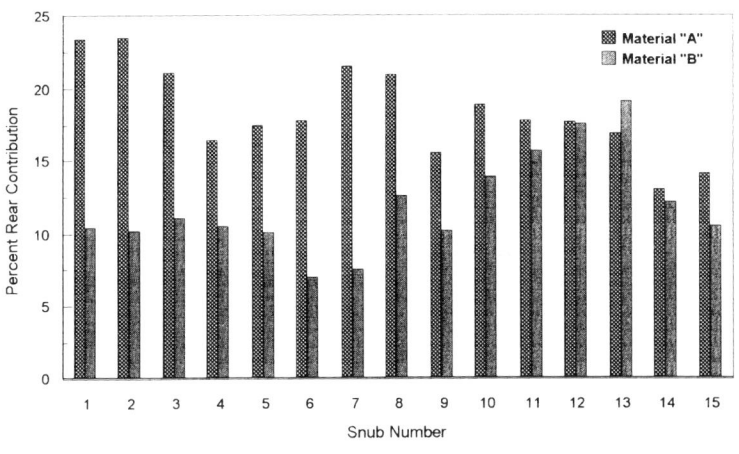

Figure 6

6 ACCEPTANCE CRITERIA IN D³EA™

A test procedure alone does not provide a sufficient basis for performance standards. Credible, vehicle relevant acceptance limits are required to determine whether the tested material passes or fails the vehicle specific requirements. Two possible approaches to this question of acceptance limits have been identified; 1) a fixed percentage deviation from original equipment, or 2) Differential Effectiveness Analysis (DEA™). The first, a fixed percentage deviation from original equipment, similar to that adopted within ECE R90 (+/-15% for disc brakes, +/-20% for drum brakes), has two basic problems. First, this approach assumes that the original equipment material provides the best compromise of vehicle performance relative to the regulatory requirements, and secondly, that the original equipment system is "centered" within an acceptance window discussed below. The second basic problem with the fixed percentage deviation from original equipment approach is that a confirming whole vehicle test is required to avoid the consequences of pairing adverse combinations of front and rear brake lining materials. It is this need for a confirmatory whole vehicle test at the replacement lining stage that makes this approach to acceptance limits distasteful to the replacement supplier.

To overcome objections to the fixed deviation from original equipment approach, Dr. Thomas A. Flaim has developed technology called Differential Effectiveness Analysis (DEA™). DEA™ is a package of computer programs and analytical methodology that translate into acceptance criteria for regressed specific torque for each specific vehicle application at each stage of the laboratory test procedure. DEA™ translates the full system performance requirements of FMVSS 105/135 or ECE R13H (separate programs and windows have been developed for each) and the design characteristics of the vehicle in question into an "acceptance window" or "effectiveness space" for regressed specific torque at the front and rear brakes. Data generated in the D³EA™ test is plotted on a graph with front regressed specific torque as its ordinate and rear brake regressed specific torque as its abscissa. The window consists of four lines, combining the stopping distance and mean fully developed deceleration equivalent effectiveness requirements of various stages of the applicable whole vehicle braking standard, including fully functional, partial failure, high temperature and other whole vehicle brake performance assessments. These boundaries thus identify the allowable ranges of front and rear brake effectiveness that will render a vehicle in compliance with the performance requirements of its respective braking standard. Figure 7 depicts the basic and competing aspects of FMVSS 135 for an exemplar vehicle.

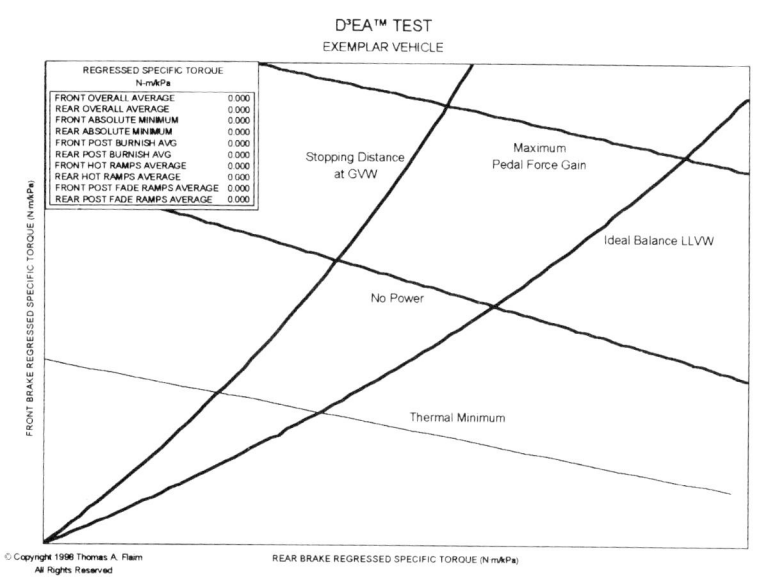

Figure 7 - D³EA™ Window

If the friction material combination (front and rear) has effectiveness values that fall within the orthogonal shaped acceptance window, the materials are capable of restoring an otherwise original equipment vehicle to an in-compliance condition. If the measured performance is not within the acceptance window, it is probable that the vehicle would not comply with the whole vehicle braking standard when fitted with that combination of friction materials.

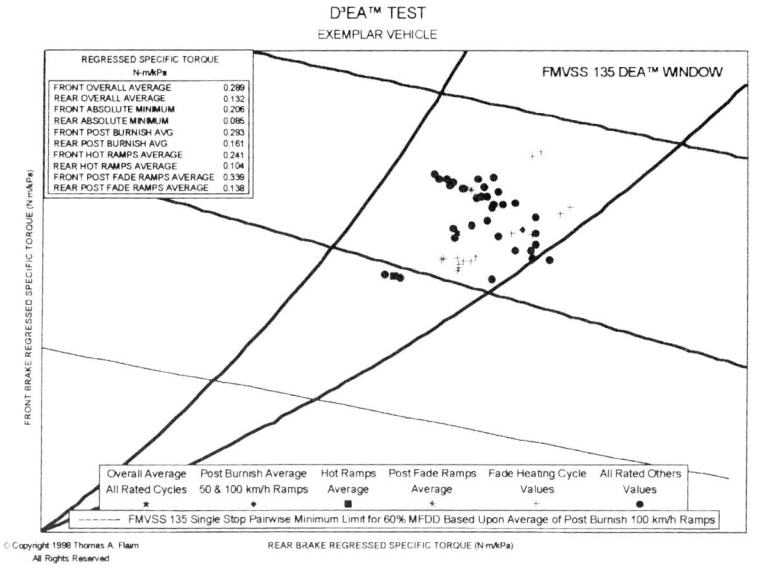

Figure 8 - D³EA™ Test Result

Figure 8 shows a test result for a full size, six passenger sedan. Selected data (such as Post Burnish Effectiveness) are averaged and plotted along with the DEA™ Window developed for the subject vehicle platform. Values for the selected test stages must lie within the window in order for the test to be judged as having an "acceptable" test result. Certification is conferred only upon materials demonstrating acceptable results in multiple test runs.

7 HYDRAULIC DISPLACEMENT

In the most recent NHTSA study of replacement brake friction materials, the hydraulic displacement of vehicle braking systems was identified as one of the two critical characteristics that must be evaluated in any objective assessment leading to a certification of consistent with Federal Safety Standards. NHTSA reported that while many replacement brake friction materials deviated significantly from OE in terms of effectiveness, they also found several that produced a displacement limited operating condition during their whole vehicle assessments.

The compressive compliance of brake friction materials can be a critical characteristic of vehicle brake systems. The hydraulic displacement capacity of vehicle brake systems is a complex function of both regulatory and consumer requirements. The hydraulic compressive compliance of brake friction materials is a nonlinear function of vehicle specific hardware characteristics and operating conditions. The interaction between material effectiveness and brake system operating pressures is a very complex interactive function of both vehicle parameters and hardware performance characteristics. Excessive hydraulic displacement due to friction material compressive compliance can produce a displacement limited operating condition in vehicles. Development of such displacement-limited conditions can have an adverse effect upon service brake pedal force gain, vehicle brake balance, stopping distance, and vehicle stability.

Objective measurements of hydraulic displacement requirements of both front and rear brake friction materials in D³EA™ tests have been implemented. Objective criteria for displacement limits are being developed for each of the platform specific D³EA™ test procedures. It is anticipated that this measurement assessment technology and criteria will be completed during 1998.

8 LABORATORY TO VEHICLE CORRELATION

Although D³EA™ is highly complex and proprietary, extensive documentation is available which demonstrates the statistical reliability and validity of its ability to predict the in-vehicle performance of brake friction materials. A total of 31 platform specific D³EA™ test configurations have been developed and validated. Across this range of vehicle configurations, current laboratory to vehicle correlation studies show a statistical reliability of from 3 to 7%, depending upon the vehicle platform and brake type, and statistical validity of less than 5% of mean. Comparable levels for whole vehicle tests are 4 to 12%. Thus, under rigidly controlled laboratory test conditions, higher levels of validity and reliability are achieved as may be expected. An example of the accuracy is shown in Figure 9. Data gathered from the Post Burnish Effectiveness stage of the D³EA™ test has been plotted against data gathered from a variety of whole vehicle tests at the same test stage.

D³EA™ to Vehicle Correlation
$R^2 = 0.993$ $S = 0.043$ With +/- 2 Sigma Limits

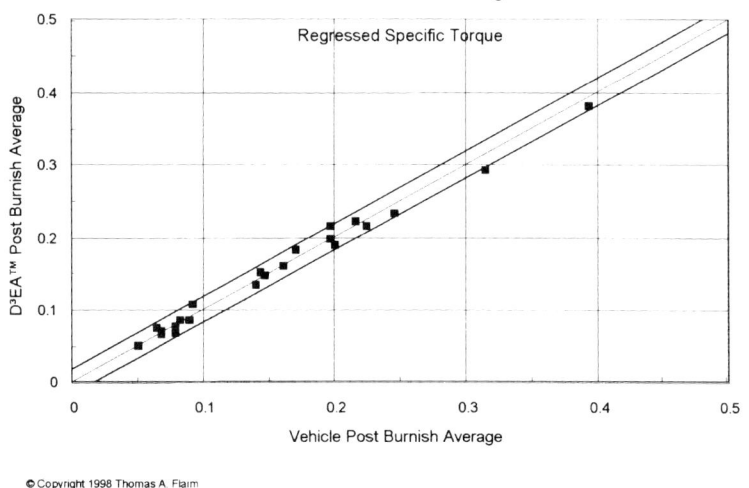

Figure 9 - Test Correlation

9 D³EA™ and ECE R90

Since D³EA™ has a substantial potential as an economically viable alternative to whole vehicle testing in the replacement market, it may be appropriate to offer comparison to replacement brake lining regulation ECE R90, enacted in January 1998.

First, the overall objective of ECE R90 is that replacement brake friction materials should meet the new vehicle performance requirements of ECE R13. In order to demonstrate this level of performance, a whole vehicle type approval test is required with the replacement brake friction material fitted to its axle and with original equipment material at the other axle. In addition, a number of laboratory tests with comparative requirements using the original equipment materials as the baseline are also required.

In particular, a whole vehicle test is required for each replacement brake friction material and for each vehicle type to which it will be fitted. In addition, replacement brake friction material suppliers are required to demonstrate conformity of production annually with a specific combination of vehicle and laboratory tests utilized. Among the laboratory test requirements included are normal and hot effectiveness as well as more extreme speed sensitivity evaluations. In each of these lab tests, requirements based upon a fixed percentage deviation from original equipment brake lining performance are applied. However, the original equipment brake friction materials are exempt from the requirements of ECE R90.

To support the identification of replacement brake friction materials meeting minimum objective performance criteria based upon FMVSS 105 or FMVSS 135, as applicable to a specific vehicle platform, a D³EA™ Certified Seal has been developed, trademarked, and applied to approved packaging. The form and content of the D³EA™ Certified Seal or approval mark was reviewed with the NHTSA prior to market introduction in 1996.

In Table 2, a general comparison highlighting features of D³EA™ and ECE R90 is provided.

Table 2
A COMPARISON OF ECE R90 AND D³EA™ CERTIFICATION

FEATURE	ECE R90	D³EA™
Scope	All New Vehicles - 1998 →	31 Platforms - 1990 →
Current Coverage	None	75 - 80% of Disc Pads
Obligation	Mandatory	Voluntary
Certification	Type Approval	Privately Held
Vehicle Tests	Yes	No
Conformity of Production	Annual	Annual
Enforcement	Technical Authorities	To Be Determined
Requirements	Vehicle + Lab Tests	Lab Test
Objective	Preserve Compliance	Preserve Compliance
Criteria	±10 to 15% of OE	Absolute FMVSS
Statistical Reliability	?	3 - 7% of mean
Statistical Validity	?	5% of mean
Hydraulic Displacement	Incorporated in Vehicle Test	Incorporated in Lab Test
Funding	Required	To Be Determined
Original Equipment Service Parts	Exempted	Testing Required for Seal

10 NEW VEHICLE BRAKE SYSTEM VALIDATION

While the entirety of this paper has focused upon the validation of replacement friction materials, it does not mean to suggest that the D³EA™ test and DEA™ analysis technology are simply limited to that area of need. In fact, the real value may not lie within in the ability to grade performance for vehicles of yesterday and today, but rather those yet to be built. These procedures and assessment technologies provide an important tool for the accelerated development and validation of automotive brake systems that will reduce lead time and expedite the certification process of new vehicles. These technologies have been successfully utilized at the original equipment vehicle level in the U.S., Canada, and Europe. The ability to validate a new vehicle braking system in the laboratory before the vehicle prototypes are available provides new opportunity for the vehicle manufacturer and supplier alike.

11 FUTURE TECHNOLOGIES

In order to bring D³EA™ laboratory validation to its full potential, additional technology for wear-life prediction in normal customer usage and thermal conductivity in sustained mountain descents will have to be developed and validated. Adding these assessment technologies to the existing D³EA™ Certification program will require the development and validation of new test procedures and vehicle platform specific criteria based upon the experience gained in effectiveness assessments conducted using dual-ended brake dynamometers. It is anticipated that these test methods and criteria will follow the development and validation of the ongoing brake fluid displacement assessments currently underway.

12 SUMMARY

D³EA™ tests and criteria represent a major step in the laboratory validation of automotive brake friction materials. The millions of brake replacements that occur each year in the U.S. are well served by the assessment of vehicle specific effectiveness and wear balance technologies available today. The implementation of hydraulic displacement assessment and criteria and the development and validation of customer-based wear and thermal conductivity tests down the road will further the utility of these laboratory methods. The lessons learned in the development of D³EA™ include the recognition and preservation of interactive conditioning between front and rear brakes, and the necessity to utilize vehicle platform specific hardware, test conditions, and performance criteria. Laboratory validation of brake friction materials is a complex, interrelated assessment process that cannot be compromised by expediency.

Correlation of scale to full-size dynamometer testing

T KURODA and **J ABO**
Nisshinbo Automotive Corporation, Michigan, USA

INTRODUCTION

There is an increasing need to evaluate performance characteristics of friction materials in a rapid and economical manner. This is a result of many factors, but is dominated by the shortened life cycle of formulations and globalization of brake systems. The customer criteria of no noise or judder combined with good life and pedal feel dictates the need to examine many developmental formulations. To characterize a friction material on a full scale dynamometer or by vehicle testing is extensive and extremely expensive.

Scale dynamometer testing was developed by Nisshinbo around 10 years ago to address the need for rapid characterization of friction material. It has been updated over the years to include an environmental chamber with fully computerized control system. Scale dynamometers have been developed for both passenger cars and commercial vehicles.

This paper will describe details of the passenger car scale dynamometer in use by Nisshinbo and demonstrate the correlation between the scaled and full size dynamometer.

SCALED DYNAMOMETER

Scaled dynamometer testing offers a rapid method to determine performance characteristics, with out the need to consider dynamometer fixtures, calipers, etc. The scale dynamometer for passenger car friction material evaluation is shown in Figure 1. It is about 2.5 meters long; 1 meter wide and 1.5 meters high.

The specifications for the scale dynamometer are shown in Figure 2. The revolutions are up to 5,000 rpm. The temperatures generated are as high as 1200 Degrees C. Basically, the scaled dynamometer for passenger vehicles is about a one-tenth the load of a full-size dynamometer. It can be operated at constant speed, pressure or temperature with different size samples. This gives extreme flexibility in characterizing friction material performance.

The size of the samples can vary, but for routine evaluations and comparative data, a sample size of 30 mm by 10 mm by 12 mm thick is used for passenger car friction materials. For commercial vehicle friction materials, a sample size of 17 mm in width with an arc of 90 degrees by 12 mm thick is employed. Two pieces are attached to a plate. This plate is pressed against a rotating disc made of the same material which is used on the vehicle brake system.

Since no caliper is used, the performance characteristics are a direct measure of the friction material and not the brake system. The output from the scale dynamometer can be in graphical format in terms of speed, friction, torque and temperature. The output response generally has a narrower distribution than that of a full-scale dynamometer since there is no influence of the brake system hardware. Of course there are limitations with the scaled dynamometers in that there is only limited measurements of noise, geometry and fluid displacement.

SCALED DYNAMOMETER TEST PROCEDURE

An overview of the scaled dynamometer procedure is shown Figure 3 and is a standard procedure used at Nisshinbo.

The procedure includes a series of 6 burnishes, 3 effectiveness and two fade and recovery's. The complete test for this procedure takes less than 12 hours. Test set-up time is around 10–15 minutes. This permits two test runs per day.

FULL SIZE DYNAMOMETER PROCEDURE

There are numerous dynamometer procedures in use by the friction material industry. Many procedures are dictated by the vehicle manufactures and many procedures have been developed internally to examine particular performance characteristics. The procedure used in this study (Figure 4) is a standard technique and is similar to that of the scale dynamometer. However, it is more detailed in certain sections. It includes extra brake sequences, such as low temperature effectiveness and water recovery. This dynamometer procedure takes slightly less than 24 hours plus set-up time. This sequence permits one test per day.

SAMPLE DESCRIPTION

The samples used in this study are shown in Figure 5. Three different friction materials were selected: A non-ferrous, a low ferrous-around 10% by volume and a medium ferrous material

– around 20% by volume. This represents most of the passenger car disc brake pad formulations.

SCALE DYNAMOMETER RESULTS

The second effectiveness usually represents the normal driving conditions for customer usage and satisfaction.

As shown in Figure 6, there is no speed sensitivity associated with the non-ferrous material. There is, however, some speed sensitivity associated with the other two materials. The aggressive type material shows the largest sensitivity to speed on the scale dynamometer. Sample C is more sensitive to pressure than the other two materials.

FULL DYNAMOMETER RESULTS

The results for the full scaledynamometer are shown in Figure 7. Sample B has the largest speed sensitivity than the other two materials. This is similar to the results obtained on the scale dynamometer. Sample C is the most sensitive to pressure changes. Sample A seems to be the most stable to pressure and speed.

Comparison – First Effectiveness
For completeness, all three effectiveness curves will be discussed. The first effectiveness curves (Figure 8) represent the green condition of the friction material. As shown Sample A and C show fairly good correlation with respect to the trends of pressure and speed. Sample B, which is the aggressive friction material, shows little correlation, except for the 100-km/hr condition.

Comparison – Second Effectiveness
Figure 9 shows the summary of the second effectiveness. Sample A shows the least sensitivity to speed and pressure, with the friction levels being similar on both the scale and full size dynamometer. The other two samples show good overall correlation.

The second effectiveness at 50 Km/hr and at 20 Kgf/ sq. cm represents the standard driving parameters used for customer satisfaction. For this case, there is excellent correlation for all the samples with respect to friction. Level,

Comparison – Third Effectiveness
The third effectiveness is shown in Figure 10. The trends between the scale and full scale dynamometer are similar for all the materials. At the 50 km/hr at 20 Kgf/ sq. cm points, there is excellent correlation.

CONCLUSION

The scale dynamometer testing is a rapid, low cost technique to characterize friction materials and is an excellent complement to full scale dynamometer testing. It is extremely flexible in evaluating the performance behavior of materials. The correlation with respect to friction and pressure trends is good between the scale and full size dynamometer. At 50 Km /hr at 20 Kgf/ sq. cm, which represents normal customer driving, there is excellent correlation.

Figure 1

Scale Dyno Specifications

Main Motor:	**3.7 kw AC Motor**
Revolution:	**0-5000 rpm**
Inertia:	**0.035--0.07 kgfms^2**
Torque:	**10 kgfm Max**
Contact Force:	**600 kg**
Temperature:	**Up to 1200 C^o**
Cooling Air	**0-13 m/s**

Figure 2

Scaled Dynamometer Test Procedure
Overview

		Sequence	
Burnish	6	20 km/h	10 - 30 (5 kgf/cm^2)
Effectiveness	3	50,100 130km/h	10 - 100 (10 kgf/cm^2)
Fade / Recovery	2	100 km/h	15 Stops (600 C)

Figure 3

Full Size Dynamometer
Modified JASO Procedure
Overview

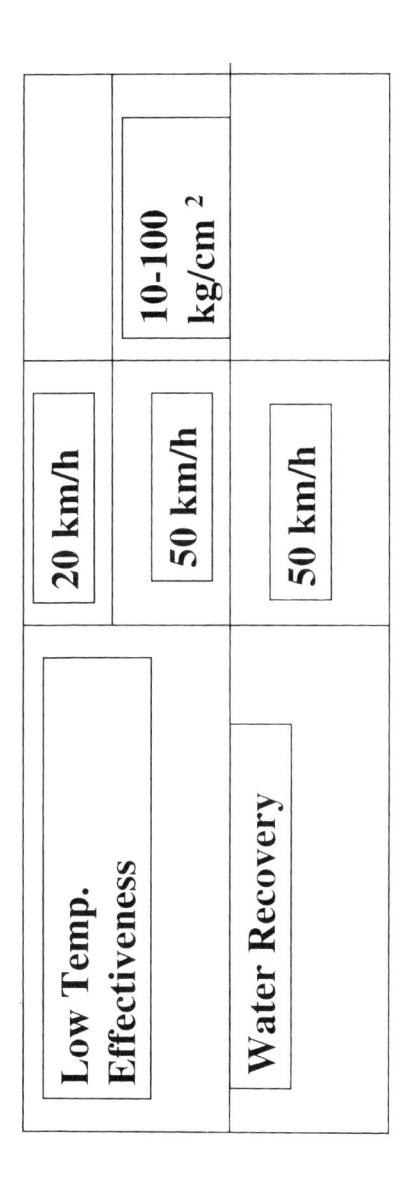

Low Temp. Effectiveness	20 km/h	
	50 km/h	10-100 kg/cm^2
Water Recovery	50 km/h	

Addition to Scaled Dynamometer Procedure

Figure 4

Sample Description

Sample A -- Non-Aggressive Type Material --- Non-Ferrous / NAO

Sample B -- Aggressive Type Material --- Medium Ferrous

Sample C -- Non-Aggressive Type Material -- Low Ferrous

Figure 5

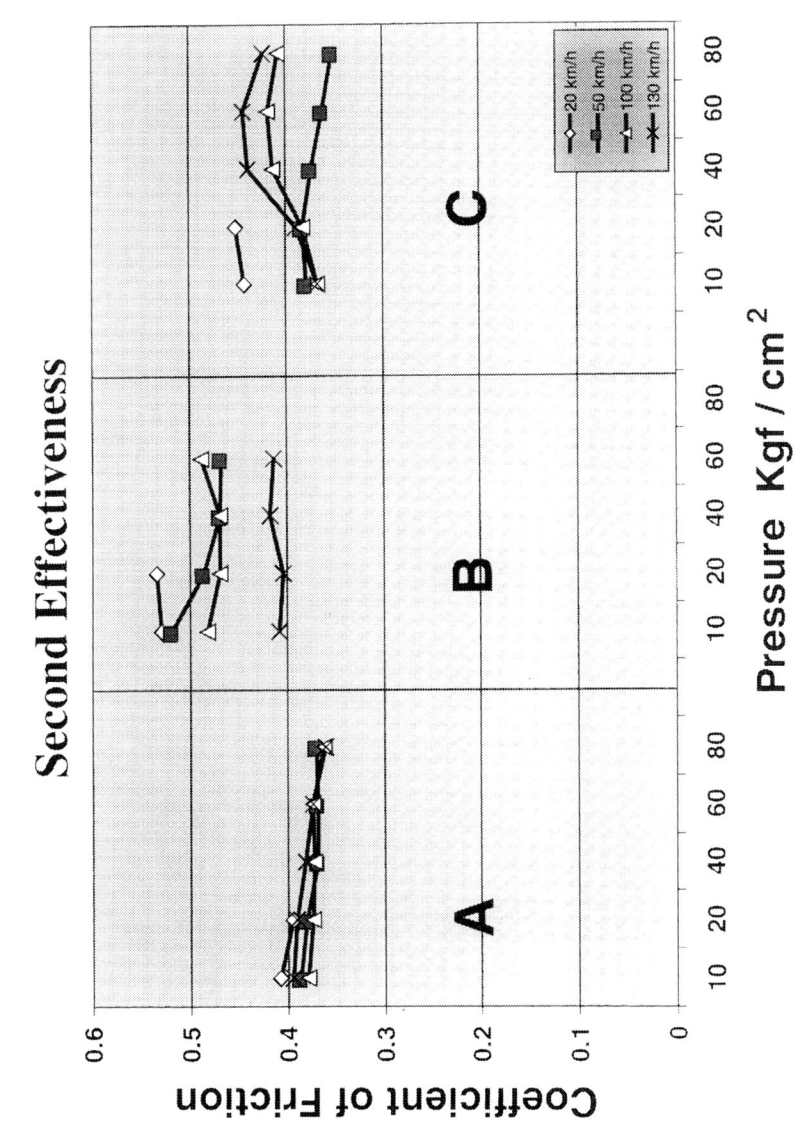

Figure 6

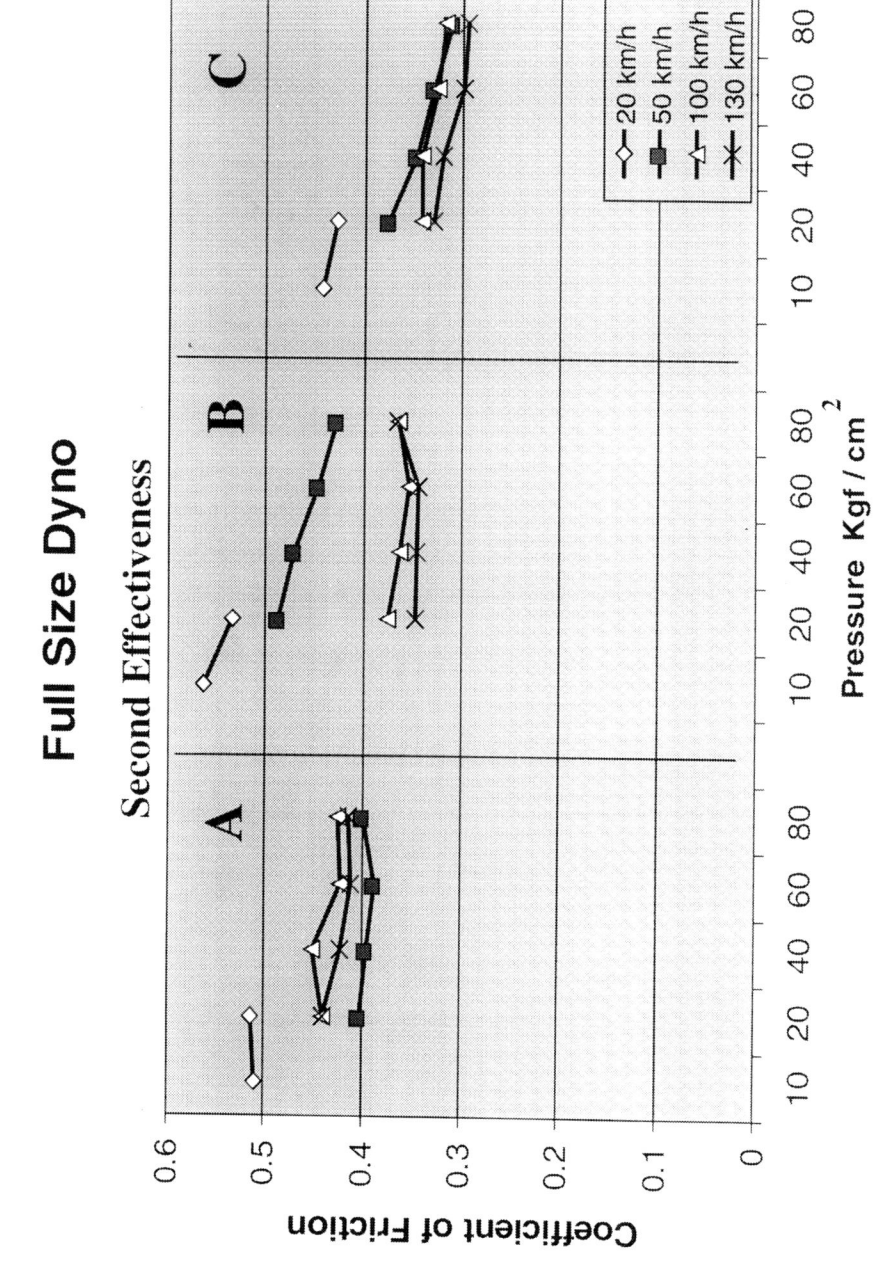

Full Size Dyno
Second Effectiveness

Figure 7

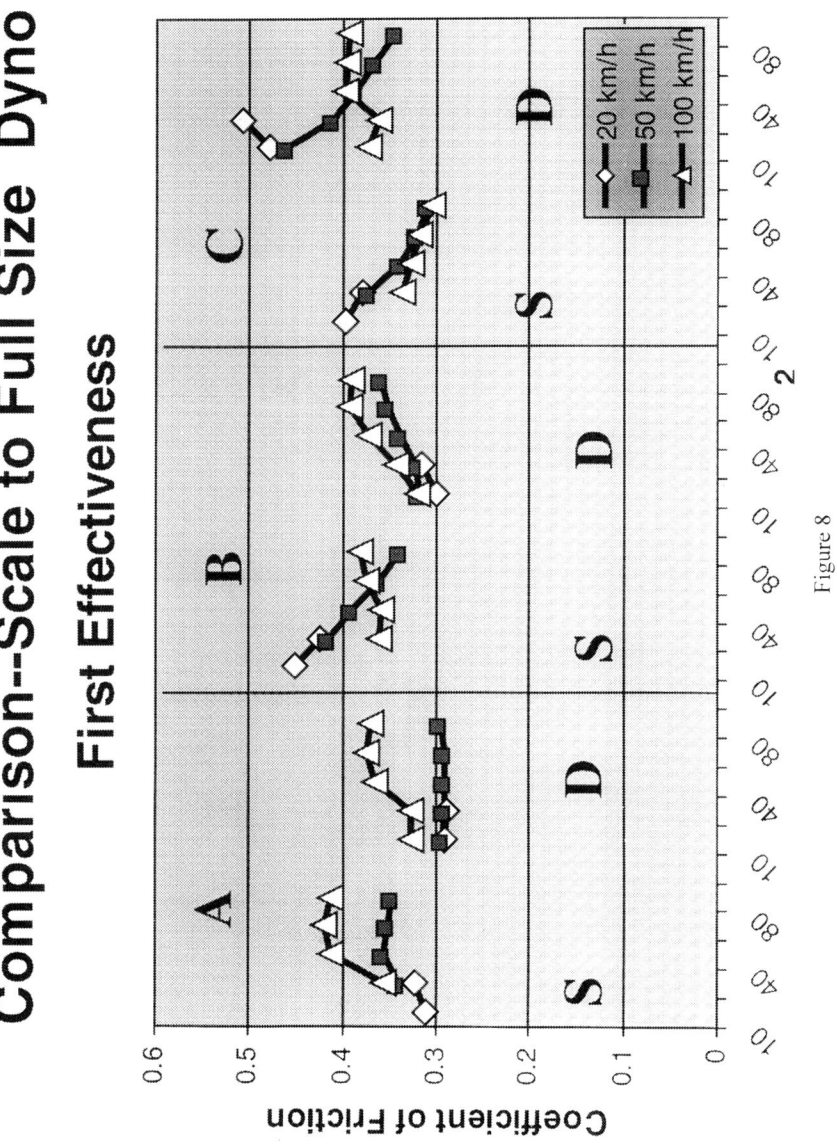

Figure 8

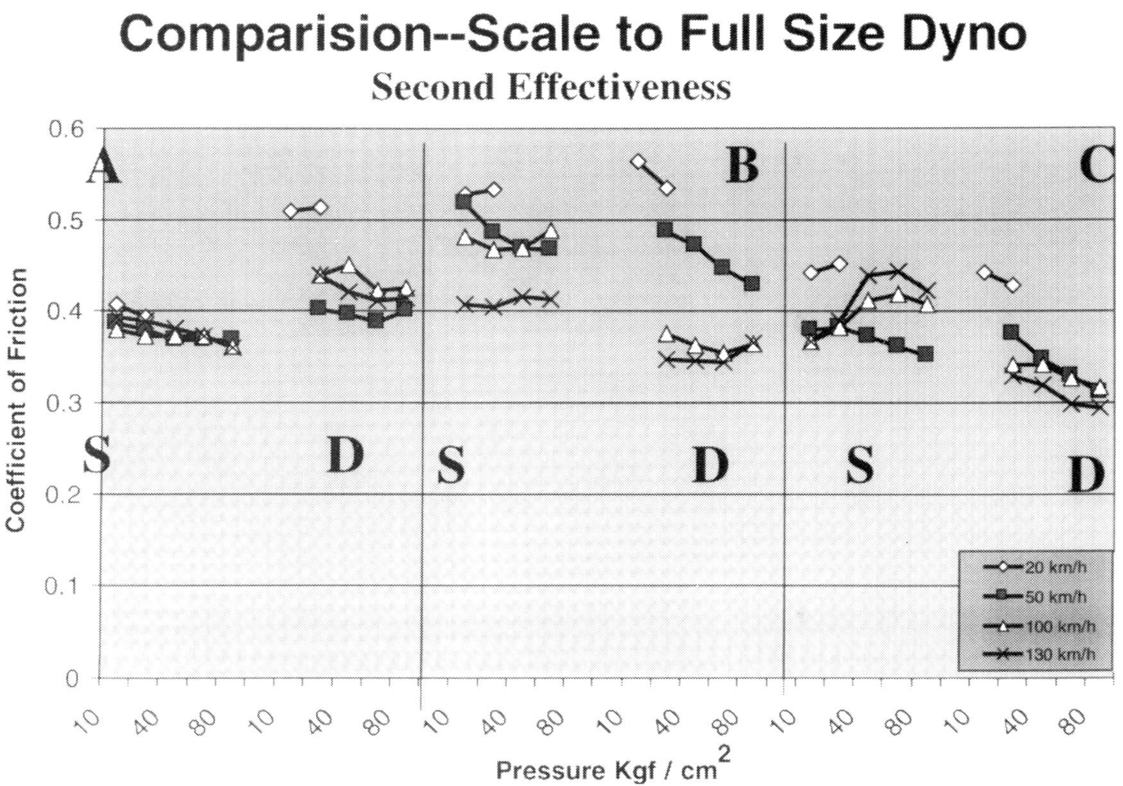
Comparision--Scale to Full Size Dyno
Second Effectiveness
Coefficient of Friction
0.6
0.5
0.4
0.3
0.2
0.1
0
A
B
C
S
D
S
D
S
D
10
40
80
Pressure Kgf / cm2
20 km/h
50 km/h
100 km/h
130 km/h

Figure 9

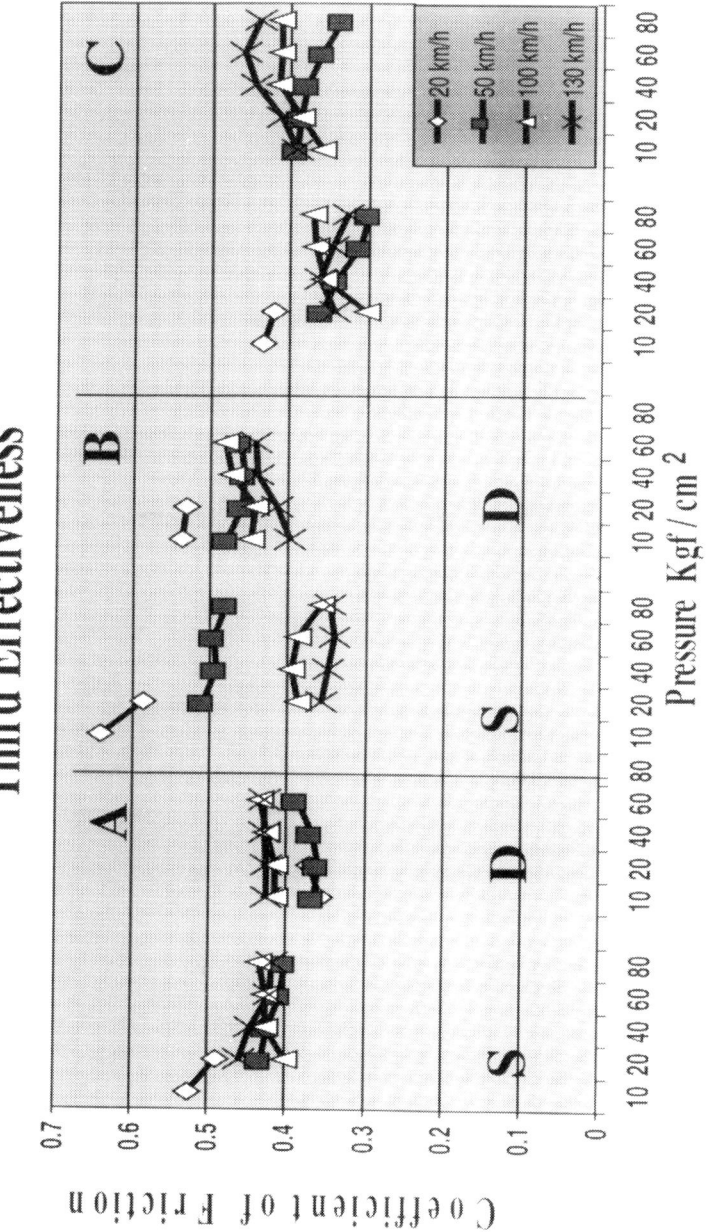

Figure 10

Characterization of automotive friction behaviour using small specimens

J W FASH, T M DALKA, D L HARTSOCK, and R L HECHT
Ford Scientific Research Laboratory, Dearborn, Michigan, USA
R KARTHIK
Varga Brakes Inc., USA

Behavior of automotive brake systems is in part dependent on the friction behavior generated at the interface between the pad (or shoe) and the rotor (or drum). Values reported for the coefficient of friction are usually determined from full scale hardware tests performed on laboratory brake dynamometers and as such integrate the effects of local pressure, temperature and speed variations into a single value. With the increased use of CAE tools such as finite element analysis in brake development there is need for representative friction behavior that is sensitive to these variations in local operating conditions. An investigation has been undertaken to observe the influence of apply pressure, sliding velocity and temperature on the friction behavior of small specimen friction samples. Three different types of production friction materials have been evaluated. Results indicate strong dependence of friction coefficient on sliding speed and temperature and to a lesser degree with apply pressure. Regression analysis has been used to model behavior over the range of test conditions to provide a quadratic equation for the friction coefficient as a function of pressure, temperature and sliding velocity. Use of improved friction models may contribute to improved screening procedures and modeling of brake performance.

INTRODUCTION

Advances in brake engineering and improved understanding of the failure modes, such as noise and judder, that occur in brake systems are driving a need for improved characterization of system parameters, in particular friction performance. Additional competitive pressures to reduce product development schedules require improved methods for screening brake performance for selection of optimum systems.

Typically, the engineering coefficient of friction (designated COF in this paper) is a fundamental parameter or value used in the engineering of a brake system. A value for the

COF is usually reported from dynamometer testing of brake hardware and is calculated from the apply system pressure, the torque output of the brake and geometric factors determined from the hardware. A series of brake stops at a variety of operating conditions are performed and the observed values of COF are averaged to produce a single value which is reported. Values obtained for the COF are, in this case, a composite average of the integrated behavior occurring over both the inboard and outboard pads as well as over the various operating conditions included in the testing. It is well established that the temperature profile varies considerably from leading edge to trailing edge and from inner radius to outer radius. These temperature variations as well as other hardware related effects, result in non-uniform pressure distributions. The sliding velocity also varies as a function of radius and with time during a stop.

The work reported here is an initial investigation to assess test methods and modeling approaches for generating specimen based friction data in order to more precisely evaluate the variation of COF over a typical range of operating parameters. Previous efforts have utilized scale or specimen testing to evaluate various friction behavior and to screen friction materials. These approaches include the Chase Test (1) and the Friction Assessment Screening Test or FAST (2). In addition, small scale dynamometers (3, 4, 5) have been used to assess more detailed friction behavior. A review of some friction related literature is given in (5).

Although much work has been done on full scale and sub-scale friction testing, there remains much opportunity to exploit the capabilities of small scale testing for friction screening and friction modeling. Results from these efforts will allow assessment of the characteristics of individual friction materials independent of the mechanical effects that result from hardware (caliper, rotor, hydraulic system etc.) effects. Appropriate modeling of friction behavior will allow improved representations of COF to be incorporated in computational models of brake system behavior, thus enabling improved correlation between analytic models and experimental brake observations. The intent being to improve understanding of brake performance and brake engineering capability.

In the work reported below, a series of test procedures have been developed to assess friction behavior. The procedures used in this series of tests were intended to provide base line data to be used in modeling, to provide data for correlation analysis of the models developed and to provide data to assess sequence effects. Data have been fit with a quadratic model to describe COF over the range of temperature, pressure and sliding velocity tested. Model predictions of COF are in good agreement with observed values for correlation tests.

EXPERIMENTAL APPARATUS

A test apparatus was developed such that a small specimen of friction material could be applied to a rotating cast iron disk for evaluation of friction behavior. A schematic of the test configuration is given in Fig. 1. An electric motor is used to control the speed of the rotating cast iron disk. Control limitations on this motor required that the speed be changed in discrete steps which, although small, result in 'stair step' ramps rather than continuously decreasing speed ramps. The effect that this causes will become evident in the discussion and results for the speed ramp test profiles. Rotational speed of the disk is measured using a 60 tooth pulse wheel mounted to the rotating shaft.

A small friction specimen is applied to the face of the rotating disk by a pneumatic cylinder. The normal force is measured with a load cell. Friction force was measured via a calibrated strain gage bridge mounted on the load arm. The temperature of the friction specimen was measured with a thermal couple embedded from the back of the pad to a distance of approximately 0.5 mm (0.020 in) from the friction surface. Temperature of the rotating disk was measured with a rubbing thermal couple at the outer radius of the disk.

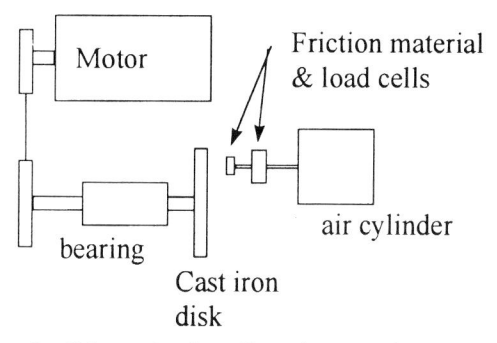

Figure 1 - Schematic of small specimen testing apparatus.

For higher temperature test conditions it was not possible to achieve the desired initial test temperatures via frictional heating. Consequently, high intensity quartz lamps were directed at the slowly rotating disk to provide radiant heating which allowed for control of the higher initial brake temperatures. During this heating the other instrumentation was shielded from the radiant heat source.

Data was collected via a Keithly Metrabyte PC based data acquisition system. Data was recorded at 100 points per second per channel in all tests. Raw data was exported to a MATLAB (9) based analysis system for evaluation and presentation.

Friction Material Specimens

Friction material specimens were cut directly from production automotive brake pads. The dimensions of the friction specimens were 12.7 mm X 25.4 mm (0.5 x 1.0 in) and included the backing plate. A small steel hemi-sphere was tack welded to the back plate of the friction specimen to allow the pad friction forces to be transferred into the load arm without need for more complex fixtures.

Three friction materials were evaluated. The materials selected were typical of the three classes of friction materials currently in common usage. These included a low metallic content material (Low-Met) typical of European style materials, a non-asbestos organic (NAO) material typical of Japanese and United States applications and a semi-metallic (Semi-Met) material. All three materials are used in production applications.

Cast Iron Disk Interface

Gray cast iron cylinders, 150 mm diameter X 300 mm length, were produced to a chemistry specification in the mid-range of the typical rotor chemistry specifications (6). The graphite flake structure was a typical type A flake of size 1-2 (7) with a pearlitic matrix. Disks for use on the tribometer were machined from these blanks and were 125 mm in diameter and 25 mm thick. The surface of the disks was a fine turned finish. A new disk was used for each friction material test.

TEST PROCEDURES

A test procedure was developed to provide information for modeling of the friction behavior as well as for correlation and evaluation of sequence effects. Four test segments were performed sequentially:

- Burnish / Conditioning
- Constant condition tests
- Speed ramp tests
- Pressure ramp tests

For each test segments a range of values for the parameters shown in Table 1 were performed. The constant condition tests were performed for the full matrix of conditions in Table 1. For the pressure and speed ramp profiles the conditions were ramped from zero to the maximum parameter value shown in Table 1. Each test segment and the analysis of the data are described below.

Table 1 - Operational conditions used in the constant condition series of friction tests. (Numerical values indicate tests used in the development of friction models.)

Pressure		0.34 mPa			1.03 mPa			2.07 mPa	
Speed (m/s)	0.08	4.33	8.67	0.08	4.33	8.67	0.08	4.33	8.67
30 C	1		5	7			19		23
95 C		3			9-12	17		21	
150 C		4			13-15	18		22	
230 C	2		6	8	16		20		24

Burnishing

A series of conditioning operations were first performed on each friction couple, similar to the burnishing or conditioning operation usually performed in conventional vehicle or dynamometer brake testing. This consisted of a series of applies under the conditions given in Table 2. The intent of this operation is to 'run in' the friction couple surface and to develop an initial surface film as would be expected from normal burnish operations.

Table 2 - Test sequence for burnish/conditioning procedure.

Number	Pressure (mPa)	Speed (RPM)	Temperature
Drag 10 min.	0.34	1000	100 C
Drag 10 min.	0.69	600	30 C
Drag 10 min.	0.34	600	100 C
25	0.34	1100 - 0 in 10 sec.	30 C
20	0.52	1100 - 0 in 10 sec.	30 C
5	0.69	1100 - 0 in 10 sec.	30 C
5	1.03	1100 - 0 in 10 sec.	30 C
10	0.69	1600 - 0 in 16 sec.	100 C
10	0.69	1600 - 0 in 16 sec.	150 C
5	0.34	1600 - 0 in 16 sec.	30 C
5	0.69	1600 - 0 in 16 sec.	30 C

Constant test conditions

Following the burnishing step, a series of applies were made to generate data under constant operating conditions. Brake applies were made for each of the test conditions shown in Table 1. The operating speed and normal force were held constant during a 10 second apply which was initiated at the desired initial brake temperature. Increases in temperature due to frictional heating were small and for the purposes of this program were considered negligible. Data was collected for 10 seconds at each of the conditions.

To assess friction level an average of 6 seconds of the data was taken from the middle of the data collection period (i.e. from 2-8 seconds). In general the variations of normal and frictional forces were small during the ten second apply. Friction coefficient (calculated as the friction force divided by the normal force) was generally very constant over the complete test window and average values over the six second window are reported.

Test conditions with numerical entries in Table 1 were used in the design of experiments analysis to develop the friction models. The remaining conditions in Table 1 were used as correlation points for the models.

Speed Ramp Tests

Two different speed ramp test profiles were developed to evaluate repeatability and sequence effects on friction behavior. The first ramp profile, Fig. 2a is called the DECEL-ACEL profile and the second speed ramp test shown in Fig 2b is the ACEL-DECEL profile. In these tests, pressure was held constant while speed was stepped by 150 RPM between 1600 and 100 RPM as shown in Fig. 2. Both ramp profiles were performed for each of the initial brake temperatures and pressure levels tested. For most of the segments the speed was held constant for 1 second before stepping to the next segment. A longer duration of 5 seconds was used at the slower speeds so that data would include several revolutions of the disk. For the speed ramp tests performed at higher apply pressures the influence of frictional heating could be observed with temperature rises of about 30-40 C for the most severe conditions.

An average friction coefficient was calculated for each step in both profiles over the time window for which the speed was held constant. These results are used to assess sequence effects and repeatability. These tests do not cover all of the possible sequence effects but give an indication of repeatability of COF relative to the range of parameters studied. Results are compared with predictions from the models of friction behavior which were developed.

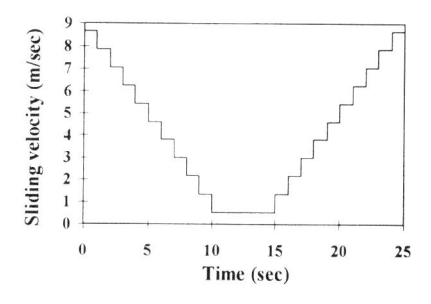

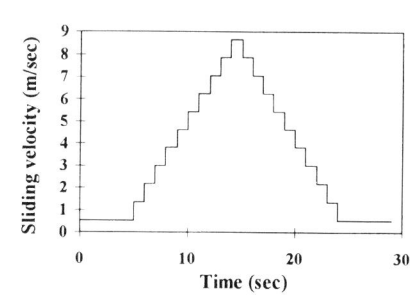

Figure 2a - DECEL - ACEL profile.   Figure 2b - ACEL - DECEL profile.

Figure 2 - Schematic of speed ramp tests.

Pressure Ramp Tests

In the pressure ramp tests, speed and initial apply temperature were held constant while pressure was ramped from zero to 2.06 mPa (300 psi) in 15 seconds. This test profile is shown schematically in Fig. 3. Tests were initiated at the desired initial brake temperature and the temperature rise over the apply time was small (less than 30 C).

For comparison with the other test conditions, COF was calculated at discrete pressure levels by averaging data over a small time window. This allowed comparison with constant condition tests even though the pressure was changing continuously.

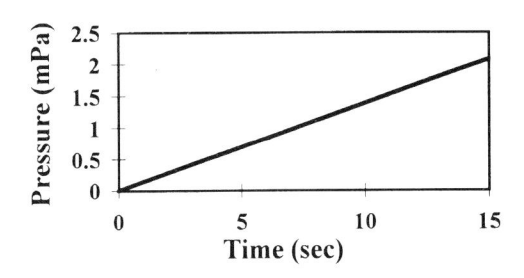

Figure 3 - Schematic of pressure ramp tests.

MODELING OF FRICTION DATA

Data for use in modeling was generated under the constant test conditions described above. A design of experiments approach (8) was used to set up the experiments and to analyze the results to generate a model of the friction behavior. A quadratic model was selected to model behavior over the range of the controlled parameters. This results in a friction model of the form;

$$COF = C0 + C1*P + C2*S + C3*T + C4*P*S + C5*P*T + C6*S*T + C7*P^2 + C8*S^2 + C9*T^2$$

where P is the pressure, S is the speed and T is the temperature.

Data generated from the experiments was fit to this form using a least squares analysis (8). In order to reduce bias in this analysis the operating conditions were first transformed to cover a range from minus one to positive one by the following equations. The model coefficients reported are for the transformed values of pressure, speed and temperature.

$$P(transformed) = (P - 1.025) / 0.806$$
$$S(transformed) = (S - 4.375) / 4.295$$
$$T(transformed) = (P - 130) / 100$$

RESULTS

Extensive test results were generated during this work. Representative examples of the temperature effects and the sequence effects observed for the three materials are shown below as a function of sliding speed.

Temperature Effects

Results showing the effect of temperature and sliding speed are taken from the initial deceleration portion of the DECEL-ACEL speed ramp tests (Fig. 2a).

Figures 4 and 5 present data for the Low-Met friction material which shows the friction behavior at various temperatures for 0.34 and 1.38 mPa respectively. Results show a strong increase in the COF with decreasing speed. There is a significant increase in COF with temperature and an increase, although less consistent, with pressure.

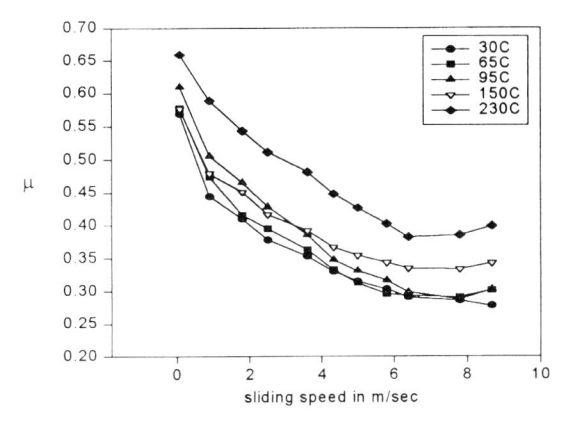

Figure 4 - Speed ramp test results for the Low-Met friction material at 0.3447 mPa (50 psi)

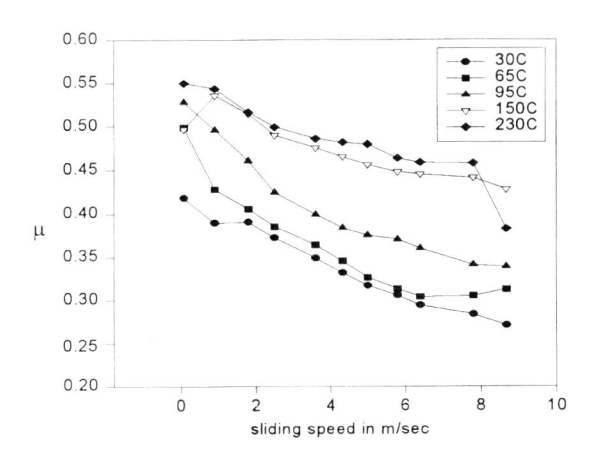

Figure 5 - Speed ramp test results for the Low-Met friction material at 1.38 mPa (200 psi)

Similar results for the Semi-Met friction material at a surface pressure of 0.69 mPa are shown in Fig. 6. At the lower temperature conditions the COF is relatively constant over most of the range of speed with a slight increase at the slowest conditions. As has commonly been observed, the COF of the Semi-Met increases substantially above some temperature, as shown by the 230C data. At the higher temperatures, the COF increases with decreasing speed.

Figures 7 and 8 show the effect of speed and temperature on COF for the NAO material at surface pressures of 0.34 and 1.03 mPa respectively. For the NAO material the COF increases only slightly with decreasing speed when compared with the Low-Met material. However, for each pressure level there is a marked increase of COF with temperature with the curves being nearly parallel up to about 150 C. For the 150 C and the 230 C conditions the behavior is nearly the same, increasing with decreasing speed but not with temperature.

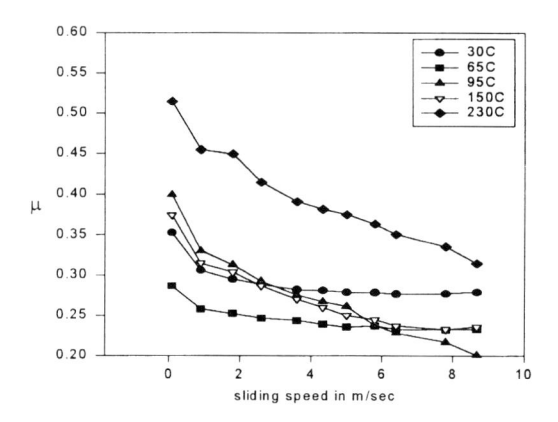

Figure 6 - Speed ramp test results for the Semi-Met friction material at 0.69 mPa (100 psi)

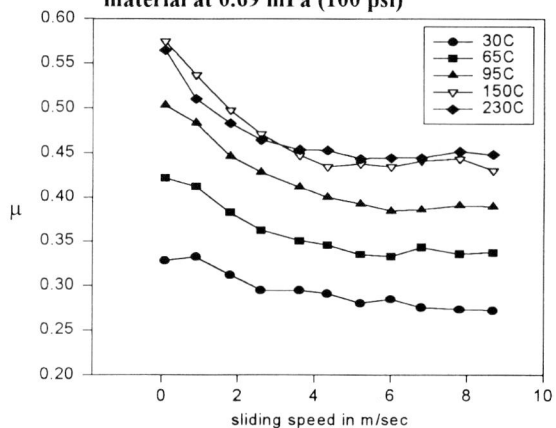

Figure 7 - Speed ramp test results for the NAO friction material at 0.34 mPa (50 psi)

Sequence Effects

Sequence effects were assessed by comparing the four individual segments of the two speed ramp profiles. For both the DECEL-ACEL and the ACEL-DECEL profiles the individual acceleration and deceleration portions of the test were analyzed at a particular pressure and initial temperature. For most of the test conditions the temperature rise due to frictional heating during the complete test profile was small with the most severe test conditions showing a rise of about 40 C.

Data for the Low-Met friction material at 0.34 mPa are shown for 65 C and 230 C tests in Fig. 9. Similar results for the Semi-Met material at 1.03 mPa pressure are shown in Fig. 10

and for the NAO material at 1.38 mPa in Fig. 11. These results show very consistent COF behavior as a function of sliding speed for the four segments of the test profiles.

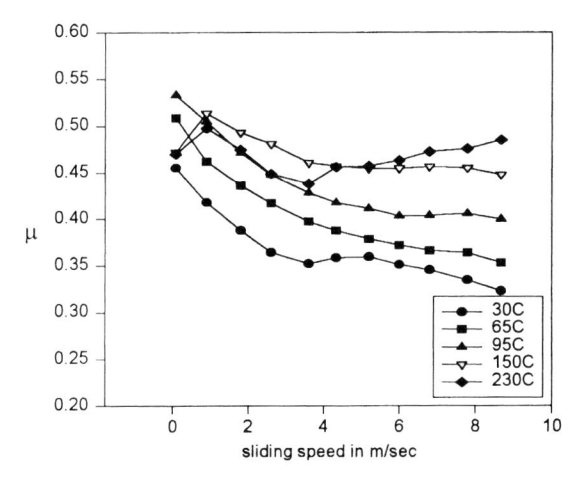

Figure 8 - Speed ramp test results for the NAO friction material at 1.03 mPa (150 psi)

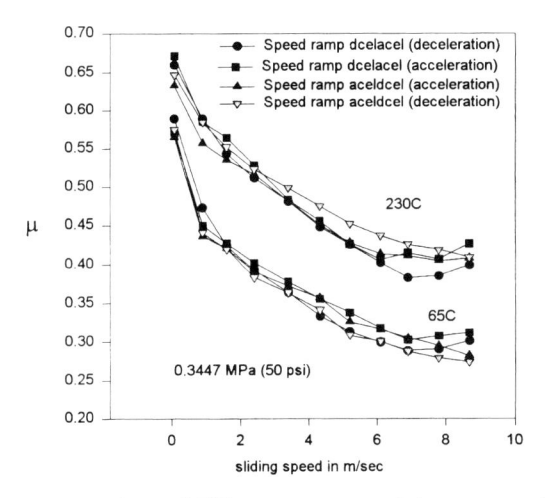

Figure 9 - Comparison of different segments of the two speed ramp tests for the Low-Met friction material at 0.34 mPa (50 psi)

Model Results

Data from the constant condition tests shown with numerical entries in Table 1 were analyzed to produce the quadratic models of the friction behavior over the operating conditions tested. Results are given in Table 3 for the transformed values of pressure, speed and temperature. A correlation coefficient, R^2, is also included indicating the correlation of the fit over the data used in the analysis. Further evaluation of the models with data not used in the development of the model is presented in the next section.

Table 3 - Model parameters for fit of friction data over temperature, pressure and sliding velocity (based on normalized parameters.)

	Low-Met	Semi-Met	Non-Asbestos Organic
Friction Model	1/COF	COF	COF
C0	2.44	0.272	0.458
C1 (P)	-0.043	-0.001	0.012
C2 (S)	0.349	-0.047	-0.033
C3 (T)	-0.340	0.043	0.069
C4 (P*S)	---	0.026	---
C5 (P*T)	0.096	---	---
C6 (S*T)	---	---	---
C7 (P^2)	---	0.025	---
C8 (S^2)	-0.322	---	0.036
C9 (T^2)	0.429	0.040	-0.056
R^2	0.89	0.85	0.92

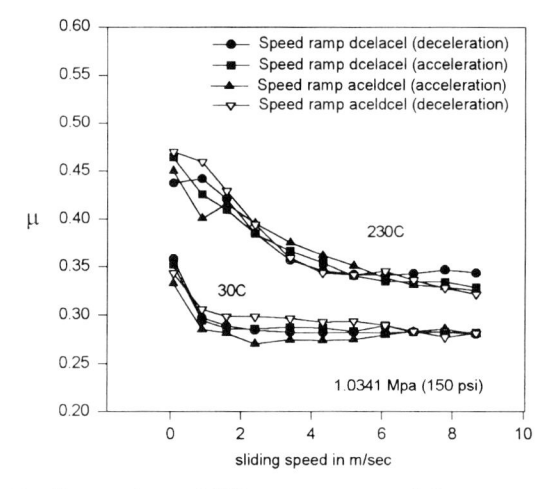

Figure 10 - Comparison of different segments of the two speed ramp tests for the Semi-Met friction material at 1.03 mPa (150 psi)

Correlation of Model

The correlation of the friction model with the data used to fit the model is shown by the correlation coefficient at the bottom of Table 3. The values of 0.85-0.92 indicate reasonably good agreement.

Evaluation of the models to predict the other test conditions not used to develop the models, such as the speed ramp and the pressure ramp test results as well as other constant condition tests, are shown graphically. Figure 12 shows results for the Low-Met material at 1.38 mPa surface pressure and two temperatures. The solid lines are the prediction of the model. The open circles are results for the speed ramp test, the asterisk indicates the constant condition tests and the cross symbol are from the pressure ramp tests. Although not perfect, the model fits follow the trend of the data very well. Similar evaluations for the Semi-Met and the NAO materials are shown in Figs. 13 and 14.

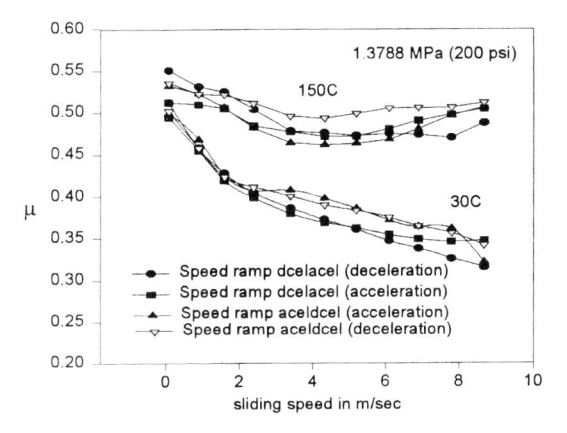

Figure 11 - Comparison of different segments of the two speed ramp tests for the NAO friction material at 1.38 mPa (200 psi).

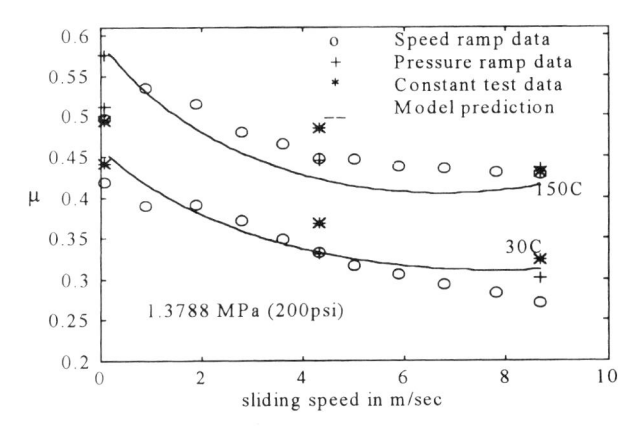

Figure 12 - Comparison of model predictions with various test conditions for the Low-Met friction material at 1.38 mPa (200 psi)

CONCLUSIONS / DISCUSSION

A series of test procedures to evaluate the dependence of friction behavior on operating conditions has been developed using a small specimen friction test rig. The results provide a means to assess and compare friction behavior for a variety of materials over a wide range of operating conditions. In general the results show behavior expected of these materials based on full scale brake dynamometer testing. Results show clear behavior difference between the materials which are qualitatively similar to observed full scale behavior. However, as this has been an initial investigation of small specimen testing, much work remains in developing meaningful small scale tests which correlate with full brake behavior. Some of the more important trends in the data and correlation's with existing brake behavior are discussed below.

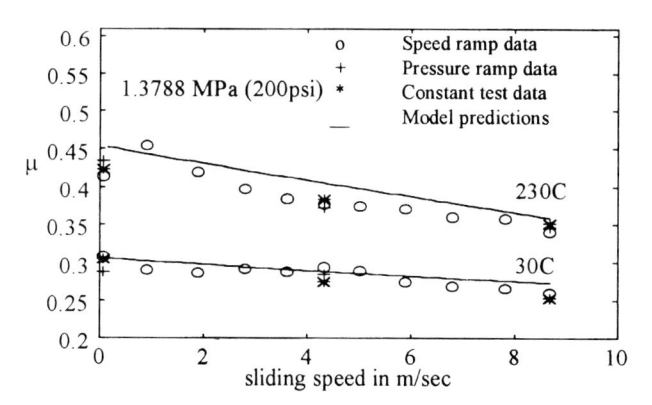

Figure 13 - Comparison of model predictions with various test conditions for the Semi-Met friction material at 1.38 mPa (200 psi)

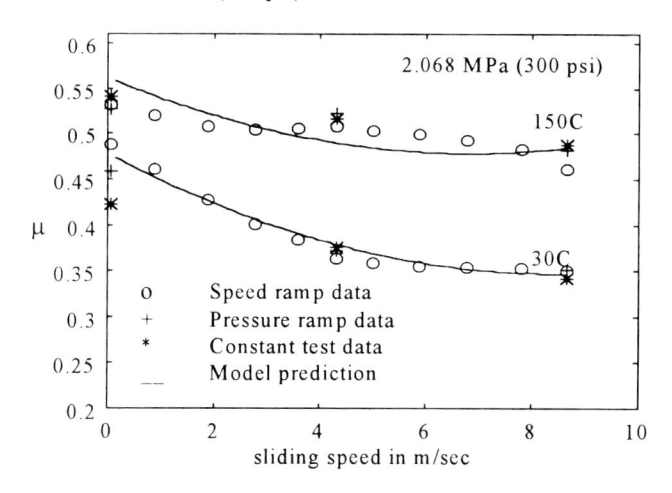

Figure 14 - Comparison of model predictions with various test conditions for the NAO friction material at 2.07 mPa (300 psi)

For the range of conditions tested, all of the materials show an increase in friction with increasing temperature. The semi-metallic exhibited lower friction behavior at low temperatures and a marked increase in the friction level for the higher temperatures, the transition occurring over a relatively small temperature range. This behavior is well known for this class of material and has been observed in vehicle and full scale dynamometer testing. This trend comes out strongly in the small specimen results. The Low-Met material shows similar changes with temperature but the trend is less consistent. The NAO material shows a relatively consistent increase in friction level with increasing temperature with the behavior becoming nearly constant for the higher temperature conditions. Fade type friction behavior was not observed in these tests as the range of temperatures evaluated was limited.

The increase in friction with decreasing sliding speed observed in all of the materials is also a known phenomenon. However, the degree of the increase observed in these tests is much stronger than was expected, particularly for the Low-Met material. The Semi-Met material showed relatively constant friction over most of the speed range at the lower temperatures. Once the increase in friction level due to temperature was reached the friction level increased nearly linearly with decreasing speed. The Semi-Met also showed a slight increase in friction level for the very slowest speed condition. This may be indicative of friction behavior which could lead to end of stop brake noise.

A variation of friction with pressure was observed, however it is a relatively small effect in these tests.

Evaluation of sequence effects based on the speed ramp tests showed surprisingly consistent behavior for all three materials. This indicates good repeatability in the measurement as well as consistent behavior in the friction material performance. This is by no means an exhaustive evaluation of the sequence and history effects that may be present due to different duty cycles. However, the results are encouraging that small specimen testing can be used to assess aspects of sequence and history dependent behavior.

Application of a quadratic curve fit to model the data provides a convenient means to concisely describe the observed behavior. This is an advantage if the friction models are to be used in more sophisticated analytic models of the full brake system such as finite element simulations. Based on this idea a general purpose friction model can be developed for such an application with the user needing only to input the coefficients from the curve fit. There remain several issues in regard to applying such a model to the complex behavior of friction. For instance, the observed dependence of the Semi-Metallic material behavior with temperature may not be reliably captured based on the selection of test conditions used to develop the model. As currently modeled, the application of the model is limited to the range of test conditions used to generate data for the model. Because of the curve fit technique, extrapolation beyond the range of the data may produce unreliable results.

Several important issues remain to be explored and clarified to extend this work. First the repeatability of performance should be understood to a much greater degree. Some work was undertaken to assess the effects of the sequence of the speed ramp loading conditions on the friction behavior. These results showed very good consistency in the observed behavior. However, there are other sequence and duty cycle events that may result in quite different friction behavior. The fade behavior of materials or the test technique have not been explored in this series of tests. Fade sequences are an important part of a typical brake effectiveness test and often show non-steady friction behavior immediately following the high temperature operations of the fade test sequence. Environmental effects, such as moisture absorption (morning effectiveness) and ambient humidity, have also been reported to effect friction performance in some systems. These types of behavior have not been captured in the testing performed here and would not be captured with the modeling approach used in this study.

Another area requiring further investigation is the correlation of the small specimen results with behavior observed in vehicle and full scale dynamometer testing. Some consistent trends in overall behavior between the two scales have been observed in these tests. However, numerical agreement with the values of friction reported has not yet been attempted. Previous researchers have at times questioned the value of small specimen tests since good agreement between full scale and small scale friction values was not achieved. In order to make this assessment all of the various assumptions that go into the testing and the analysis of the test results to obtain friction values must be examined. Some of the lack of correlation between the two scales is not surprising when one considers that the values obtained from full scale hardware on a dynamometer, for example, are an average of the behavior occurring over the full contact surface on two brake pads. This remains a difficult issue to be addressed.

Finally, the test apparatus used in this preliminary investigation had limitations in test control which would ideally be overcome in future specimen studies. The speed control on the motor control system did not allow a continuous speed ramp to be applied and consequently the speed ramp tests were a series of discrete speed steps. Improvement in the design of the load apparatus to improve resolution of both the normal and friction force would improve the accuracy of these measurements. Possibly most significant is the measurement of temperature. In this study, temperature measurements were made with typical brake testing techniques using thermal couples. The rotor temperature was measured with a rubbing thermal couple about 7 mm from the friction track. Clearly, this temperature is an approximation to that occurring directly at the friction surface. Improvements in the measurement of temperature will allow better characterization of the role of temperature on friction behavior but will likely introduce more complexity in the test procedures and interpretation.

REFERENCES

1) "Brake Lining Quality Control Test Procedure", Society of Automotive Engineers Procedure J661.
2) Anderson, A. E., Gratch, S., Hayes, H. P., "A New Laboratory Friction and Wear Test for the Characterization of Brake Linings", SAE Paper # 670079.
3) Wilson, A. J., Belford, W. G., Bowsher, G. T., "Testing Machine for Scale Vehicle Brake Installations", The Engineer (London), Vol. 225, P.317, 1965.
4) Oldershaw, R. M., Prestidge, A. F., Birkmyre, R. C., "Brake Road Testing in the Laboratory", SAE Paper # 730563.
5) Karthik, R., "Investigation of the Friction Behavior of Automotive Brake Materials", MSc. Thesis, May 1997.
6) Hecht, R. L., Dinwiddie, R. B., Porter, W. D., Wang, H., "Thermal Transport Properties of Gray Cast Iron", SAE Paper # 962126.
7) ASF - ASTM A247-67, "Standard Test Method for Evaluating the Microstructure of Graphite in Iron Castings".
8) RS/Discover, Analysis Software and Statistical Manual, BBN Software Products Corporation, 1989.
9) MATLAB Software, The MathWorks Inc., Natick Mass.

ACKNOWLEDGMENTS

The authors would like to acknowledge Mr. R. Baer for his assistance in the DOE methods employed both in testing and data analysis for this program.

Hayes' increased airflow rotor design

A R DAUDI and **W E DICKERSON**
Hayes Lemmerz International Inc., Romulus, Michigan, USA
M NARAIN
Optical Inc., Novi, Michigan, USA

Abstract - Hayes Lemmerz has evaluated and compared the performance of airflow through different brake rotors using CFD techniques and Hayes experimental airflow machine. Hayes has patented a unique rotor with 50%-50% inboard and outboard air entry design. This increased airflow rotor design was used successfully to cool Al MMC rotors currently being developed for automotive applications. The CFD predictions of pressure, velocity and rate of airflow matched the experimental data obtained from airflow machines. The objective of this study is to compare airflow through different design rotors using TASC-FLOW software. The assignment was also to evaluate the HAYES COOL ROTOR design analytically and experimentally.

1. INTRODUCTION

In brake rotor design, fin configuration is altered to maximize airflow for effective heat removal during braking. In Fig-1, 2 and 3 three rotor designs shows air entry through inboard side, outboard side and 50%-50% from both inboard and outboard side.

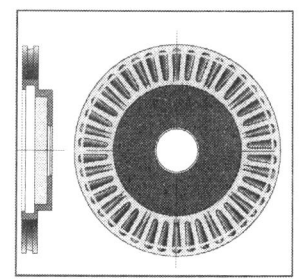

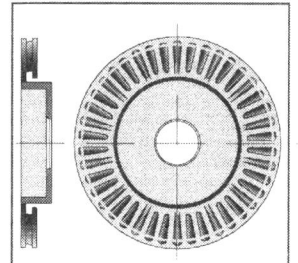

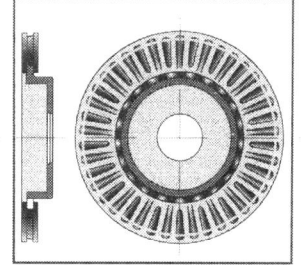

Figure 1 (PN42886)　　Figure 2 (PN-----)　　Figure 3 (PN43136)
Three rotor designs shows air entry through inboard side, outboard side and 50-50 from both inboard and outboard

In these designs the fins are straight with one or two half fins in between the full length fins. In Fig- 4, 5 and 6 three rotor designs shown have similar air entry paths but the fins are curved with one or two half fins in between the full length fins. This study has also analyzed the airflow when the number of fins increase from 30,36,42 and to 48 and finally to 72

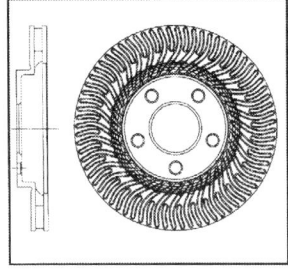

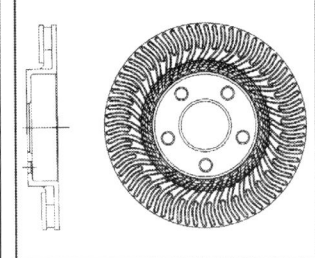

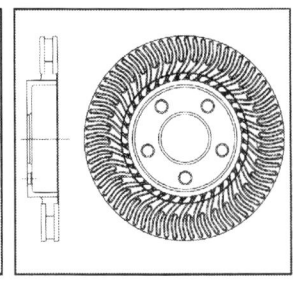

Figure 4 (X43166 in)
100% Inboard
Air Entry

Figure 5 (X43166 Out)
100% Outboard
Air Entry

Figure 6 (X43166)
50% - 50% Inboard
and Outboard
Air Entry

In brake rotor design flow separation regions are avoided in the brake rotor to eliminate hot spots on the rotor surface. Computational fluid dynamics (CFD) techniques were used to map the airflow in each section of the rotor. The straight fins were altered in length and changed in entry and exit fin radii, to remove flow separation regions. When straight fins are curved and fin number increased from 42 to 72, there is an increase in airflow velocity, but the lack of rotor symmetry makes the rotors not interchangeable from both the left and right side wheels and this adds a part number and increases manufacturing cost. This study (Part-1) analyzes flow through rotor designs. These designs are further verified for maximum heat transfer and low distortion of rotor in Part-2 of this study to be presented later.

Hayes Lemmerz uses an experimental set-up to evaluate airflow through different brake rotor designs. The airflow measurements were made on cast iron rotors and Ren-wood models of rotors. Ren-wood models are economical and quick way to verify experimentally a high airflow design. The airflow were measured by two methods : (1) Pitot tube and manometers (2) Air duct around the rotor and flowmeter. The airflow experimental measurements were extremely close to the analytical predictions of TASC-FLOW software. CFD is extremely valuable tool to quickly analyze brake rotor designs and obtain performance data to compare with other brake rotor designs performance data. The advantage of CFD technique is quick turnaround time and design completed with low development cost.

2. SUMMARY

1) Airflow through two brake rotor designs (Fig-1 to 3 straight fin design versus Fig-4 to 6 curved fin designs with different fin configurations using CFD techniques are compared.
2) First rotor (#42886) has 14 long straight fins with 2 short straight fins between each long fin (see Fig-1). Second rotor (#43136) has 21 long straight fins with 1 short straight fin in between each long fin (see Fig-3). Third rotor (# X43166 IN) has 36 long curved fins with 1 curved short fin in between each long fin (see Fig-4). Fourth rotor (#X43166) has same number of long and short curved fins as third rotor but has 50%-50% inboard and outboard air entry (see Fig-6).
3 Flow is assumed to be steady, incompressible, isothermal and turbulent.
4) CFD finite volume mesh was generated in ICEM-CFD and CFD analysis was done using TASC-FLOW (Ref-1).techniques.

5) Rotating frames of reference capability of TASC-FLOW was used to model the rotating brake rotor.
6) Hayes airflow machine (see Figure 11), was used to experimentally measure pressure, velocity and rate of airflow.
7) CFD show that first rotor straight fin design (PN 42886) has higher velocities and less flow separation than second rotor (PN 43136). CFD show that third rotor curved fin design (PN X43166-IN) also has higher velocities and less flow separation than fourth rotor (PN X43166).
8) Comparison of the flow results between the CFD and experiments are reasonable

3. OBJECTIVES

1) Evaluate and compare the performance of airflow through different brake rotors using CFD techniques.
2) Maximum velocities and regions of flow separations through brake rotor will be used to evaluate rotor designs.
3) To compare performance data for both the brake rotor designs, its important to keep the same operating conditions.
4) Use Hayes airflow machine to experimentally verify the CFD predictions of pressure, velocity and rate of airflow.

4. METHODOLOGY

4.1 Geometrical Modeling

Figure 7 and 8 show the three dimensional computation domain for the CFD analysis for the brake rotor designs. The computational mesh for the CFD analysis was created from the IGES data provided by Hayes Lemmerz. Only one pair of fins was considered for this analysis, because all the fins have the same shape and the flow was assumed to be steady state. Inlet and outlet sections are added upstream and downstream of the rotor, where the boundary conditions were imposed as shown in Figure 7 and Figure 8.

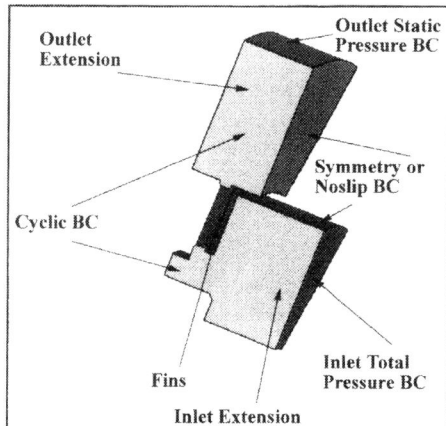

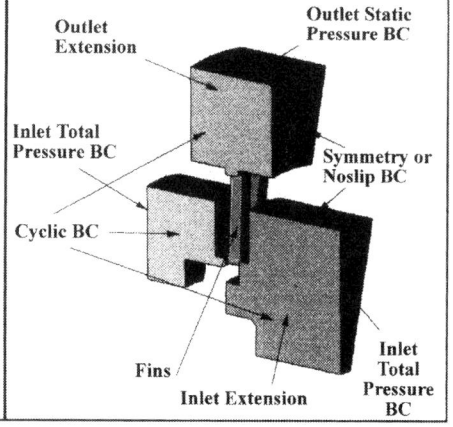

Figure 7 Part No. 42886
Computational domain and boundary conditions

Figure 8 Part No. 43136
Computational domain and boundary conditions

The CFD mesh on both rotor designs in the region of the fins, are shown in Figure 9 and 10.

Figure 9 Part # 42886
Computational mesh in the region of the fins

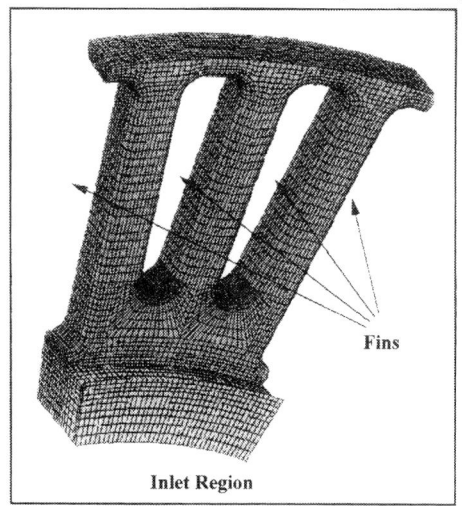

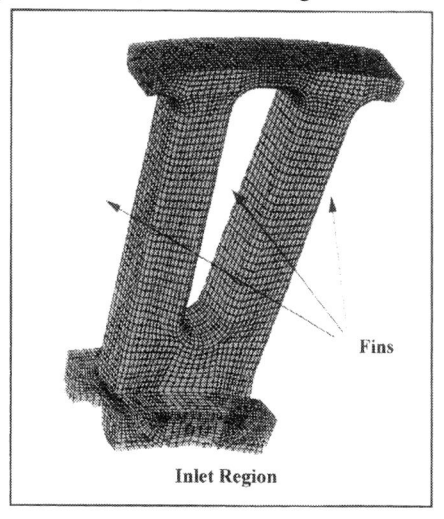

Figure 10 Part # 43136
Computational mesh in the region of the fins

4.2 CFD Mesh generation
The CFD mesh for this study has been created using ICEM-CFD, a grid generation package developed by Control Data Corp. The module used (MULCAD) allows for interactive generation of hexahedral cells.

The CFD mesh had approximately 160,000 cells for each rotor. To ensure good quality of results only hexahedral (or brick) cells were used in the CFD model. The average determinant of cells in the model is 0.9. The minimum determinant of cell is 0.4. Determinants less than 0.1 are usually not acceptable to get good results. Fine cells were used around the fins to predict the flow separation.

Mesh size and quality was monitored at all times and commonly accepted norms of mesh quality for CFD meshes were adhered to whenever possible. This is necessitated by the fact that an ill built mesh would result in poorly converged (or in extreme cases even non-physical) results. Numerical accuracy restricts grid characteristics such cell aspect ratios, warpage and skewness. All portions of the grid were developed to hold these constraints, as much as possible without generating an unacceptably large grid.

4.3 CFD Modeling
A commercially available CFD software, TASC-FLOW (Ref-1) was used for this CFD analysis. TASC-FLOW numerically solves the generalized Navier-Stokes equations. The Navier-Stokes equations describe most of the fluid flow and heat transfer phenomenon. TASC-FLOW is a finite volume based CFD software. In this CFD analysis the air flow through the brake rotor assumed to be steady, incompressible and turbulent. k-ϵ turbulence model was used in this CFD analysis. To obtain the 3D velocity field and pressure field in TASC-FLOW, the continuity, momentum and turbulence equations were solved. The governing equations were solved in rotating frames of reference.

4.4 Boundary conditions

Boundary conditions need to be provided at the faces of the computational domain. Refer to Figure 7 and Figure 8 for the location of the boundary conditions. Atmosphere pressure boundary conditions were applied at both the inlet and outlets. At the inlet and the exit a turbulence intensity of 5% and mixing length scale of 0.01 meter was used. No-slip velocity boundary condition was used at the surface of the brake rotors. Periodic boundary conditions are used at either side of the wedge of the brake rotor computational domain. Symmetry boundary conditions were used for the remaining regions.

It is extremely important that the boundary conditions be imposed far away from the region of interest. In this case the region of interest is around the brake rotor, therefore care is taken to impose the boundary conditions away from the rotor. Imposing the boundary conditions (BC's) away from the brake rotor insures that the BC's do not have an undesirable effect on the flow around the rotor. This is one of the main reasons for selecting such a big computational domain, even though the flow around the rotor is of the most importance.

The boundary conditions applied to both the brake designs are defined in Table 1

Boundary Condition	Values
Inlet	Total Pressure = 101325 Pa Turbulence intensity = 5%
Outlet	Static Pressure = 101325 Pa Turbulence intensity = 5%

TABLE 1
Boundry Conditions Applied to Brake Rotor Designs

Three operating speeds of 699,466 and 233 rpm are analyzed for each rotor design.

4.5 Material properties
Standard properties of air were used for this CFD analysis.
Density = 1.164 Kg/m^3
Viscosity = 1.824e-5 Pa-s

5. RESULTS AND DISCUSSION

5.1 Experimental and CFD results correlation
Hayes Lemmerz conducted experiments on four designs of the brake rotors. Hayes Lemmerz airflow machine is shown in figure 11.

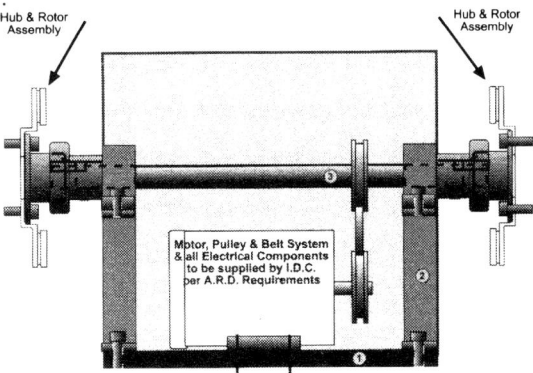

Figure 11
Hayes Lemmerz airflow machine

Rotor designs are cut into renwood models using CNC machines. Two rotors are mounted on airflow machines and with Pitot tube and digital readout of pressure and velocity head, and are measured at exit of the brake rotor. The air resistance around rotor is created in the airflow machine by duplicating the environment around the rotor in a vehicle. The hub, knuckle and wheel can be mounted. In the experiment first a Pitot tube was used to measure the velocity head at the exit of the brake rotor. The velocities obtained by the experiments are quantified in table 2. The velocities obtained by CFD are also shown in the same table.

PART # 42886

RPM	Exp. (in. of water)	Exp. Velocity (m/s)	CFD Velocity (m/s)	CFD flow rate thru rotor (kg/s) $\times 10^{-4}$	Exp. Flow Rate Kg/s $\times 10^{-4}$
699	0.215	9.4	9 - 10	206.08	229.53
466	0.09	6.1	5.5 - 6.5	116.76	153.71
233	0.02	2.9	2.5 - 3.0	60.76	65.58

PART # 43136

RPM	Exp. (in. of water)	Exp. Velocity (m/s)	CFD Velocity (m/s)	CFD flow rate thru rotor (kg/s) $\times 10^{-4}$	Exp. Flow Rate Kg/s $\times 10^{-4}$
699	0.225	9.6	9.25 - 10.5	194.10	175.20
466	0.1	6.4	6 - 7	122.64	118.87
233	0.025	3.2	2.5 - 3.0	59.03	52.26

PART # X43166 IN

RPM	Exp. (in. of water)	Exp. Velocity (m/s)	CFD Velocity (m/s)	CFD flow rate thru rotor (kg/s) $\times 10^{-4}$	Exp. Flow Rate Kg/s $\times 10^{-4}$
699	.320	11.5	10.5 - 11.5	360.72	368.9
466	.138	7.6	7.0 - 8.0	232.45	235.7
233	.031	3.6	3.4 - 4.0	106.81	106.57

PART # 43166

RPM	Exp. (in. of water)	Exp. Velocity (m/s)	CFD Velocity (m/s)	CFD flow rate thru rotor (kg/s) $\times 10^{-4}$	Exp. Flow Rate Kg/s $\times 10^{-4}$
699	.194	9.0	9.0 - 10.5	244.84	215.19
466	.087	6.0	6.0 - 7.0	162.94	144.48
233	.023	3.1	3.4 - 3.9	76.79	62.50

Table 2
Flow Rate and Velocity Through the Four Brake Rotor Designs

It can be seen from Figure 12 and Figure 13, that the velocities at the exit of the rotor is not the same but varies over a range and this range can be determined from the color legends in these figures.

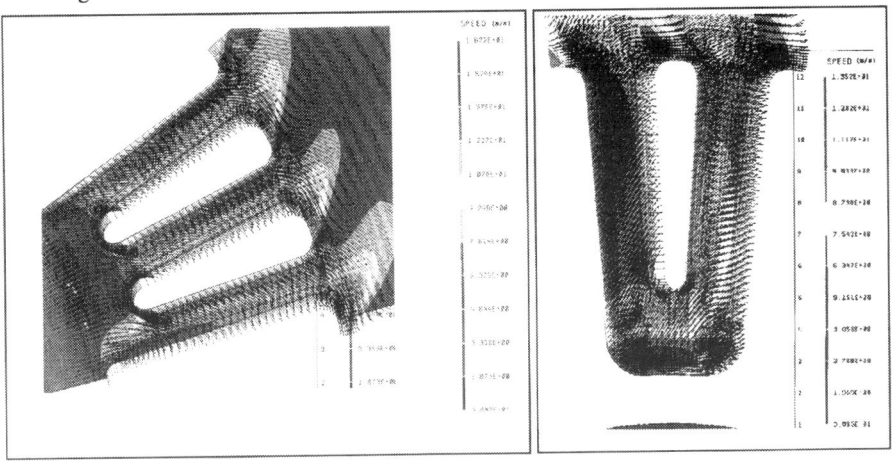

Figure 12
Part # 42886
699 RPM
Absolute Velocity mid-plane

Figure 13
Part # 43136
699 RPM
Absolute Velocity mid-plane

In experiments only an average (or smeared) velocity at the exit of the rotor is measured. Therefore, in table 2, a range of velocity is given for correlation purpose. The table also includes the total flow rate (CFD) through the rotor. It is not possible to compute the total flow rate with the current experimental setup, as such the experimental setup was altered as shown in figure 14, which shows a convolute duct built around the rotor to capture air exiting from rotor.

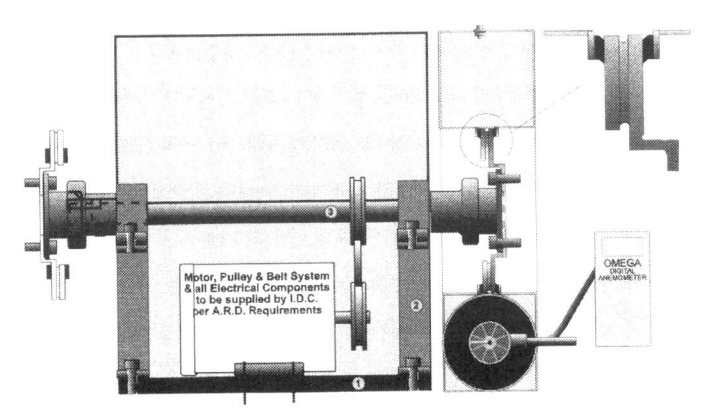

Figure 14
Hayes Lemmerz airflow machine with convolute duct

The rotor rotates with duct stationary and interface is a stationary lip seal shown in Figure 14. It can be seen from the table that velocities at the exit of the rotor and the flow rates for both the designs are similar in the CFD study and the experiments.

5.2 CFD Results
The CFD results will be primarily shown on four or three planes. The mid-plane passes through the middle of all the fins as shown in Figure 15 and 16. The pass1, pass2 and pass3 cut planes goes through passages 1, 2 and 3 respectively as shown in figure 15 and 16.

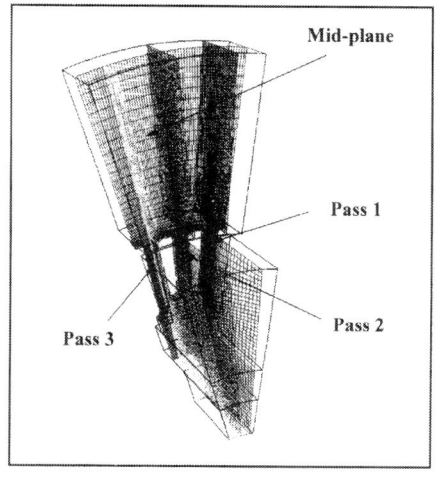

Figure 15 Part No. 42886
Cut planes description for CFD results

Figure 16 Part No. 43136
Cut planes description for CFD results

The following quantities will be plotted on these planes at different speeds:
   1. Absolute Velocity Vectors and Magnitudes: This is the flow pattern in a stationary frame of reference. In other words the velocities observed by person standing away from the rotor. This information is used to correlate with the experimental velocity measurements.
   2. Relative Velocity Vectors and Magnitudes: This is flow pattern in the rotating frame of reference. In other word, this is the flow pattern observed by the person standing on the rotor and rotating with it. This information is most valuable as it gives an indication of the effectiveness of heat removal. It is the relative motion of the air with respect to that to that of the rotor that removes the heat and not the absolute value of the velocity itself.
   3. Absolute Total Pressure: This is the sum of the velocity head and the static head in an absolute frame of reference. This provides information on the head rise.
   4. Relative Total Pressure: This is the sum of the velocity head and the static head in a relative frame of reference. This provides information on the head loss or energy loss.
   5. Absolute Velocity Head: This is the head measured by the Pitot tube experimentally.

5.2.1 CFD results of airflow through rotor with different fin spacings (30,36,42 and 48)
The objective of this test is to determine the affect of number of fins on airflow through the brake rotors. For this test PN 42886 was selected and four fin spacings 30,36,42 and 48 were studied. The CFD methodology, mesh generation, boundary conditions and properties were same as detailed above. The relative velocity vectors on mid-plane through all the fins are shown in figure 17, figure 18, figure 19 and figure 20 for 48,42,36 and 30 fins respectively. Table 3 shows the comparison of flow rates and exit velocities, through the fin passages for all the cases. From the tables and the figures it can be seen that the flow rate through the brake rotor increases as the number of fins decreases. The average exit velocity at the end of the fins for all the cases in around 5 m/s. A higher flow rate through the brake rotor does not necessarily lead to increased heat transfer rates. The heat transfer rate depends on the surface area and the local temperature gradients at the surface of the rotor. Less the number of fins, less will be the surface area available

for heat transfer from the rotor. It is highly recommended that we run these airflow designs through thermal and distortion software to find the heat transfer effects through the rotor surface.

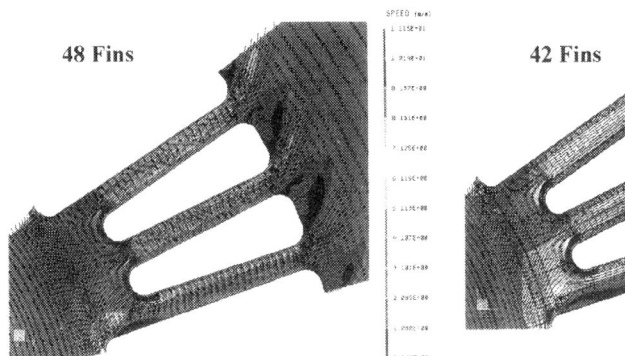

48 Fins

42 Fins

Figure 17 Part No. 42886
466 RPM Relative velocity mid-plane

Figure 18 Part No. 42886
466 RPM Relative velocity mid-plane

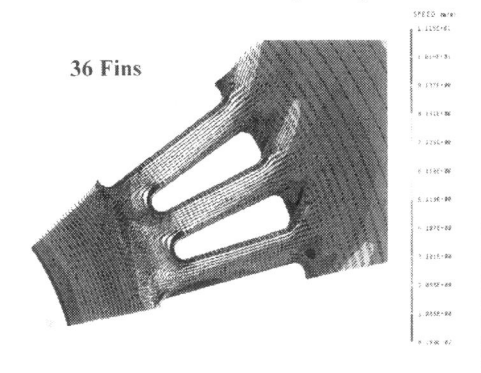

36 Fins

30 Fins

Figure 19 Part No. 42886
466 RPM Relative velocity mid-plane

Figure 20 Part No. 42886
466 RPM Relative velocity mid-plane

# of Fins	Mass flow thru' a section of the brake rotor	Flow Area at the exit between the fins	PASSAGE 1			PASSAGE 2			PASSAGE 3			Mass flow through the rotor
			Flow Rate	Ave. Velocity at the exit	% of Total	Flow Rate	Ave. Velocity at the exit	% of Total	Flow Rate	Ave. Velocity at the exit	% of Total	
	kg/s*10⁻⁴	m²*10⁻⁴	kg/s*10⁻⁴	m/s	%	kg/s*10⁻⁴	m/s	%	kg/s*10⁻⁴	m/s	%	kg/s*10⁻⁴
48	6.96	3.96	2.50	5.42	35.9	2.37	5.14	34.1	2.09	4.53	30.0	111.36
42	8.34	5.76	3.18	4.74	38.1	2.94	4.39	35.3	2.22	3.31	26.6	116.76
36	13.59	8.16	5.44	5.73	40.0	5.09	5.36	37.5	3.05	3.21	22.5	163.08
30	16.85	11.5	6.89	5.15	40.9	6.35	4.74	37.7	3.61	2.70	21.4	168.50

TABLE 3
Flow Rate and Velocity Through the Brake Rotor

5.2.2 CFD results for first straight fin brake rotor design (PN 42886)

Figure 12 shows the absolute velocity vectors. From this figure it can be seen that the velocities at the exit of the rotor are around 9-10 m/s, which was also obtained by the experiments.

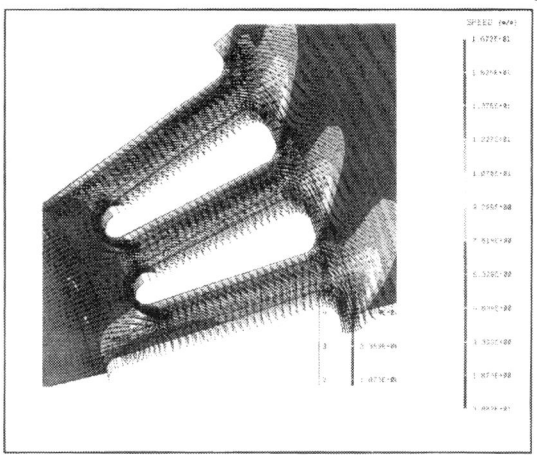

**Figure 12
Part # 42886
699 RPM
Absolute Velocity mid-plane**

Figure 21 shows the velocity head in inches of water. The velocity head is in the region of the exit of the rotor is around 0.23-0.20 inches of water.

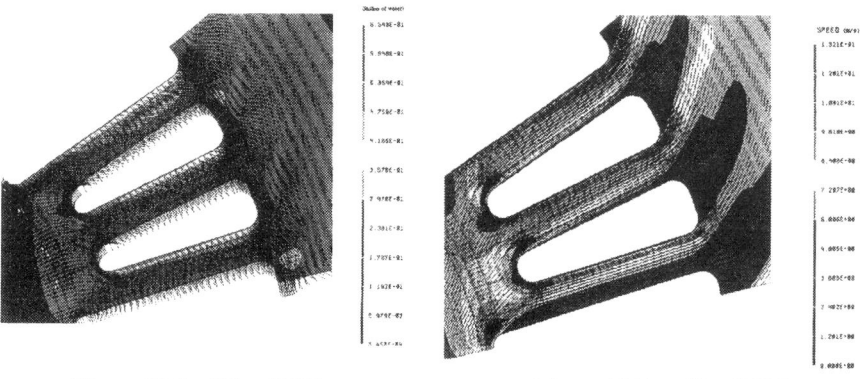

**Figure 21 Part No. 42886
699 RPM Velocity head mid-plane**

**Figure 22 Part No. 42886
699 RPM Relative velocity mid-plane**

Figure 22 shows the relative velocity on the mid-plane. From this figure a big separation zone (or recirculation zone) can be seen near the bottom long fin. The reason for this separation is that the air is not able to negotiate the sharp bend to get into the passage 1. A pocket of air is residing in this region and it will not remove any heat. This will lead to temperature hot spot. The flow through passage one is the least and the flow increased from passage 1 to passage 3.

Figure 23 shows the total head rise in the rotor. Figure 24 shows the losses in the rotor. The losses (relative total pressure drop across the fins) are the greatest for passage one and minimum for passage 3.

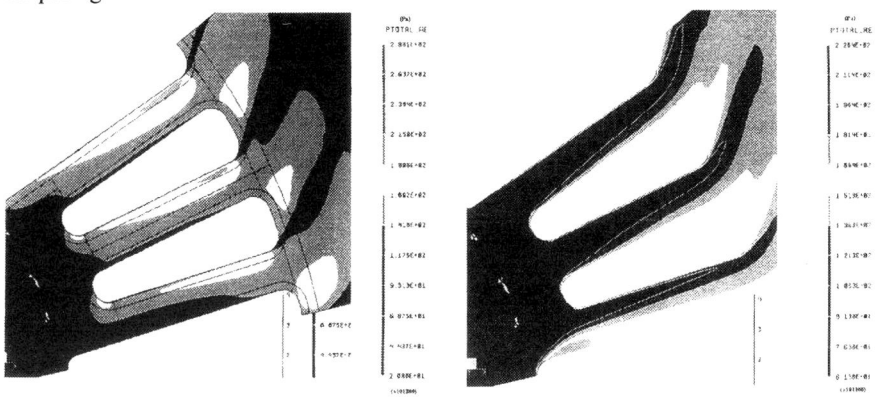

Figure 23 Part No. 42886
699 RPM Absolute total pressure
mid-plane

Figure 24 Part No. 42886
699 RPM Relative total pressure
mid-plane

Figure 25 through 30 show the relative velocity vectors and relative total pressure drops on pass1, pass2 and pass3.

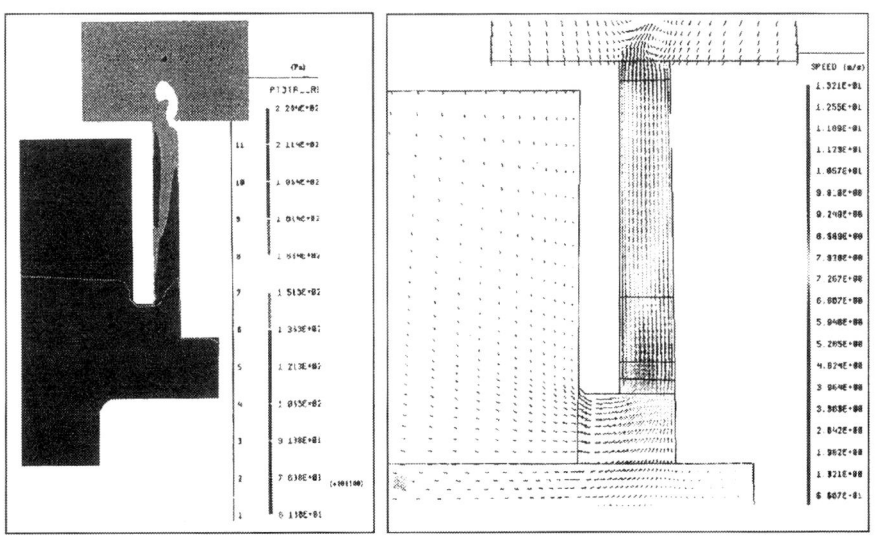

Figure 25 Part No. 42886
699 RPM Relative total pressure
PASS 1

Figure 26 Part No. 42886
699 RPM Relative velocity
PASS 1

137

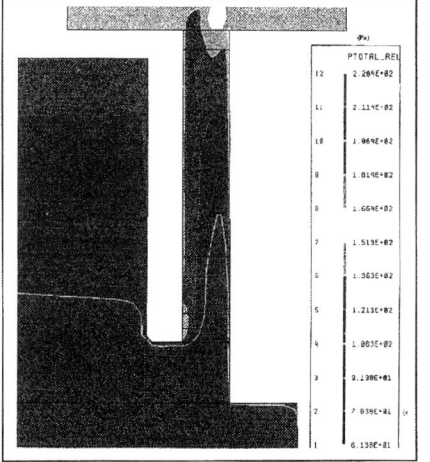

Figure 27 Part No. 42886
699 RPM Relative total pressure
PASS 2

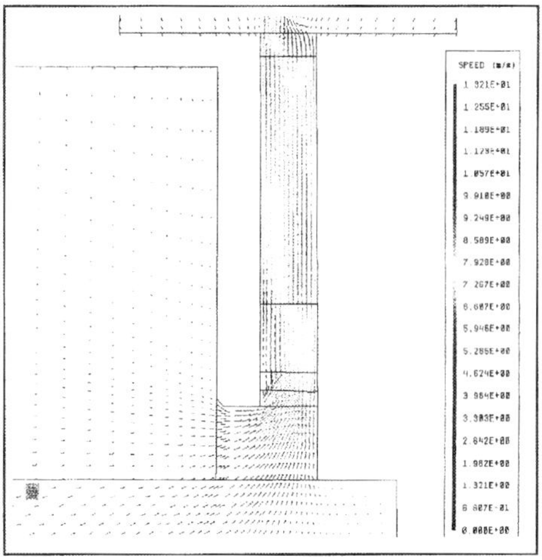

Figure 28 Part No. 42886
699 RPM Relative velocity
PASS 2

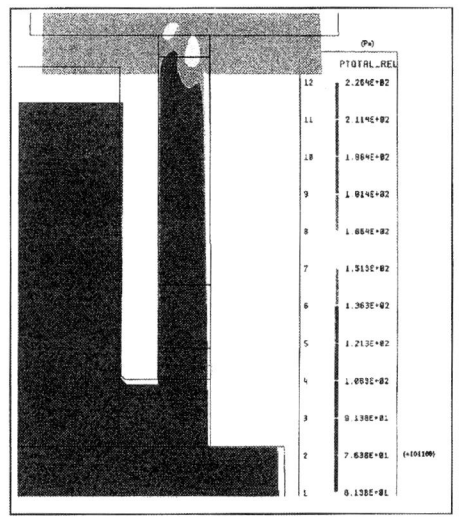

Figure 29 Part No. 42886
699 RPM Total pressure
PASS 3

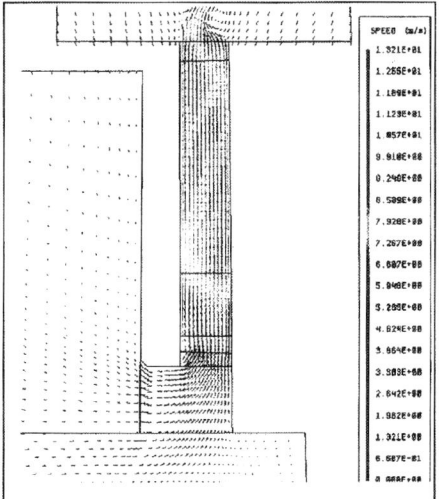

Figure 30 Part No. 42886
699 RPM Relative velocity
PASS 3

5.2.3 CFD results for second straight fin brake rotor design (PN 43136)

Figure 31 shows the absolute velocity vectors. From this figure it can be seen that the velocities at the exit of the rotor are around 9.25-10 m/s, which was also obtained by the experiments.

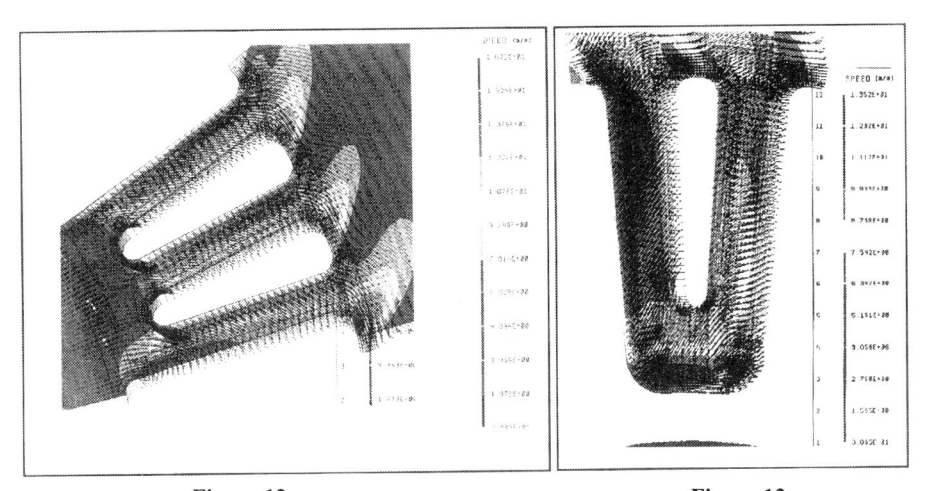

Figure 12
Part # 42886
699 RPM
Absolute Velocity mid-plane

Figure 13
Part # 43136
699 RPM
Absolute Velocity mid-plane

By comparing figure 12 and 13 it can be seen that the difference in the maximum velocity at the exit of the rotor is negligible.
Figure 32 shows the velocity head in inches of water. The velocity head is in the region of the exit of the rotor is around 0.24-0.21 inches of water. Figure 33 shows the relative velocity on the mid-plane. From this figure it can be seen that the most of the flow is entering the second passage. This will lead to poor heat rejection rates around passage 1.

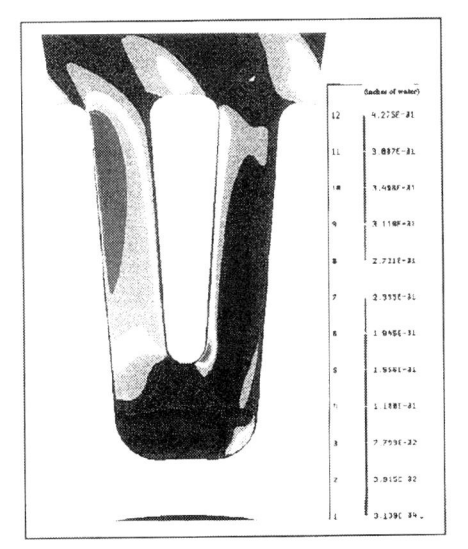

Figure 31 Part No. 43136
699 RPM Velocity head mid-plane

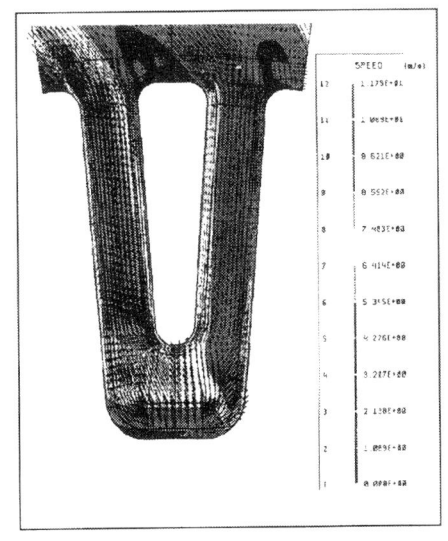

Figure 32 Part No. 43136
699 RPM Relative velocity mid-plane

The reason for this phenomenon maybe a large radius at the base of the first long fin. The absolute total pressure on the mid-plane is shown in figure 33. Figure 34 shows that most of the losses are across passage 1.

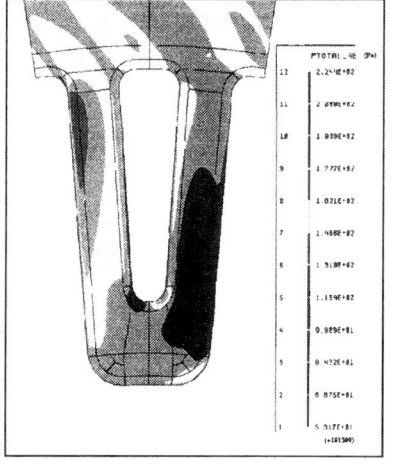

**Figure 33 Part No. 43136
699 RPM Absolute total pressure
mid-plane**

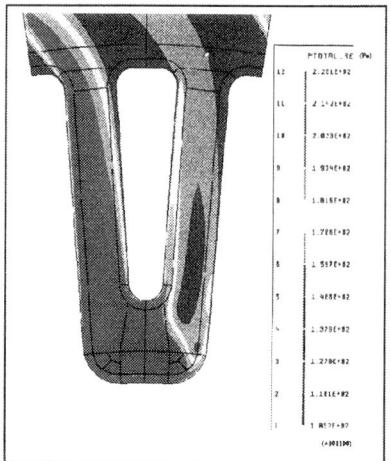

**Figure 34 Part No. 43136
699 RPM Relative total pressure
mid-plane**

The relative velocity vectors on pass1 are represented in figure 35. It is interesting to notice that more flow enters from the right side of the rotor than the left side of the rotor. Also from this figure it can be seen that this design has two separation zones instead of one (PN 42886, figure 26) in the axial direction. This could lead to more hot spots and higher losses.

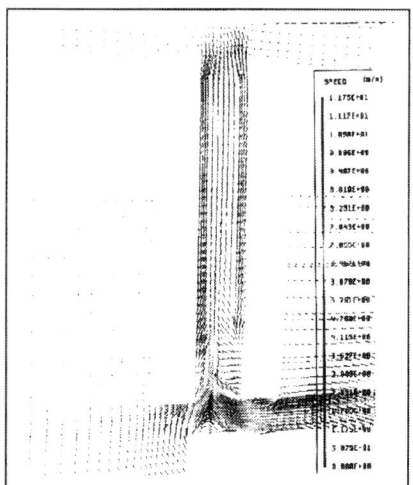

**Figure 35 Part No. 43136
699 RPM Relative velocity
PASS 1**

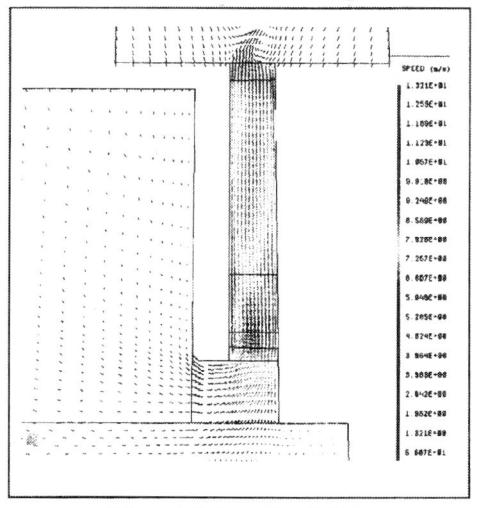

**Figure 26 Part No. 42886
699 RPM Relative velocity
PASS 1**

For brevity we have not included figures for 466 and 233 rpm. The CFD results for lower rpm's are shown only on the mid-plane. Higher speeds (699 rpm) are considered for detailed study because they represent the worst case scenario.

5.2.4 CFD results for third curved fin rotor design (PN X43166 IN)

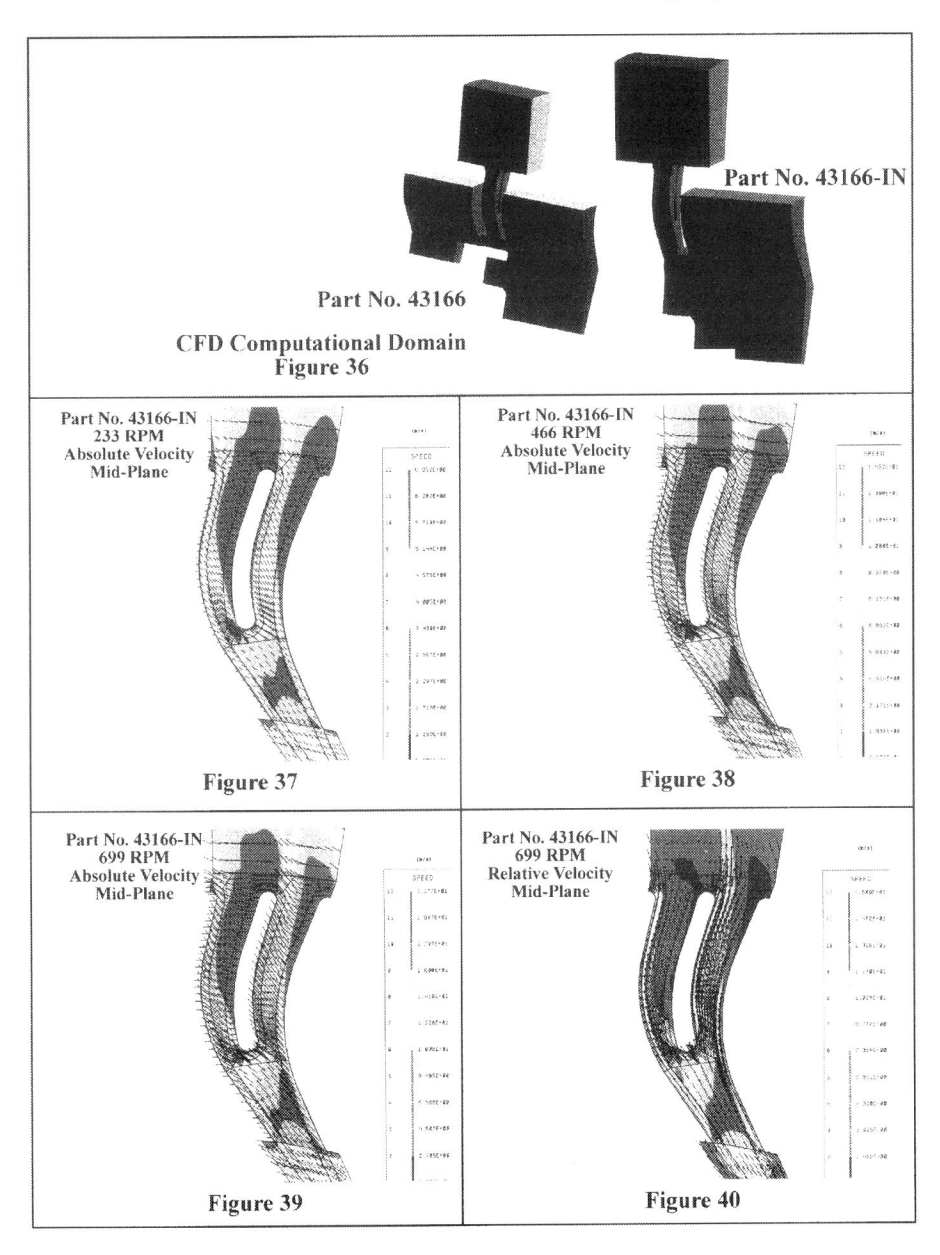

Part No. 43166-IN

Part No. 43166
CFD Computational Domain
Figure 36

Part No. 43166-IN
233 RPM
Absolute Velocity
Mid-Plane

Figure 37

Part No. 43166-IN
466 RPM
Absolute Velocity
Mid-Plane

Figure 38

Part No. 43166-IN
699 RPM
Absolute Velocity
Mid-Plane

Figure 39

Part No. 43166-IN
699 RPM
Relative Velocity
Mid-Plane

Figure 40

5.2.5 CFD results for fourth curved fin rotor design (PN X43166)

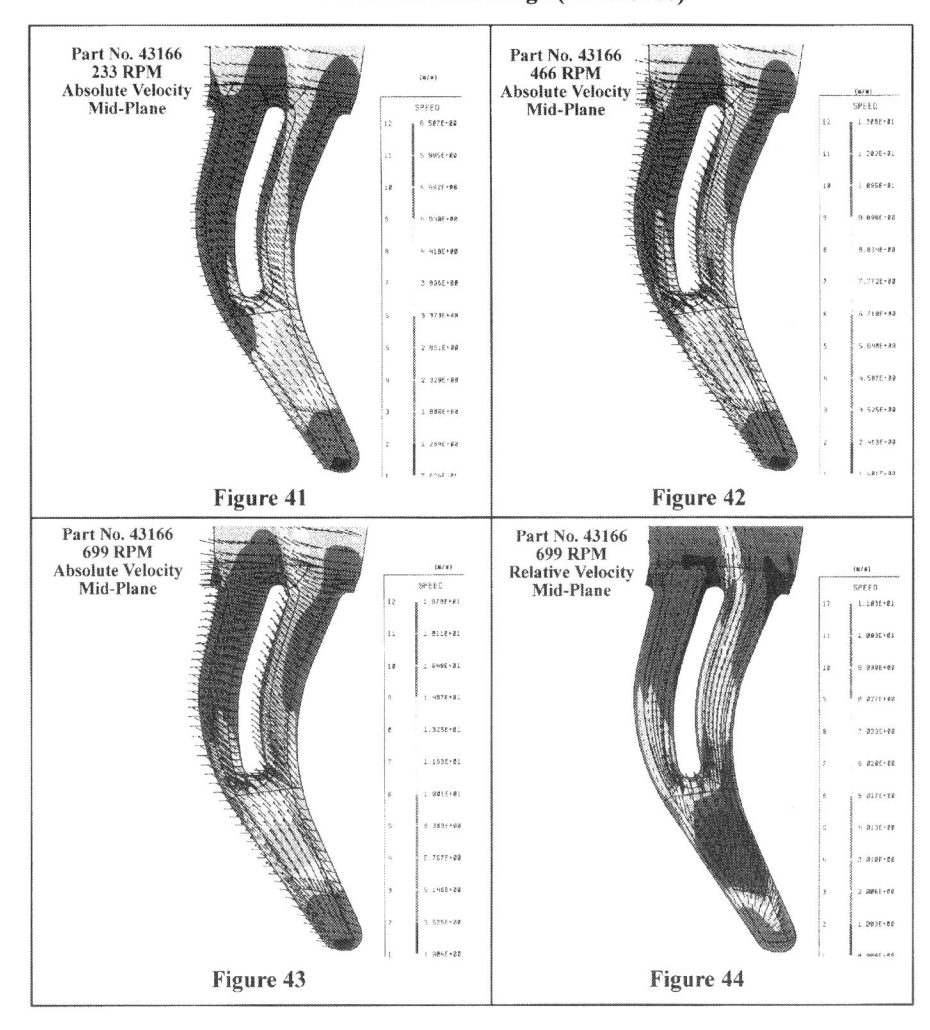

Figure 41

Figure 42

Figure 43

Figure 44

The mass flow rate through the rotor PN 43166-IN is 360.72 kg/s and the velocities at the exit of the rotor are in the range of 10.5 to 11.5 m/s (Table 2) for 699 rpm. Among all the rotor designs considered, the flow through this design is about 25-35% better than the other designs based on the operating conditions. These results compare very well with the experimental data. From Figure 40, it can be seen that the flow through the left passage is more than in the right passage. There is also a region of low velocity to the left of the middle fin. By changing the fin profile at the lower region, the flow uniformity and distribution through the channels can be further improved.

For the rotor PN 43166, the flow rate and the velocity at the exit of the rotor are slightly higher than the experimental measurements. The velocity distribution in this rotor is similar to that of rotor PN 43166-IN, except that the magnitudes are lower in the rotor PN-43166.

Since the fins are curved to guide the flow, there are no flow separation regions at the entrance of the rotor and as seen in earlier designs

The other advantage of the 72 fin rotor over the 42 fin rotor is the increased surface area which will improve heat removal from the rotor.

6. CONCLUSIONS

Computational fluid dynamics (CFD) techniques have been successfully used to evaluate and compare the performance of four rotor designs with different fin configurations. CFD results have also compared very well with the experimental data from Hayes Lemmerz airflow machine. CFD results show that the maximum velocities and flow rates are similar for both the rotor designs for all the rpm's considered. The first rotor with part # 42886 has less regions of flow separation than the second rotor design with part# 43136. Therefore it is likely that the second rotor (# 43136) will have more temperature hot spots. But further thermal study (Part 2) needs to be done using CFD techniques to determine the temperatures in the rotors and resulting distortion.

Temperature distribution in the rotor depends on the effectiveness of heat removal by the airflow through the fins. Heat transfer through rotor depends on (1) Total surface area available for heat transfer (2) Local temperature gradient at the metal-air interface. Local temperature gradient depends greatly on the airflow patterns in the rotor and the bulk fluid temperature. From the above we note that (a) higher airflow velocity leads to increased heat transfer (b) recirculation zones (flow separation region) decreases heat transfer (c) Less bulk fluid temperature leads to increased heat transfer. Airflow studies give only partial information to successful brake design. Good airflow rotor designs have to be run through rotor thermal transfer and distortion software to arrive at the best robust brake rotor design.

The new straight 42 fin rotor design (PN 43136) has 5% (*) more airflow than current rotor design (PN 42886). The new rotor design runs cooler than current design. Also the new curved 72 fin design (PN X43166-IN) has more airflow than straight 42 fin design. Rotor temperature and thermal distortion analysis of these designs using the airflow computer models and the experimental verification of rotor temperature and rotor distortions is being done presently and would be reported later.

(*) Velocity Head Difference $(.24-.20)/(.20) \times 100\% = 20\%$ as seen from Figure 21 & 31

$$\text{Velocity} = \sqrt{20} = 4.47\% \perp 5\%$$

Reference:
1- TASC-FLOW an AEA Technology, Advanced Scientific Computing developed software used (under lease)by Manish Narian of OPTIMAL Inc. Novi, Michigan 48375 USA.

Determination of water content in brake fluid by refractive index

T E RYAN
Leica Microsystems Inc., New York, USA

ABSTRACT

Water concentration in Brake Fluid can play a key role in the safety of the brake system. This paper examines the use of refractive index measurements as an indicator of water content in brake fluid. The boiling point of brake fluid is significantly reduced as water content increases. High water content and brake overheating can combine to create a compressible gas phase in the system causing loss of brake function. Traditionally water content in brake fluid is determined by Karl Fischer Titration, a technique not easily adaptable as a routine service test. Refractive index measurements are quick and can be performed with simple, inexpensive instruments. The formulation of brake fluid within a class (e.g. DOT 3, DOT 4) may vary. Therefore, to be a viable technique the absolute refractive index (nD), the slope of nD vs. % water, the temperature coefficient and the dispersion of the available formulations must be consistent.

The refractive indexes, with and without added water were determined for a range of commercially available DOT 3, High Temperature DOT 3 and DOT 4 brake fluids. The results are compared with Karl Fischer % water content and actual boiling points of the fluids. Similar tests were conducted on fluids selected from a random sample of 129 "in use" road vehicles. Data from these vehicles is consistent with data obtained with new clean fluids with and without added water.

The results indicate that measurement of the refractive index of brake fluid can be correlated to % water content for the range of fluids and road vehicles tested. Brake fluids within a class exhibit little difference in absolute refractive index and identical changes due to water content.

1. INTRODUCTION

Over the service lifetime of brake fluid the percentage of water dissolved in the fluid increases (1). This is due to the extremely hygroscopic nature of glycol ether based brake fluids. As water content increases the boiling point of the brake fluid decreases, the viscosity at low temperatures increases and the fluid becomes more corrosive (2). This can result in damage to cylinder bores and pistons. Ashley, Kumar and Firey (3) have recently demonstrated that high disk and pad temperatures cause a gradual heating of brake fluid. The measured fluid temperatures are high enough to boil brake fluid containing significant levels of moisture. The compressible gas phase produced by the boiling fluid in the closed hydraulic system leads to loss of braking (4).

A number of methods exist for determining water content in brake fluid. The Department of Transportation Motor Vehicle Safety Standard No. 116 part 571 section 6.1 describes a method for equilibrium reflux boiling point (see also American Society of Testing and Materials Method D1120) (5). Water content can also be measured by Karl Fischer titration as described in American Society of Testing and Materials method D 1364 (6). Our previous work (7) has demonstrated that volumetric Karl Fisher titration provides excellent results with DOT 3 and 4 brake fluids. A method relating dielectric constant to boiling point (8) and an infrared technique (9) have also been described. All of these techniques require expensive instrumentation, skilled laboratory technicians and significant set up and run time. They are not easily adapted for use in a motor vehicle service facility.

Refractometers have long been used as field instruments for a variety of applications. They have proven useful in determining sugar content in grape juices, water content in honey, the concentration of water soluble metalworking lubricants and protein levels in blood serum. American Society of Testing and Materials Standard Practice D 3321 (10) details use of a refractometer for determining the freeze point of engine coolants with far greater accuracy than traditional hydrometer methods. Refractometers require a very small sample size, they are relatively inexpensive and they are fast and easy to use.

There is no standard formula for brake fluid. Commercial brake fluids must meet viscosity specifications, wet and dry boiling point specifications and an evaporation specification. The proportion of various brake fluid components is not specified. To be useful as a field test the absolute refractive index of a population of fluids which can vary in makeup must be consistent. The change in refractive index as water content increases must also stay fairly constant. We have previously shown that the refractive index of clean, unused brake fluid decreases as water is added to the fluid (7). The slope of this change is identical for both DOT 3 and DOT 4 brake fluids. Preliminary data has demonstrated that the absolute refractive index of a population of DOT 3 and DOT 4 brake fluids is consistent enough that this measurement could prove feasible as a field test for water in brake fluid. The variation in refractive index caused by temperature changes and the optical dispersion of the fluids studied are also very similar for clean, unused brake fluids.

A final test to determine the efficacy of refractometry in measuring water content in brake fluid is that testing of "in service" fluids produces results similar to clean new fluids. Used brake fluid can contain rubber particulate, corrosion products and may have undergone compositional changes as a result of exposure to high temperatures.

The purpose of this study is to determine if Refractometry is a viable field test method for determining water content and boiling point of DOT 3 and 4 brake fluids. Commercially available brake fluids were purchased and analyzed by Refractometry, Karl Fischer titration and boiling point determination. Brake fluid samples from 125 operating vehicles were also collected and analyzed.

2. EXPERIMENTAL

Commercially available DOT 3 and 4 Brake Fluids were purchased from automotive supply dealers at locations throughout the U. S. All the containers remained sealed until the start of experiments. Samples of brake fluid (5 to 50 ml) were removed from the master cylinder of the test vehicles and stored in air tight polyethylene bottles. All testing was done within 3 hours of sampling. Sodium D-line (589nm) refractive indexes measurements were read to 0.00001 at 20 Deg. C. and 24 Deg. C. using a Leica AR600 automatic refractometer. The refractometer was calibrated with standard oils traceable to the National Institute of Standards and Technology. Karl Fischer water concentration was determined on 1 ml samples of brake fluid with an Orion AF8 automatic volumetric titrator. Pyridine free reagents were used for all titrations. Boiling points were determined following the method outlined in ASTM D 1120. A calibrated electronic thermistor was substituted for the glass thermometer described in the method.

3. RESULTS

3.1 The Relationship Between Refractive Index, Water Content and Boiling Point of Brake Fluid.

The relationship between refractive index and water concentration for DOT 3, High Temperature DOT3 and DOT 4 brake fluids is shown in Figure 1. 1% to 8% water (wt./wt.) in increments of approximately 1% were added to each new fluid to develop this data. Refractive indexes were measured at approximately 20 and 24 Deg. And were corrected to 20.0 Deg. C. It is interesting to note that all three types of brake fluid exhibit excellent correlation between refractive index and water concentration despite the fact that the composition from fluid to fluid within a category may vary. The curve fitted data in figures 1B and 1C are almost identical. This demonstrates that there is no appreciable difference between the refractive index and water concentration correlation for the High Temperature DOT 3 and the DOT 4 fluids.

The boiling points of a smaller number of fluids were also determined over a 1% to 6% water concentration range. The relationship between boiling point and water as determined by Karl Fischer titration is shown in Figure 2. The results from Figures 1 and 2 were used to determine average water concentration and refractive index at the specified wet and dry boiling points for DOT 3 and DOT 4 fluids. These results are presented in Table 1 along with the average boiling point, % water and refractive index for fluid fresh from the container. All of the fluids studied greatly exceeded the minimum wet and dry boiling points specified by Society of Automotive Engineers J1703 and Department of Transportation Motor Vehicle Safety Standard No. 116.

Table 1
Water Concentration and Refractive Index
New Fluid and at Wet/Dry Boiling Points

	DOT 3, N = 7		
	Boiling Point	% H2O wt./wt.	Refractive Index, nD
Fresh Boiling Point	252 C.	0.23	1.4435
	+/- 7 C.	+/-0.14	+/-0.0004
Range	240-262 C.	0.06-0.42	1.4428-1.4440
Minimum Dry Boiling Point	205 C.	1.15	1.4429
		+/- 0.38	+/-0.0004
Range			
Minimum Wet Boiling Point	140 C.	4.4	1.4406
		+/-0.3	+/-0.0004
Range			
	High Temp DOT 3, N = 6		
Fresh Boiling Point	281 C.	0.14	1.4457
	+/- 18 C.	+/-0.07	+/-0.0004
Range	263-305 C.	0.07-.26	1.4449-1.4460
Minimum Dry Boiling Point	205 C.	1.56	1.4447
		+/- 0.35	+/-0.0004
Minimum Wet Boiling Point	140 C.	4.65	1.4425
		+/-0.39	+/-0.0004
	DOT 4, N = 9		
Fresh Boiling Point	259	0.18	1.4452
	+/- 18 C.	+/-0.08	+/-0.0005
Range	248-264 C.	0.06-0.30	1.4448-1.4462
Minimum Dry Boiling Point	230 C.	1.11	1.4447
		+/- 0.28	+/-0.0005
Minimum Wet Boiling Point	155 C.	5.1	1.4418
		+/-0.46	+/-0.0005

3.2 Water Content and Refractive Index Measurements From Brake Fluid in Operating Vehicles.

The master cylinder brake fluid water content of 131 vehicles was analyzed. The reservoir cap of all the vehicles tested recommended DOT 3 Brake fluid. A histogram of the results is shown in Figure 3. The average water concentration in the vehicles tested was 2.6% +/-1.5%. The vehicles ranged in year from 1976 to 1997. Average age of the vehicles was 8 years +/- 4 years. The median water content was 2.4%. In 60% of the vehicles the measured water content was greater than 2%. 25% of the vehicles had a water content greater than 4%, near the wet equilibrium reflux boiling point of typical DOT 3 brake fluid.

The physical appearance of the fluids tested varied greatly from sample to sample. The representative range of color of the samples is depicted in Figure 4. A subjective analysis of the color of the samples is shown in Figure 5. There was no indication that the color or darkness of the samples contributed to any variability in the analysis. Calculated correlation coefficients for refractive index versus % water were identical even when the darkest samples were excluded.

The data shown in Figure 6 includes all samples, regardless of color. Figure 6 compares refractive index readings to water content for 125 "in use" road vehicles. No attempt was made to differentiate between standard DOT 3 and High Temperature DOT 3 used brake fluids. A curve for the used brake fluid data falls between the curve fit lines for clean standard DOT 3 and high temperature DOT 3 fluids as shown in Figure 7.

3.3 Temperature Coefficients of New Clean and Used Fluids

Temperature has a large effect on refractive index measurements and must be controlled or corrected for very carefully. Aqueous solutions generally exhibit a temperature coefficient of -0.0001 to -0.0002. The temperature coefficient of organic liquids is approximately -0.0004. This corresponds to about 0.5% change in concentration per degree centigrade. Our previous work demonstrated that the temperature coefficients of new DOT 3, High Temperature DOT 3 and DOT 4, brake fluid were identical. The average temperature coefficient of a sampling of used DOT 3 brake fluid is also identical. Temperature coefficient data is shown in table 2.

Table 3

Temperature Coefficients of DOT3 and DOT4 Brake Fluids

Water Concentration % wt./wt.	DOT3 N=9	DOT4 N=7
New	-0.00038 +/-0.00002	-0.00041 +/-0.00004
2%	-0.00038 +/-0.00003	-0.00040 +/-0.00004
4%	-0.00039 +/-0.00002	-0.00039 +/-0.00003
6%	-0.00037 +/-0.00003	-0.00038 +/-0.00002
Used Fluid, N = 101	-0.00037 +/- 0.000014	

4. DISCUSSION

Refractive index is a unique physical property of chemical compounds. It is often specified as an indicator of purity and is used to determine concentration for a wide range of commercial solutions. Measuring refractive index to a high degree of accuracy (+/- 0.00002) is fairly easy with an automatic laboratory refractometer. The accuracy of concentration as determined by refractive index is very dependent on temperature corrections. Temperature must be measured accurately and the entire system including sample, measuring prism and environment must be in equilibrium (11). Based on a comparison of methods between Karl Fischer titration and temperature corrected refractive index measurements this work has demonstrated that refractive index can be used to determine water concentration in brake fluid. The measurement can be applied to a population of different brands of brake fluid within a class of brake fluid. This is significant in that each brand of brake fluid within a DOT classification can vary with respect to type and concentration of glycol ether, glycol derivatives, alkylamines and alkyl aromatics. The correlation coefficients and standard error for the comparison of methods are shown in Table 3.

Table 3
Determination of Water Concentration in Brake Fluid
Correlation Coefficients and Standard Error

	Used	DOT 3	HT DOT 3	DOT 4
R Square	0.85	0.94	0.93	0.92
Standard Error	0.72	0.57	0.58	0.64
Observations	125	103	100	135

To be useful as a field test the results from brake fluid which has been in service for a period of time must be similar to results obtained on new, clean fluid. In service fluid is typically darker and may contain particulate contamination. The absolute refractive index, the change in refractive index as water concentration changes, and the temperature coefficients for a population of fluids should be consistent for a large population of samples. The data from used fluid samples is very similar to the data for the clean fluids. Contaminants in the used fluid samples appear to have little effect on the Karl Fischer titration, the refractive index measurement and the temperature coefficient. The per cent water versus refractive index curve fit of the used samples falls between the two types of DOT 3 fluids as expected.

The safety of a motor vehicles braking system is directly related to the water content of the brake fluid. As water content increases, the boiling point decreases. Water content in brake fluid increases 1% to 2% per year (1). A recent study (3) has shown that temperatures of brake fluid in the wheel cylinders of vehicles can reach high enough levels to boil brake fluid containing significant amounts of water. Our study on used brake fluid confirms an earlier study that showed 25% of samples vehicles had brake fluid with water content at or below the wet equilibrium reflux boiling point (11). Figure 3 of this study shows 36 of 131 (27%) vehicles had water content exceeding 4%. The brake fluid in these vehicles could reach the boiling point under high braking load for short periods or light but continuous braking such as encountered when descending from elevation.

Regular checking of brake fluid by measuring the refractive index could eliminate the safety hazard represented by vehicles with high moisture content brake fluid. If 2% water as measured by Karl Fischer titration was chosen as a cutoff point for our used vehicle sample the curve fit line relating refractive index to water in Figure 6 would have correctly predicted that 74 of 125 vehicles (59%) exceeded 2% water, and 37 of 125 (30%) had less than 2% water. The refractive index determination would have placed 7 vehicles of 125 (5.6%) in the OK range while the actual water concentration exceeded 2%. The highest Karl Fischer water determination in this group was 3.1%, still significantly below the wet equilibrium reflux boiling point for DOT 3 fluids. The refractive index would have indicated an additional 7 vehicles (5.6%) exceeded 2% water while the Karl Fischer titration indicated they were below 2%.

REFERENCES

1. Radlinski, R. W., R. J. Forthofer and J. L. Harvey. 1971. Operating Performance of Motor Vehicle Braking Systems as Affected by Fluid Water Content. Society of Automotive Engineers paper No. 710253.

2. Society of Automotive Engineers J 1703. June 1991. Motor Vehicle Brake Fluid.

3. Emery, A. F. P. Kumar and Joseph C. Firey. 1997. Experimental Study of Automotive Brake System Temperatures. WA-RD 434.1. Final Report Washington State Department of Transportation, Planning and Programming Service Center, in cooperation with the U.S Department of Transportation, Federal Highway Administration.

4. Morgan, S. and R. W. Dennis. 1972. A Theoretical Predication of Disc Brake Temperatures and a Comparison with Experimental Data. Society of Automotive Engineers paper No. 720090.

5. American Society of Testing and Materials (ASTM). Method D 1120. Standard Test Method for Boiling Point of Engine Coolants. Annual Book of Standards. Vol. 15 p 7.

6. American Society of Testing and Materials (ASTM). Method D 1364. Standard Test Method for Water in Volatile Solvents (Fischer Reagent Titration Method).

7. Ryan, T. E. and T Hinz. 1997. Analysis of Water Content in Brake Fluid. Part 1. Method Comparison: Karl Fischer Titration Versus Refractive Index. Society of Automotive Engineers paper No. 973023. In Proceedings of the 15th Annual SAE Brake Colloquium, SAE P-319.

8. Hara, M., T. Hizikata and I. Miyagaw. 1984. Measurement by Relative Dielectric Constant of Boiling Point o Brake Fluids in an Extremely Small Quantity. Transactions of the Institute o Electrical Engineers o Japan, Section E., Vol. 104. No. 11-12. p. 239-246.

9. Haines, E. L. 1972. Determination of Water in Brake Fluid by Near Infrared. Society of Automotive Engineers paper No. 720057.

10. American Society of Testing and Materials (ASTM). D 3321. Standard Practice for Field Test Determination of the Freezing Point of Aqueous Engine Coolants. Annual Book of Standards. Vol. 15. p 107.

11. **Ryan, T. E.** (1993). Effects of Temperature on the Accuracy of Aqueous Engine Coolant Freezing Point Determination by Refractometer. Society of Automotive Engineers Technical Paper Series. SP-960 pp 48-54.

12. National Highway Traffic Safety Administration. 1973. Technical Report R-11, Field Evaluation of Brake Fluid Moisture Pickup in Vehicles.

**1A. DOT 3 Brake Fluid
18 New Fluids**

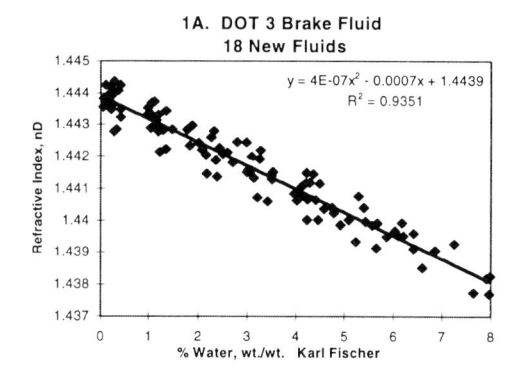

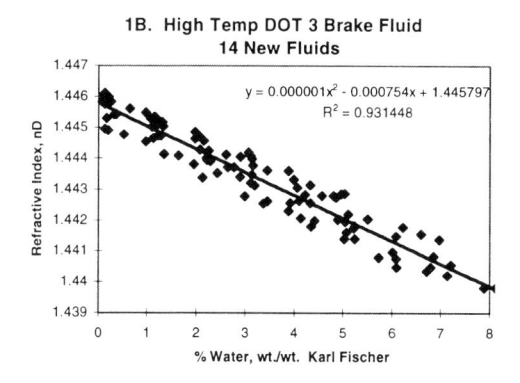

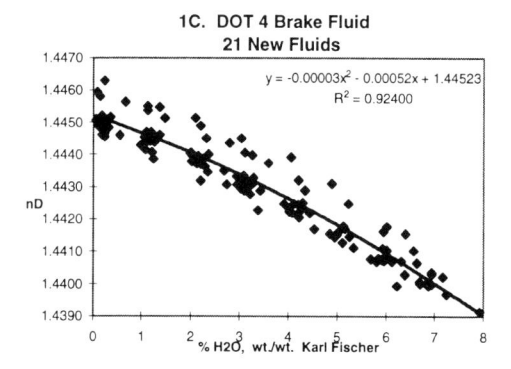

$y = 4\text{E}{-}07x^2 - 0.0007x + 1.4439$
$R^2 = 0.9351$

**1B. High Temp DOT 3 Brake Fluid
14 New Fluids**

$y = 0.000001x^2 - 0.000754x + 1.445797$
$R^2 = 0.931448$

**1C. DOT 4 Brake Fluid
21 New Fluids**

$y = -0.00003x^2 - 0.00052x + 1.44523$
$R^2 = 0.92400$

Figure 1. The relationship between refractive index and water concentration for three different types of brake fluid

Figure 2. The relationship between boiling point and % water as determined by Karl Fischer titration. DOT 4 N = 9, DOT 3 N = 7, High Temp. DOT 3 N = 6

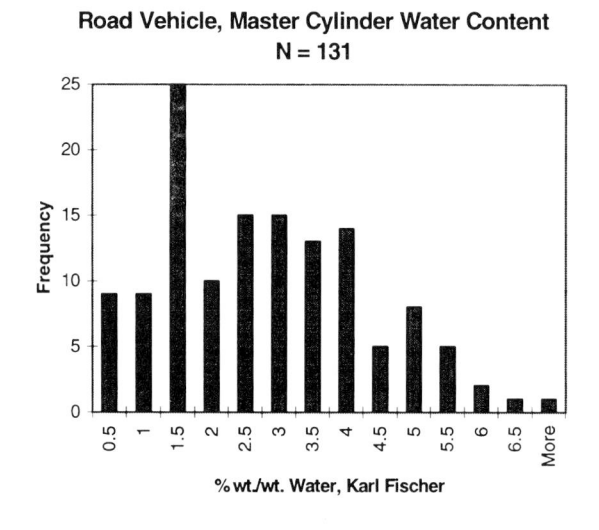

Figure 3. Histogram showing water content of a random sample of operational vehicles. DOT 3 Brake Fluid was specified for all vehicles

Figure 4. Range of appearance of the 131 in service brake fluids tested

Road Vehicle Brake Fluid Color

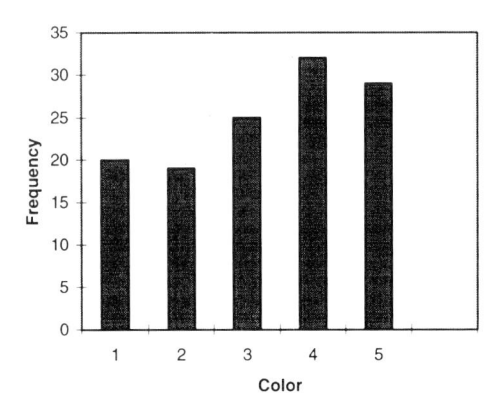

Figure 5. Color intensity of the in service brake fluid samples. Numbers the horizontal axis correspond to the representative samples shown in Figure 4.

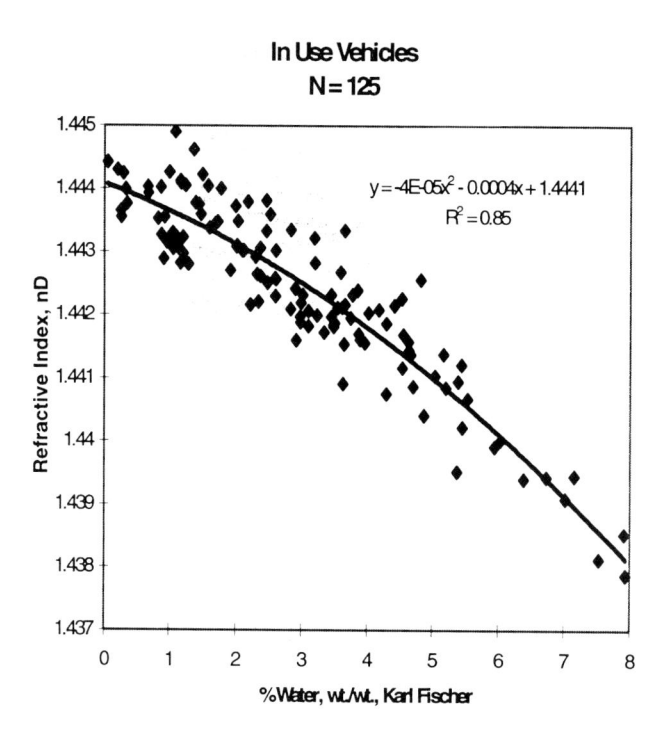

Figure 6. The refractive index versus water concentration of 125 in service vehicles. Samples were taken from the master cylinder. DOT 3 brake fluid was specified in all the vehicles.

High Temp, Regular and Used DOT 3 Brake Fluid

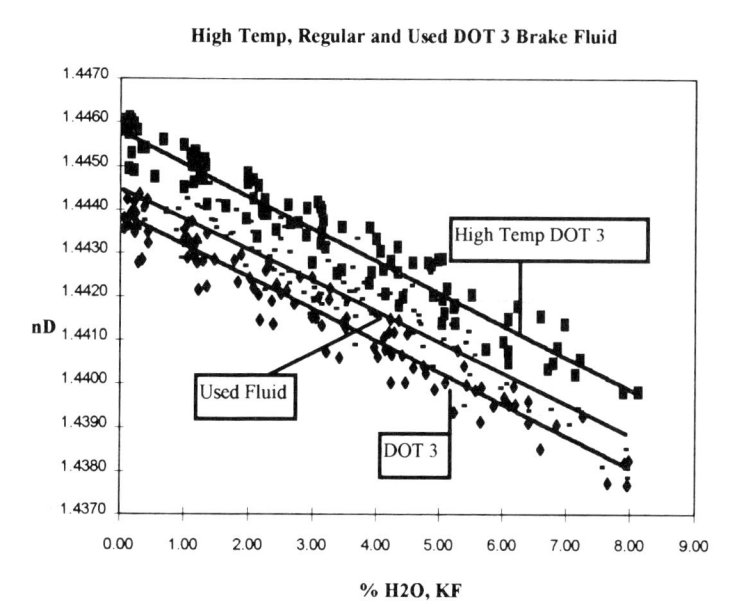

Figure 7. The used brake fluid samples depicted in Figure 6 are compared to the standard DOT 3 and High Temperature DOT 3 data shown in Figures 1A and 1B.

Automotive braking – a patent perspective

T H LEMON
The Patent Office, Newport, UK

Synopsis

During the last two decades considerable progress has been made in vehicular braking technology. Nowhere have these advances been recorded in greater detail than in patent literature. Patents represent, by far, the largest single source of technical knowledge, and, increasingly, they are a source of growing importance of commercial intelligence. Analysed patent data collected from over thirty issuing authorities are presented identifying organisations and individuals addressing actively the braking performances of vehicles. Details are discussed on the approaches adopted by the more active organisations, with trends indicating on a global basis projected levels of future developments. Such intelligence will help reduce wastage of vital resources and minimise the risk of infringement actions that potentially may impede proper exploitation of innovative vehicle-brake constructions.

1 BACKGROUND

There have been a number of studies reported both in the general press and in specialised, financial periodicals on the performance of European organizations in comparison with those in the other two major trading blocs, the Americas and Asia-Pacific. Not all those reports are complimentary about the European performance, although there are distinct signs that Europe is starting to adopt practices that have led to foreign successes. One of the many aspects that has produced improved performances is the manner in which organizations have used information technology and, in particular, the way in which those organizations have embraced information as a management tool to increase competitiveness.

The smaller and medium sized companies are considered to be one of the major sources of economic progress in Europe through to the next millennium. A high proportion of those companies will be concerned with technical innovation, including an increasing number from enterprising British academics eager to establish their own companies to commercialise their discoveries. It is to be hoped that the intellectual property system, and, in particular patents, will be invoked in a suitable manner to provide protection for these innovations. It has to be said, however, that many of these new organisations will be disillusioned, believing initially that they have patentable innovations only to discover after considerable expenditure of valuable resources that this indeed not the case. One of the principal causes of this wastage arises from inadequate literature surveys which, if technical innovation is involved, will have to include within its scope a scan of the global collection of published patent documentation.

1.1 Patents and Information

Virtually all machines, products and processes across the entire industrial spectrum are patentable provided they satisfy three criteria: they must be new, they must be inventive and they must be capable of industrial application. A patent is one of a number of legal rights known as 'Intellectual Property Rights', administration of which is undertaken in the UK by The Patent Office, an Agency within the Department of Trade and Industry.

A patent is a bargain between the State and the inventor; as long as the patent is maintained by paying annual renewal fees, the inventor has the right to prevent others from using, selling or importing the technology which is the subject matter of patent protection. In return for this 'monopoly' the inventor must file a full, detailed description of the invention which is published at the appropriate time. It is these patent publications that are overlooked so frequently by industrial and commercial operations hoping effectively to exploit technical innovation.

The collective, world-wide publications of patents currently represent the largest single body of technical information. There are over 36 million patents published to date, and each year over a million new patent records become available (Figure 1). Most patents are published eighteen months after first filing and this publication frequently represents the first substantive public disclosure of a new piece of technology. Independent studies have shown that much of this information, over 80% in some technical areas, is unique to patents and will never be published elsewhere. Perhaps what is not so well appreciated is that only a small proportion of the total number of patents are still in force, providing the monopolistic rights mentioned above; the remaining majority represents 'free technology' - available information that may freely be used by anyone for whatever purpose with no fear of infringement.

It is not surprising, therefore, that patents are viewed by progressive companies as a fertile source of technical information. Furthermore, during the last three decades patents have been viewed as a source of increasing importance of commercial information, since most patents documents display on the front page the name(s) of the inventor(s) and corporate ownership alongside the technical description and other data relating to the invention for which protection is sought (Figure 2).

Patent information, therefore has a role to play in virtually every stage of a project - from the conception of an idea through to the exploitation of the final product (Figure 3). Before research projects are initiated, a scan of the published literature can save a great deal of time pursuing fruitless lines of investigation, and in doing so help preserve vital capital by avoiding unnecessary duplication of basic investigatory work. It has been estimated by the European Commission that around 30% of all research and development carried out across member states is merely duplicating earlier work, the results of which have been prior published in patent literature. According to the Commission this amounts to a wastage in academic and industrial operations of around £20 billion each year.

Detailed knowledge of problems and solutions evolved in a field of technology can stimulate further ideas with innovative spin-offs often culminating in successful production lines. Unforeseen difficulties in production may have been encountered elsewhere and solutions recorded in patent literature; time saved with this information could determine whether or not a vital foothold is gained in a competitive market. An indication of the size of a market can be obtained by evaluating the global patent practices of organisations operating in the same technical field. At the same time, periodic reviews of the most recent published literature will help keep abreast of the latest developments and offer an early warning of emerging markets.

2 PATENTS AND BRAKE TECHNOLOGY

The vast majority of patents have their technical content classified according to an internationally agreed system; thus those documents relating to automotive braking, for example, carry a variety of codes typically beginning with the characters F16D (Figure 2). Attention is confined to this so-called sub-class of the International Patent Classification scheme in the following analyses conducted on data culled from the Derwent World Patent Index database which covers over thirty worldwide patent issuing authorities.

Looking at the patenting trend with time in this area of technology it can be seen that there has been a general, steady decline in activity from the year 1974 until very recently (Figure 4). A closer inspection of the data shows that the bulk of the increased activity since the mid-1990s has arisen from an increase in patenting from the Far East, and in particular, from Japan (Figure 5). Over the last twenty years or so the patenting frequency in braking technology originating from Europe exemplified by the number of applications filed by the United Kingdom and by Germany has shown a steady downward drift (Figure 5). Nonetheless, cumulatively over the entire period from 1974 to date, Germany is the origin of the majority of patents relating to braking technology (Figure 6).

The original set of data within the F16D classification area can be reduced to isolate those patents relating specifically to innovations concerned with reducing or removing noise during the braking operation. This exercise was undertaken by introducing into the original search patterns keywords such as 'squeal', 'judder' or 'vibration' to pick out patents addressing problems of the unwanted noise accompanying braking.

Analyses of the smaller set of data show quite a different picture with Japan clearly being the major country of origin of patents concerned with brake-noise reduction (Figure 7); those companies most active in this technical area are shown in Figure 8. Inspection of the patents in the name of, say, Akebono Brake Co. Ltd., indicates that their approach to the problems of reducing brake noise has changed from one of invoking innovative materials, often in laminar formations, as brake pad components to one proposing novel constructions for braking assemblies. On the other hand, the constructive approach is one which has consistently been adopted by the company Toyota to solve braking problems.

An outline of where globally major markets exist for a technology can be obtained from analyses of the patent holdings of the most active companies in that area. In the case of braking technology, for example, the top eight regions where major manufacturers seek patent protection, and consequently where they see their main markets, are shown in Figure 9.

It is possible from patent literature to identify with some accuracy 'competitors', namely companies or organisations that are operating along similar lines of investigation to solve similar technological problems. As seen from Figure 2, the front page of patent documents identify earlier publications or citations that have addressed the same problem in a closely-related manner to that being proposed in the patent document under review. Such citations arise during the normal patent granting procedure in order that the patentability of an application may properly be assessed. Thus, taking as examples patent holdings of Toyota and of Akebono, databases can be interrogated to determine which later patents are citing those by either of these two companies. Analyses of the proprietors' names of the later patents identify organisations whose activities have similarities in technical approach to specific problem solving as those adopted by either Toyota or by Akebono. That patents originating from Toyota or from Akebono are consistently being cited against those from the same company (eg., Allied Signal, Bendix etc., see Figure 10) is an indication of just how closely technologies to overcome brake noise are developing in seemingly rival organisations; further inspection of the data also shows that there a number of cooperative ventures occurring by, for example, among the companies Toyota, Sumitomo and Aisin Seiki.

Finally, the names of inventors, displayed on the patent front page may be analysed to identify key research workers operating in technical areas of interest. Figure 11 lists the names of three of the most prolific individual inventors in the field of brake noise along with their respective affiliations and the patenting policies of the corresponding assignees.

3 CONCLUSIONS

As shown above in the case of automotive braking, a great deal of knowledge about a technology, the manner in which it is being developed and exploited where that activity globally is occurring and by whom, may be obtained from publicly available, patent-based information. It is often said that we are now living in a knowledge-based society and knowledge is one of the keys to industrial competitiveness and in particular to the efficient and effective investments in research and development. In today's competitive environment, with increasing demands being made on apparently dwindling resources it makes little economic sense to ignore the ready availability of such information; information that is inexpensive to collate and to interpret and which potentially can have such a profound effect on the manner in which technical innovation is conducted and eventually exploited. And, of course, the higher the innovative content in a programme the more compelling becomes that argument.

- **Huge Information Source**
  - 36 million patents
- **Unique Information**
  - 80% only Published in patents
- **First Published Information**
  - 18 months from priority date
- **Free Technology**
  - 85% of all UK Patents not in force

Fig 1 Patents as a source of technical and commercial information

Kind of document
Country Code

(12) UK Patent Application (19) GB (11) 2 311 107 (13) A

(43) Date of A Publication 17.09.1997

Application number
Filing date
Priority application number date and country code
Name/address applicant

(21) Application No 9705362.3
(22) Date of Filing 14.03.1997
(30) Priority Data
(31) 08087533 (32) 15.03.1996 (33) JP

(71) Applicant(s)
Akebono Brake Industry Co Ltd

(Incorporated in Japan)

19-5 Nihonbashi Koami-cho, Chuo-ku, Tokyo, Japan

Names of Inventors

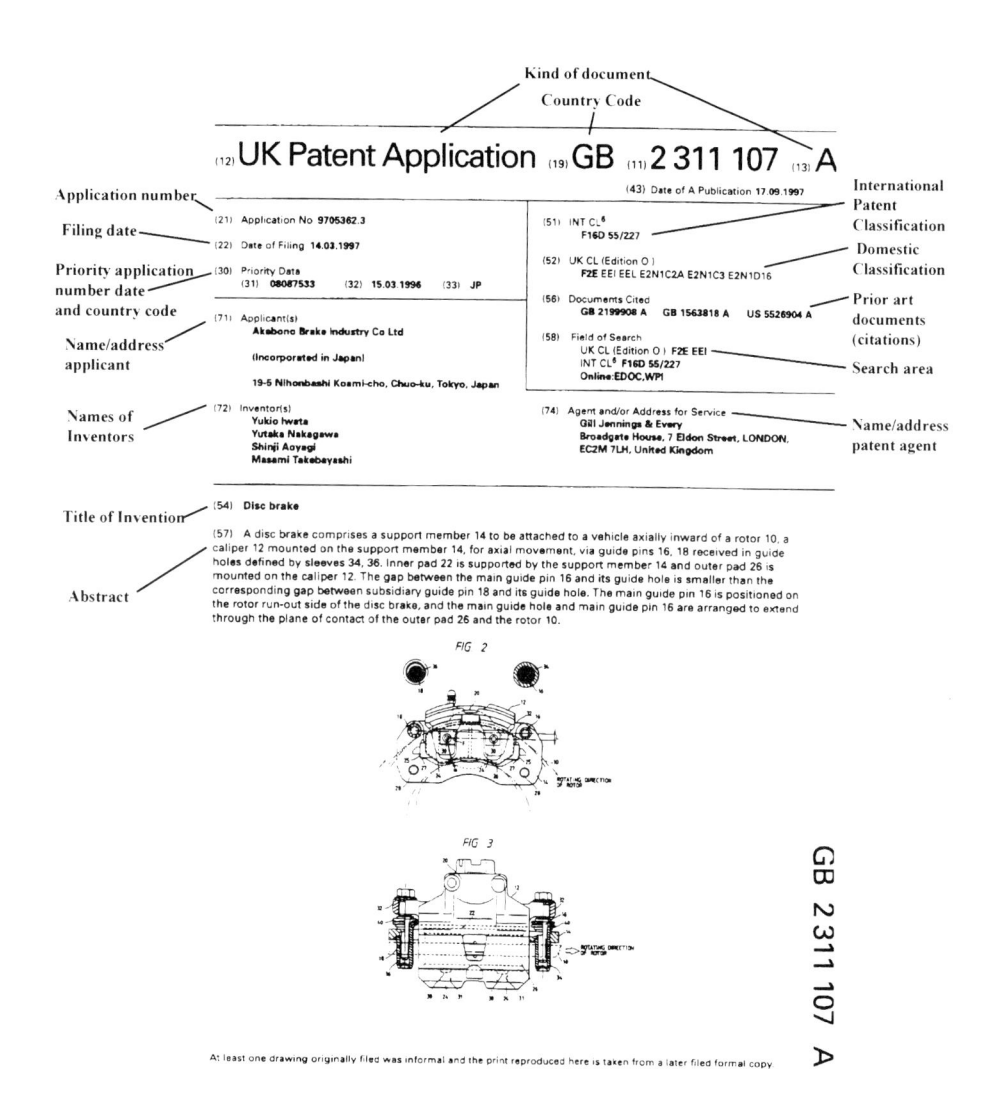

(72) Inventor(s)
Yukio Iwata
Yutaka Nakagawa
Shinji Aoyagi
Masami Takebayashi

(51) INT CL[6]
F16D 55/227

(52) UK CL (Edition O)
F2E EEI EEL E2N1C2A E2N1C3 E2N1D16

(56) Documents Cited
GB 2199908 A GB 1563818 A US 5526904 A

(58) Field of Search
UK CL (Edition O) F2E EEI
INT CL[6] F16D 55/227
Online:EDOC,WPI

(74) Agent and/or Address for Service
Gill Jennings & Every
Broadgate House, 7 Eldon Street, LONDON,
EC2M 7LH, United Kingdom

International Patent Classification
Domestic Classification
Prior art documents (citations)
Search area
Name/address patent agent

Title of Invention
Abstract

(54) Disc brake

(57) A disc brake comprises a support member 14 to be attached to a vehicle axially inward of a rotor 10, a caliper 12 mounted on the support member 14, for axial movement, via guide pins 16, 18 received in guide holes defined by sleeves 34, 36. Inner pad 22 is supported by the support member 14 and outer pad 26 is mounted on the caliper 12. The gap between the main guide pin 16 and its guide hole is smaller than the corresponding gap between subsidiary guide pin 18 and its guide hole. The main guide pin 16 is positioned on the rotor run-out side of the disc brake, and the main guide hole and main guide pin 16 are arranged to extend through the plane of contact of the outer pad 26 and the rotor 10.

FIG 2

FIG 3

At least one drawing originally filed was informal and the print reproduced here is taken from a later filed formal copy.

GB 2 311 107 A

Fig 2 Front page of a published GB patent application

164

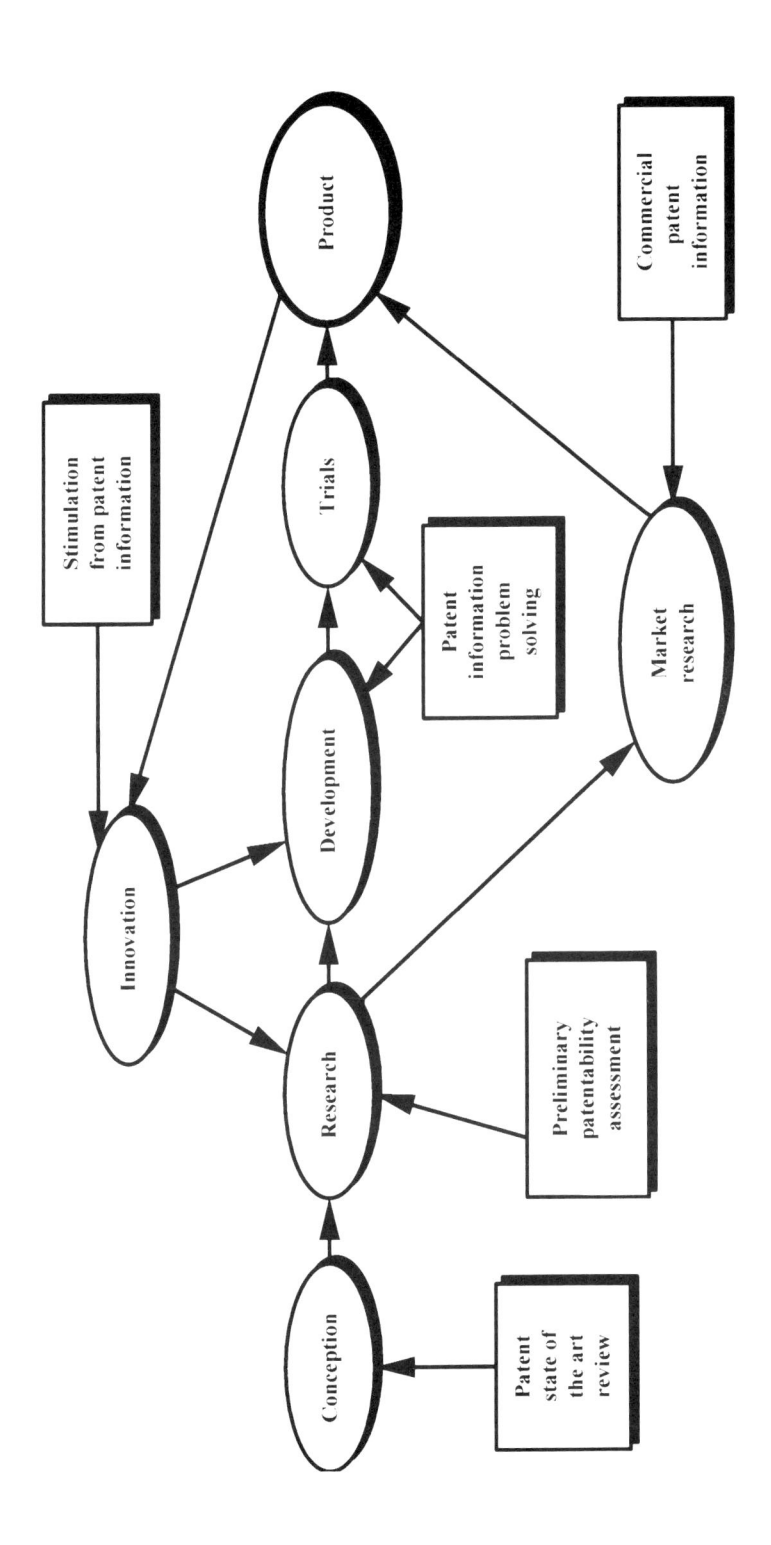

Fig 3. Industrial uses of patent information

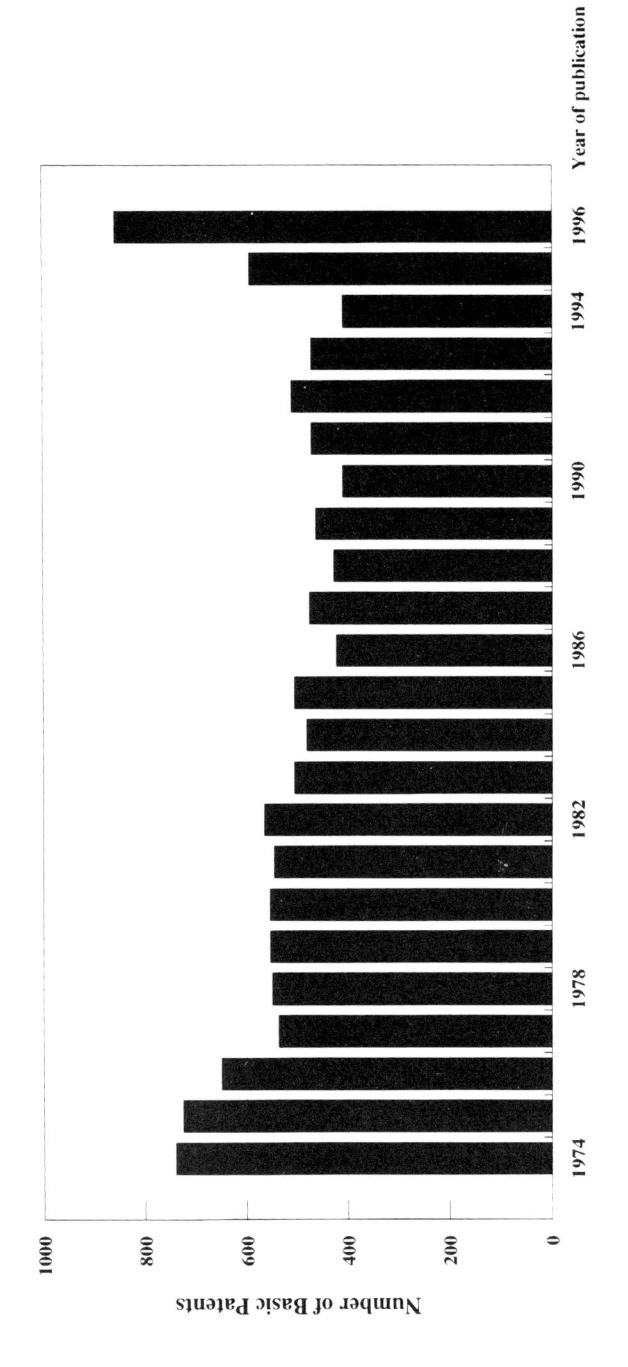

Figure 4. Patenting activity in the area of braking technology (F16D)

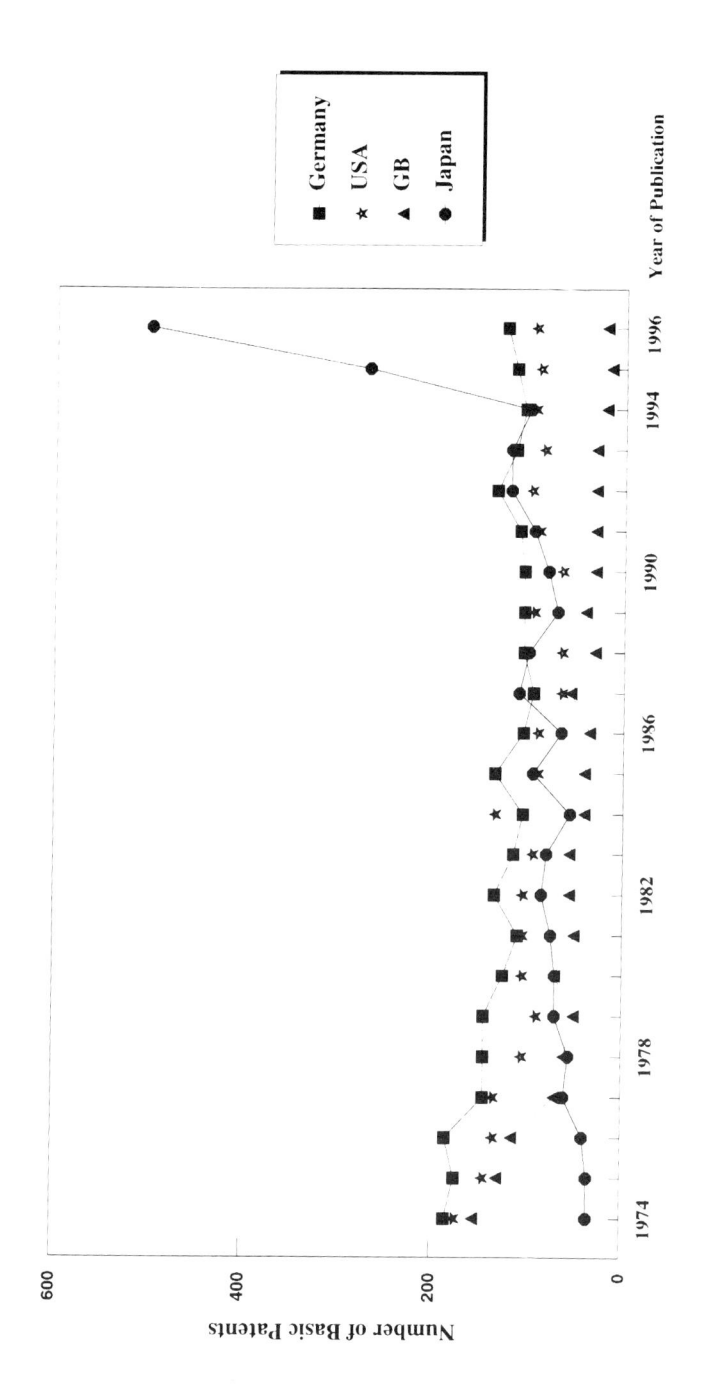

Figure 5. Patenting activity by country in the area of braking technology (F16D)

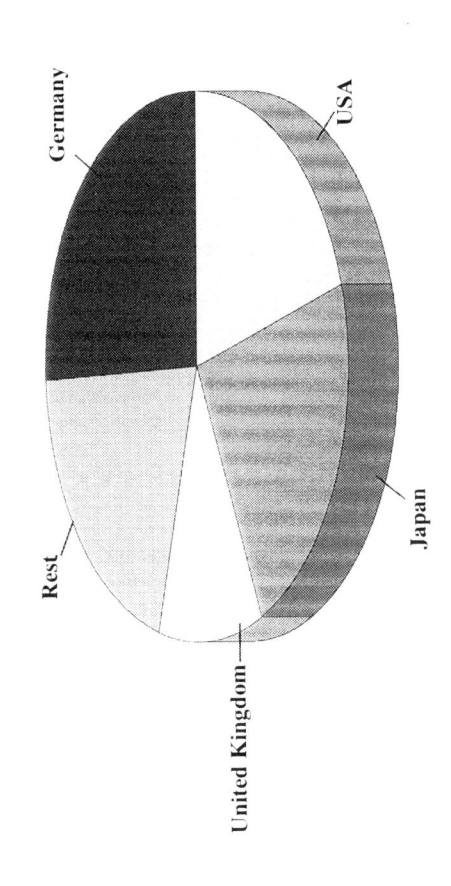

Figure 6. Total patenting activity by country of origin in the area of braking technology (F16D) from the year 1974 to date.

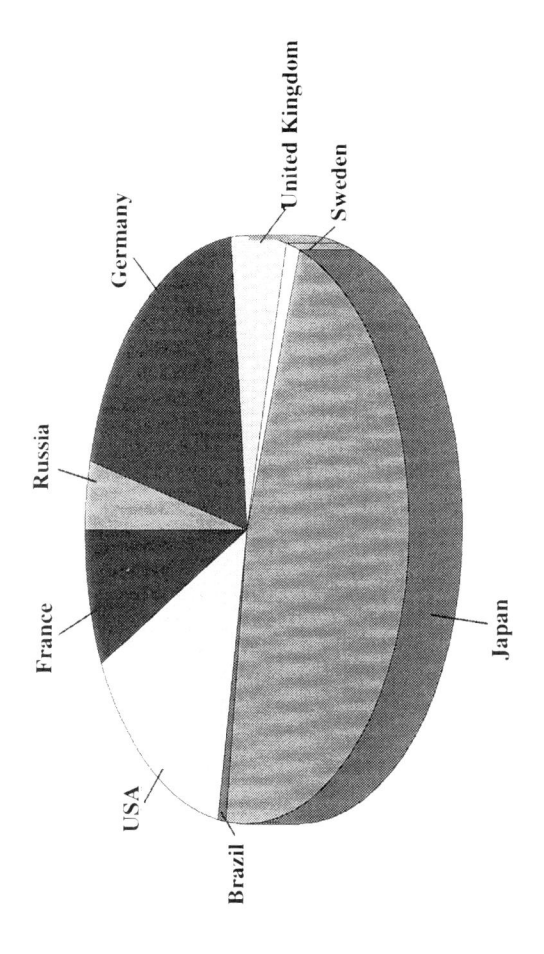

Figure 7. Total patenting activity from the year 1974 according to the country of origin in the field of brake noise

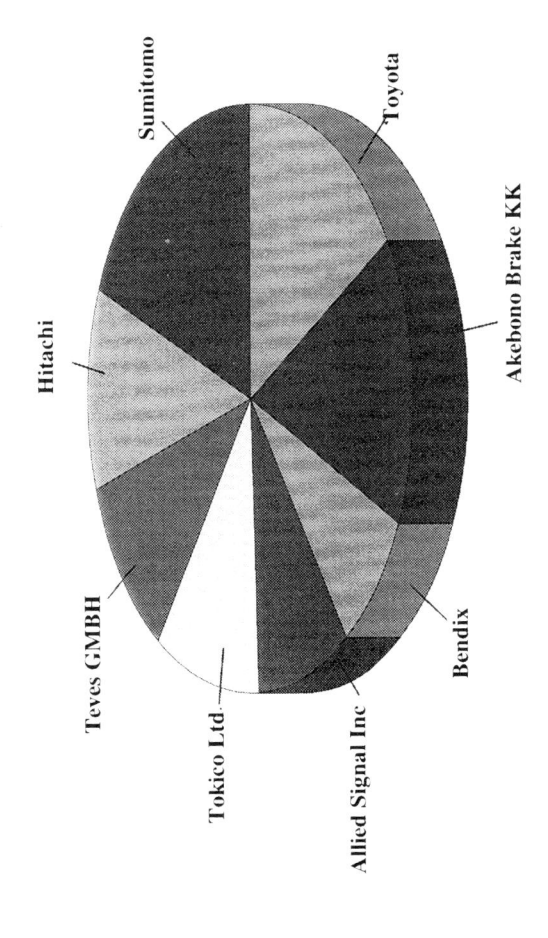

Figure 8. Major patenting organizations in the field of brake noise

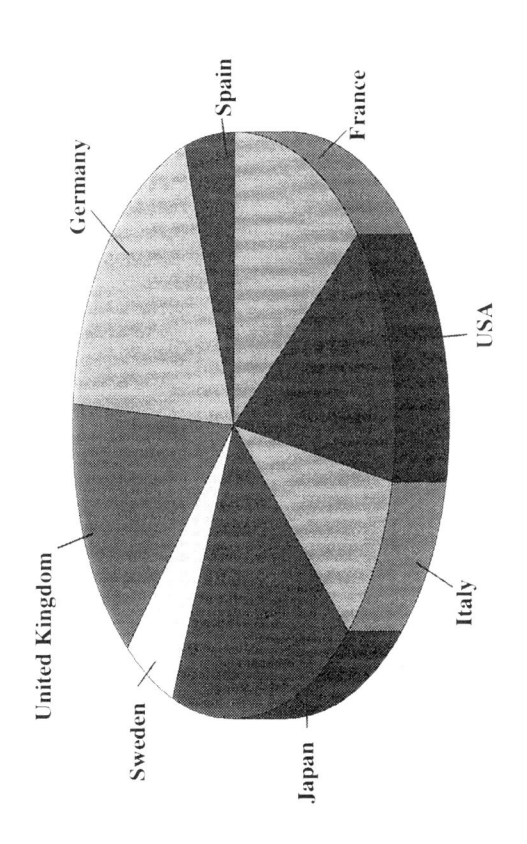

Figure 9. Major countries where protection is being sought for patents relating to brake noise

Fig 10. Major "competitors" of Toyota and Akebono in the field of brake noise from analyses of patent citations.

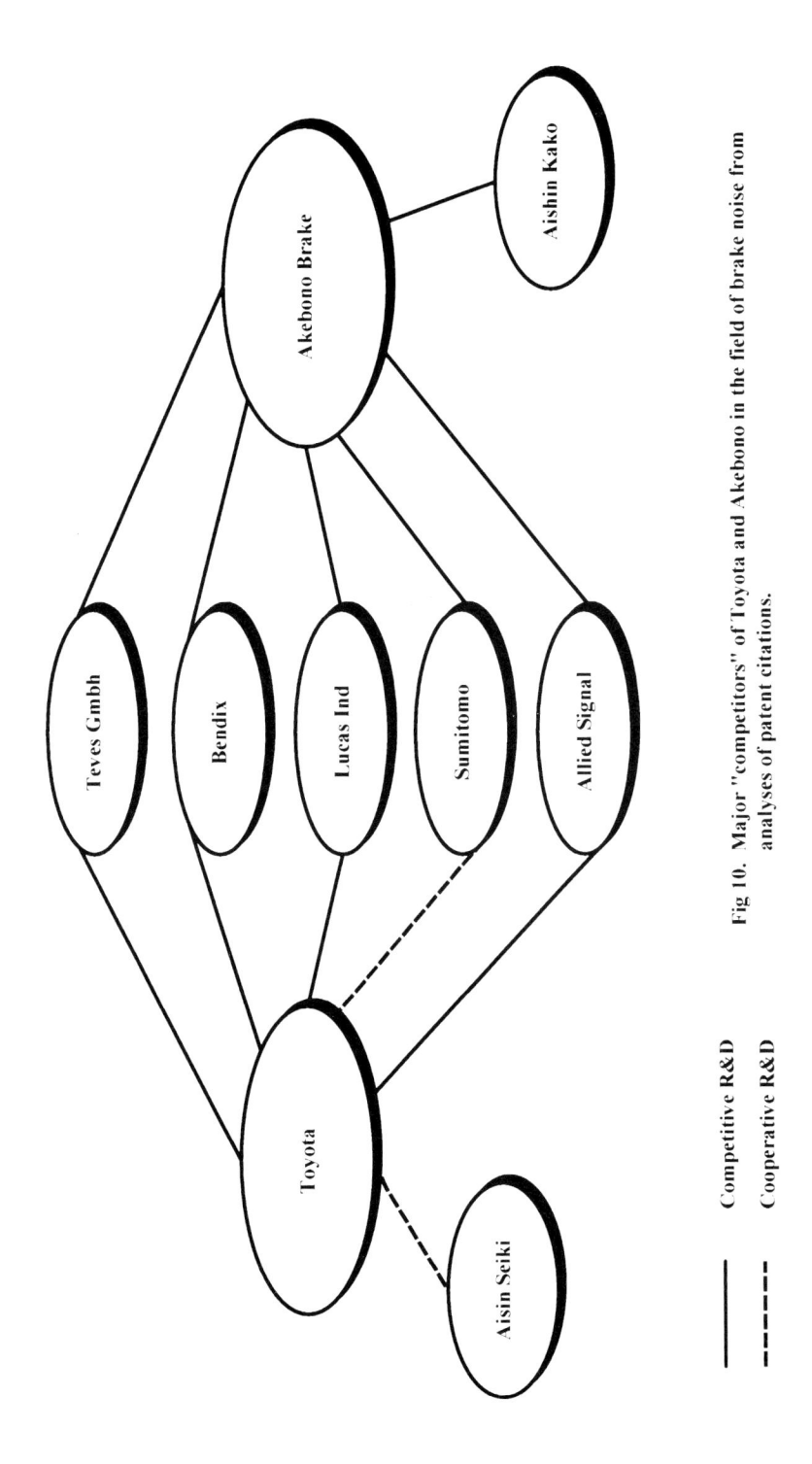

―――― Competitive R&D
------ Cooperative R&D

```
┌─────────────────────────────┐
│         J L GERARD          │
└─────────────────────────────┘
```

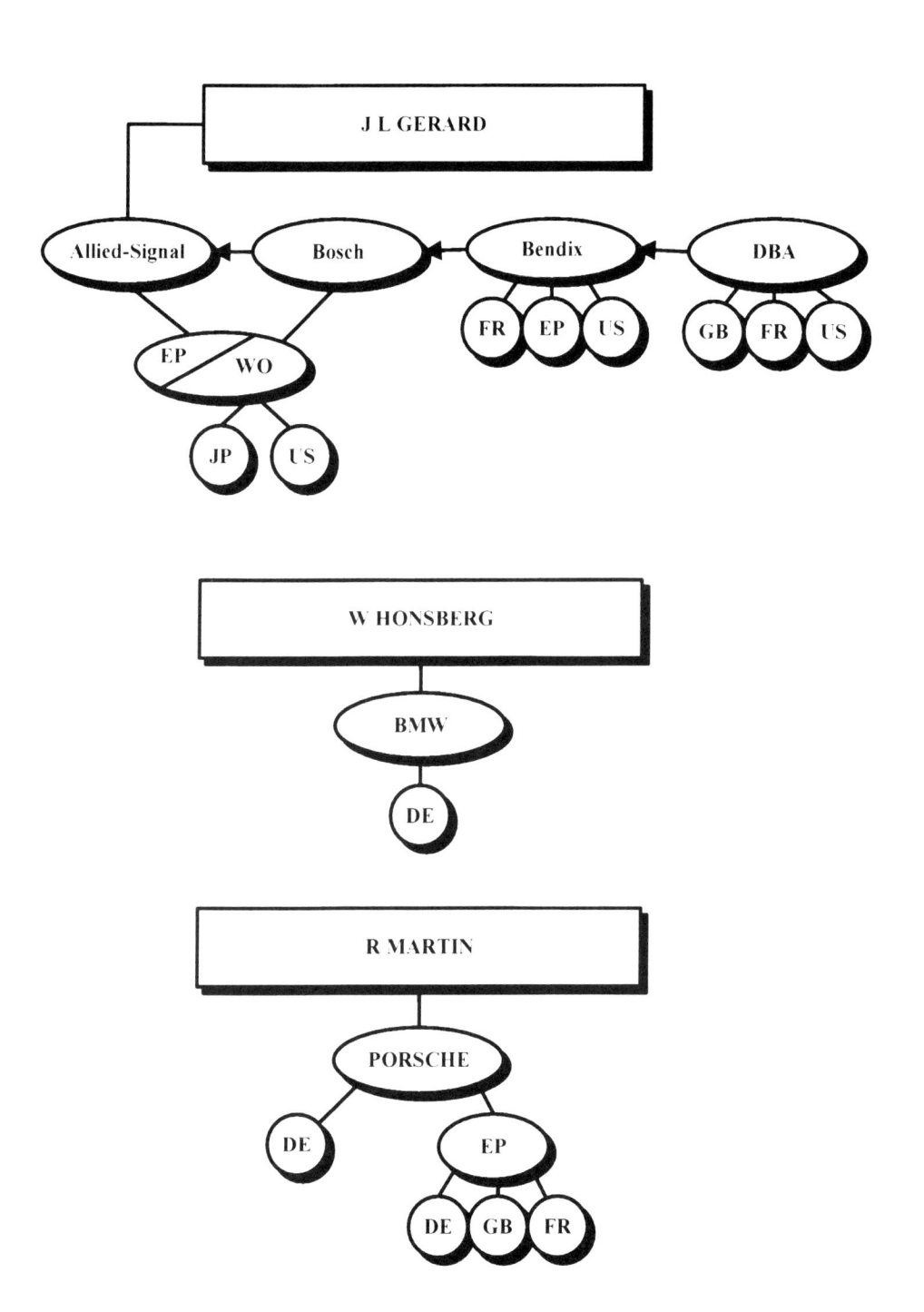

- J L GERARD
  - Allied-Signal ← Bosch ← Bendix ← DBA
    - Bosch/Allied-Signal: EP / WO
      - JP, US
    - Bendix: FR, EP, US
    - DBA: GB, FR, US

- W HONSBERG
  - BMW
    - DE

- R MARTIN
  - PORSCHE
    - DE
    - EP
      - DE, GB, FR

Materials and Commercial Vehicle Brakes

Cast iron – a brake disc material for the future?

M P MACNAUGHTAN and **J G KROSNAR**
Precision Disc Castings Limited, Poole, UK

SYNOPSIS.

The vast majority of both production road car, and high performance racing vehicle brake discs have, since the introduction of the brake disc in the early nineteen fifties, been produced in Grey cast iron. This paper introduces the various cast iron materials used in industry today, and discusses their effect on disc performance

Precision Disc Castings have made significant investments, both in new plant and equipment, and in a research and development programme, a radical move for an automotive foundry. The view is put forward, that by working with the designer, a go ahead and innovative company such as Precision Disc Castings, can help influence future thinking on brake disc design, materials and production methods, and ensure that there is a future for grey cast iron, an extremely versatile and cost effective material.

1.INTRODUCTION.

Precision Disc Castings, part of the European Automotive Components group, located at Poole, in Dorset, is one of Europe's leading ferrous foundries specialising in the production of ventilated brake discs for the automotive market. The production of some 4.5 million discs a year, with 70% of these being exported to Europe, has led to international recognition. The company employs 200 people, with a turnover of £19.0 million and a capacity of 40,000 tonnes per annum. Over the years, the gaining of ISO 9000 and QS 9000, both internationally recognised quality standards, has led to the expansion of the companies supply base to a position where it now extends to all the major European countries.

The companies first involvement with the automotive industry came in 1970, with the production of clutch pressure plates, this was followed in 1972 by solid brake discs. In 1979 the first ventilated discs were produced for the Rover Metro and Montego, since then the companies portfolio has expanded to 300 different part numbers, which cover some 600 vehicles. Precision Disc Castings also has a major involvement in the production of discs for racing and sports cars, with approximately 90% of the total market share.

The foundry facility at Poole, in Dorset, is currently melting in the region of 1,200 tonnes of wet metal a week to produce to some 100,000 finished components. Grey cast irons are produced to satisfy various customer specifications ranging from grades 150 to 250, in both alloyed and non-alloyed grades. Raw materials are melted in four ABB six tonne medium frequency furnaces. Metal is poured through Calamari automatic pouring furnaces to three Disamatic 2013 moulding lines. Each pouring furnace is controlled by a state of the art, Selcom laserpour system.

The EURAC group also has two other companies associated with the automotive brake disc business, High Precision Machining, located in Birmingham, and Prazisions-Brems-Scheiben located at Lemgo in Germany. Both facilities offer a full machining service for the original equipment and aftermarket sectors of the industry.

2 HISTORICAL REVIEW

Brake discs as we know them today, were first used by the Conze brothers at the 1951 Indianapolis 500 race held in the United States, these discs were originally developed for aircraft use (1). The new discs were first used at a Grand Prix in Europe in 1953, when the T24 Cooper Alta of Stirling Moss raced in the Italian Grand Prix held at Monza. In the same year, Jaguar won the Le Mans 24 hour race with its C-type sports car fitted with solid brake discs. Jaguar again won the race in 1955 with their D-type, also fitted with brake discs.

It was Ferrari who introduced some of the first ventilated brake discs in 1964. The discs manufactured by Dunlop, were 12 inches in diameter, ½ inch thick and radially drilled with 60 equally spaced 3.5 mm diameter holes drilled to a depth of 45 mm and thereafter 3 mm. The calipers were reported to be cast iron fitted with Ferodo DS11 pads, 22 mm thick (2). Unfortunately the disc material is not known, but was thought to be cast iron.

Ventilated cast iron brake discs remained standard equipment on all Grand Prix cars until the advent of the 1000 bhp turbo machines introduced in the early 1980's. This rapid increase in power and hence braking requirement lead to the development of the first carbon composite brake discs, and the rapid phasing out of cast iron as a disc material. However grey iron discs have become standard equipment for the majority of mass produced family vehicles, especially on the front brake system, and still remain in use in the majority of all other motor sport formulae. Indeed grey cast iron would appear to be most suitable material for use in braking applications that require stability at operating temperatures of between 400 and 600 degrees centigrade.

3 BASIC MATERIAL PHILOSOPHY

A review of disc materials used by the automotive industry today, will show that there are two basic material philosophies. The first, used for family sized vehicles, operates on the principle of small diameter, high strength discs with sufficient inherent strength to resist any tendency towards the formation of thermal cracking, and distortion, at high operating temperatures. These discs whilst having good strength properties, have relatively low thermal conductivity.

The second principle, that of large, weaker, low strength discs with high thermal conductivity, has been applied more commonly, to the larger high powered type of vehicle where space constraints are not so critical, and as a consequence, a larger diameter thicker disc can be employed.

3.1 BRAKE DISC REQUIREMENTS.
The role of the brake disc, is to react with the brake pads within the braking system and convert, by means of friction, the kinetic energy of motion into thermal energy, or heat. The amount of heat produced depends on the vehicles weight and speed at the moment the brakes were applied. A vehicle weighing say 1500 kg, will convert some 240 kW of kinetic energy into heat if arrested from 90 mph to 0 mph in a few seconds. Such large amounts of heat input means that the disc can, for short periods, reach temperatures of 800 degrees centigrade, this can result in a very steep thermal gradient between the friction surface and the core of the disc, sometimes in excess of 500 degrees centigrade.

Clearly therefore, the disc material must be able to resist the high thermal stresses resulting from repeated thermal cycling experienced during use. The large amounts of heat generated during braking, must be absorbed and dissipated as quickly as possible. The material must also be reasonably strong, easy to machine, as well as being light, cheap and easy to cast.

Grey cast iron is both cheap and easy to produce in high volumes, to tightly controlled specifications. It is reasonably light, strong and easy to machine in high volumes and most importantly, it possesses good thermal conductivity and thermal diffusivity. Another advantage of cast iron is that the specific heat of the material increases with temperature, thereby improving the ability of the component to absorb more heat during normal operating conditions. Cast iron also possesses a low coefficient of thermal expansion.

4. CAST IRON AND THERMAL CONDUCTIVITY.

The performance of cast iron discs, in the main, relies on the inherent thermal conductivity of the basic material. This in turn can be related to the type of cast iron employed. It is therefore pertinent to discuss this point in some detail.
There are three graphite morphologies present in cast iron
- Flake graphite, or Grey cast iron
- Compacted, or as it is sometimes referred to quasi iron.
- Nodular, or spheroidal graphite cast iron.

4.1 FLAKE GRAPHITE CAST IRON.
Graphite in grey cast iron is present in the form of thin elongated flakes. The material is generally classed as low to medium strength, with no elongation at fracture. It is a brittle material with no inherent ductility.

4.2 COMPACTED GRAPHITE CAST IRON.

Reducing the sulphur content of grey iron and treating the melt with alloys such as cerium, magnesium or titanium, produces a modification in the graphite shape and the flakes become shorter and thicker. This change results in an increase in strength and the introduction of a limited amount of elongation at fracture, or ductility.

4.3 NODULAR GRAPHITE CAST IRON.

A further modification to the treatment, removing the titanium, and lowering the sulphur content to less than 0.02%, will produce a totally nodular or spherical graphite form. Magnesium is the preferred treatment alloy for this process. The resulting iron has high strength and in some cases as much as 20% elongation at fracture. Why therefore, are these modified materials not used in the manufacture of brake discs ? The answer to this question can be best illustrated by the information in table 1.

TABLE 1. THERMAL CONDUCTIVITY COMPARISON GREY VS S.G. IRON

400/18 SG at 300 degrees centigrade = 36.2 W/mK.
220 Grade grey iron at 300 degrees centigrade = 48.1 W/mK.

As reported by the Castings Development Centre (3) & (4)..

It can be clearly seen that grey iron, has an inherently higher thermal conductivity than nodular iron. Studies have shown (5) that it is the high surface area and shape of the graphite flakes that allows for a rapid transfer of heat through the matrix of the structure and hence the component. Experience has shown that discs produced in nodular iron will always tend to crack sooner than the equivalent disc produced in grey cast iron The effect of graphite on thermal conductivity can be taken one stage further with the information illustrated in table 2, here the thermal conductivities of a range of grey irons are tabulated. Examination of the table will clearly show the benefit of adding graphite to grey cast iron, and hence the increasing popularity of high carbon grey irons.

TABLE 2. THERMAL CONDUCTIVITY FOR GREY CAST IRONS.

Grade	Thermal conductivity W/mK at 300 Centigrade
350	47.7
250	45.4
200	48.1
150	50.5

As reported by the Castings Development Centre (3).

5. BRAKE DISC MATERIALS

A review of disc materials used today by the automotive industry, will show that there are three distinct classes of material.

5.1 MEDIUM CARBON HIGH STRENGTH IRONS.
These materials posses good resistance to distortion and cracking, and are found in either alloyed on non-alloyed form. Non-alloyed discs are usually fitted to the vast majority of 'standard' passenger vehicles. Alloyed variants are only employed where brake system requirements dictate that resistance to cracking and associated thermal problems are marginal. Such discs will usually be found on 'hot hatch' versions of standard vehicles. Medium carbon high strength discs are normally small in size, and are used where space considerations are of paramount importance.

5.2 HIGH CARBON LOW STRENGTH IRONS
The development and use of low strength high carbon materials for racing applications, has led to the introduction of a range of similar materials for the prestige, large saloon/sports car market. Such vehicles tend to be large in size, and high in power output. High carbon irons have improved thermal conductivity and as a consequence can, if the discs are large enough, prevent the on-set of thermal cracking. The addition of alloys can improve thermal properties in marginal conditions.

5.3 ALLOYED IRONS.
Alloy additions such as molybdenum, chromium, nickel and copper can be added to all grades of cast iron. The use of such alloys will improve tensile strength, and resistance to problems, such as the on-set of thermal cracking. Unfortunately, the addition of molybdenum and chrome will adversely affect the castability of cast iron, so should always be used with caution. The addition and control of titanium was reported in detail by Chapman and Hatch (6) & (7).

5.4 METALLURGICAL AND PHYSICAL PROPERTIES.
The mechanical properties of grey iron can vary considerably according to the type and amount of any alloy addition, this is particularly relevant when considering material tensile strength, as table 3 shows.

TABLE 3. MECHANICAL PROPERTIES.

	Medium carbon alloyed	Medium carbon non-alloyed	High carbon alloyed	High carbon non-alloyed
Tensile strength N/mm2.	250	220	200	150
Hardness BHN	190 - 240	180 - 230	180 - 230	180 - 220

Table 3 also shows that material hardness is relatively unaffected by either additions, or changes in tensile strength. Indeed, it should be appreciated, that the relationship between tensile strength and hardness is only of a very general nature, and will vary according to component section thickness and cooling rate.

Variations in the microstructure of the various materials under discussion are illustrated in table 4. It will be seen that the nature of the matrix i.e. the amount of pearlite and ferrite are not affected by alloy content, it should be noted however, that the density of the pearlite may change and will only be noticed by a change in the machinability of the component. Molybdenum will however affect the matrix and should be used with caution. The graphite type, when evaluated according to ASTM 247, should not alter with alloy addition, the preferred graphite type is always type A. The major change in microstructure will be found in the graphite size, as carbon content increases so does the size of the graphite flakes, with a corresponding decrease in the tensile strength, assuming of course that no alloy additions are made.

TABLE 4. METALLURGICAL STRUCTURES.

	Medium carbon alloyed	Medium carbon non-alloyed	High carbon alloyed	High carbon non-alloyed
Graphite	Flake type A	Flake type A	Flake type A	flake type A
Size	4 -6	4 -6	3 - 5	3 - 5
Matrix	Pearlitic	Pearlitic	Pearlitic	Pearlitic
Ferrite	5% max	5% max	5% max	5% max

5.5 GENERAL CHEMICAL ANALYSIS.

Base chemical composition, with the exception of carbon and silicon content remains the same for all the alloys under discussion. The five main constituent elements, and the most common alloy addition ranges are illustrated in table 5.

6. DISC MATERIAL PERFORMANCE.

6.1 LOW CARBON IRONS.

Very few true low carbon materials are in use today, those that are, tend to be related to older specifications compiled before the move towards higher carbon materials. Low carbon cast irons have high strengths, are relatively hard and are therefore used where resistance to wear, thermal cracking and distortion are thought to be necessary. Their application is limited to the smaller diameter disc typically found on passenger vehicles. From a foundry point of view, low carbon cast irons have several distinct disadvantages.

TABLE 5. GENERAL CHEMICAL ANALYSIS.

	C	Si	Mn	S	P
Low carbon	3.0 - 3.4	1.9 - 2.2	0.5 - 0.8	0.10 max	0.10 max
Medium carbon	3.3 - 3.5	1.9 - 2.2	0.5 - 0.8	0.10 max	0.10 max
High carbon	3.6 - 3.9	1.8 - 2.1	0.5 - 0.8	0.10 max	0.10 max

Alloys.
- Molybdenum - 0.3 - 0.5%
- Chromium - 0.2 - 0.4%
- Nickel - 0.1 - 0.3%
- Copper - 0.5 - 0.8%
- Titanium - 0.02 - 0.04%

Note. Alloy additions can be added individually or in combination to any of these base grades.

The materials tend to have a narrow freezing range, this can lead to a reduction in the structural homogeneity of the component, especially in areas of dramatic section thickness change, where defects such as interdendritic shrinkage may be found. Problems with microstructure segregation can also occur. The fairly rapid freezing inherent in these irons can also induce cooling stresses in the casting, it is often the case that these will require relieving, by a further thermal heat treatment, if the component is to be machined to tight dimensional tolerances.

It is the firm belief of the author that there is no need for low carbon high strength materials for brake discs application today. If strength is required, it is far better to strengthen a higher carbon material with one of the many alloy elements available.

6.2 MEDIUM CARBON IRONS.
The bulk of discs used today, are cast in medium carbon irons conforming to commonly accepted grades such as GG 20 (Germany), Gh 190 (Italy) and Grade 220 (UK). Brake discs produced in this material, are used for the majority of standard passenger vehicle applications. The increased carbon content offers improved castability, and higher thermal conductivity. Medium carbon irons are good to machine to an excellent standard of surface finish, and have good wear resistance.

6.3 HIGH CARBON IRONS.
Brake discs produced in high carbon irons have optimal thermal conductivity. Foundry casting performance is excellent, with good structural integrity and minimal microstructural problems. As a result of their good thermal conductivity, high carbon discs have an inherently good resistance to distortion and thermal cracking. In spite of their high carbon content, the hardness levels of high carbon irons are not much lower than those of the low-medium carbon materials. Consequently, wear properties are only marginally worse, but are however, significantly more homogeneous. The need to resort to alloy additions for improvements in thermal properties is

rarely required, except in extreme circumstances, such as high performance motor sport applications. Even when alloys are required, machinability is excellent.

The thermal properties of high carbon irons have been reported by Keiner & Werning (8), who note that at low temperatures, the thermal stresses of high carbon irons are approximately 35% lower that that of an alloyed medium carbon iron. At higher temperatures the difference was found to be as much as 45%. It was also found that the resistance to thermal cracking of high carbon irons can be as much as 70% higher than that of an alloyed medium carbon material. Brecht & Egner-Walter (10) have investigated the effect of material performance on the stresses found in ventilated brake discs, they also concur with the view that high carbon materials have the best resistance to thermal related performance problems. They suggest that resistance to thermal stress is related to the Young's modulus, the thermal expansion coefficient and the temperature difference within the disc. For high carbon materials both the Young's modulus and the thermal expansion coefficient are significantly lower than low carbon materials and hence extremely beneficial.

6.4 ALLOYED IRONS.

Alloys can, as stated earlier, be added to all base grades of grey cast iron, and offer improved resistance to thermal cracking and distortion. However there are sound practical reasons, from a foundry point of view, why alloying should, if possible be avoided.

Additions of chromium and molybdenum markedly affect the cooling characteristics of grey cast iron to such an extent that production rates usually have to be reduced in order to avoid the formation of undesirable bainitic structures.. Problems with structural integrity are exacerbated, and machinability is reduced. Additionally, discs may require stress relieving before machining if dimensional stability is to be maintained.

Copper on the other hand, is an extremely user friendly element from the foundry point of view. Additions of up to 1% can easily be made to maintain adequate strength at high carbon levels. Tensile strength is maintained without increasing hardness and thus affecting machinability. The element is relatively inexpensive, and there are no detrimental production implications. Nickel would also fall into the category of benign additives, but is extremely expensive at effective levels. It should also be appreciated that most alloy elements with the exception of copper and molybdenum reduce thermal conductivity.

Titanium can be considered as an alloying element in grey cast iron and its effect in cast iron has been reported in detail by Chapman and Hatch (6) & (7). The use of titanium to control the coefficient of friction of cast iron, has only been accepted to a limited degree, principally by Ford and Rover in the U.K. It is believed that the impetus behind the work was more as a result of problems with particular Ferodo friction materials rather than any inherent problem with cast irons, and that pad material development has in effect by-passed the problem. Wirth and Whitaker briefly alluded to work at BBA Friction (9) which tends to support this view. The Ferodo work has been totally ignored on the continent where titanium is very rarely, if ever, listed in proprietary specifications. It is certainly accepted that a vehicle may slew under braking, if discs having dramatically varying titanium contents are fitted to the front brake system, however it is common practice to replace discs in pairs and for reputable suppliers to only source discs from one or two foundries. It is the authors experience that discs examined from sources throughout Europe very rarely if ever contain harmful levels of titanium.

7. DISC PERFORMANCE RELATED PROBLEMS.

In the context of this paper, disc performance related problems fall into four main subject headings,
- Machinability
- Corrosion
- Judder
- Squeal and groan

7.1 MACHINABILITY
Machinability of cast iron, especially at the rates necessary to machine brake discs economically, is dependant on carbon content. Increasing carbon content improves machinability by virtue of slightly reduced hardness levels. Machinability is also improved by the better structural integrity and homogeneity of discs produced in high carbon irons. Alloy additions in the main, reduce machinability. Dimensional accuracy is crucial for sound performance, with some cast iron materials, especially low carbon and alloyed materials, a stress relieve may be required before machining. The quality of the machined surface of the disc is important for good disc/pad performance. It has been found (11) that the coefficient of friction between the disc and pad is reduced by a coarse surface finish. The type of finish on turned surfaces will also be affected by the type of tool used in the finishing operation, i.e. carbide or ceramic tipped (12).

7.2 CORROSION
Catastrophic failure due to corrosion is seldom, if ever a problem. However light surface corrosion can be considered a problem with cast iron, as it is with all ferrous materials unless highly alloyed. Localised rusting may cause problems, especially with effects such as low speed groan and squeal.

7.3 JUDDER.
Brake judder is defined as a first order excitation of the vehicle suspension, and therefore, the vehicle structure during braking. Any rotating components in the brake system can therefore contribute to the on-set of the phenomenon. However from the brake disc point of view, judder related problems fall into two categories:-
- material specification and structural integrity
- dimensional stability.

The use of high carbon disc materials can, by virtue of the fact that discs run cooler and are therefore less likely to distort, help reduce judder. Foundry experience has shown that high carbon discs possess a more homogeneous structure, with a reduced tendency to suffer from micro porosity, such factors will reduce the tendency to create hot spots at the disc/pad friction interface and therefore judder.

Dimensional variations within the disc can be markedly influenced at the casting stage. An understanding of such variables is important. Factors such as mould and core stability, and core location within the mould are all important variables to monitor and control. Machining can, if not correctly controlled, introduce further variations through factors such as, run-out and disc thickness variation. Disc distortion can also be influenced by the build-up of internal stress within the component. Such problems can be minimised by choosing high carbon disc materials and introducing thermal stress relief treatments into the machining cycle. The use of two stress relief treatments, one before the first rough cut, and the second before the final finishing operation, have proven beneficial in the reduction of stress, and therefore distortion induced during both casting and machining operations.

Other causes of judder such as chemical reactions between disc and pad, and the formation of the third layer interface are influenced by the both the disc and pad materials. Indeed there is some evidence, as yet unconfirmed that high carbon disc materials prove beneficial from the point of view of the formation of the third layer interface.

7.4 SQUEAL AND GROAN.

Vehicle noise of any kind is rarely a safety issue, although in today's quality driven automobile environment where vehicle refinement is a major issue, any kind of noise is important. Squeal can be related to the coefficient or friction between the disc and pad. The higher the coefficient of friction, the higher the tendency to squeal. Groan is a low frequency noise, generated at low speed with light brake applications.

Both squeal and groan can be reduced by employing high carbon irons, whose damping capacity especially at very high carbon levels can be quite marked. Miller (13), reported that the most pronounced effect was with very high carbon irons having a low chromium content, it was also reported that pearlitic irons, with damping capacities of sufficient magnitude to significantly reduce squeal have been produced.

8. RESEARCH AND DEVELOPMENT PROJECT

As part of its investment for the future, Precision Disc castings embarked in early 1996, on a major research and development project. The project was managed by The Motor Industry Research Association (MIRA), and had as partners, European Friction Industries (EFI) and The Castings Development Centre (CDC). The first stage of the project was essentially a bench marking exercise, comprising of a vehicle disc survey, and a dynamometer testing exercise to evaluate four of the most common grey cast iron materials in use today.

8.1 DISC SURVEY

Discs from 49 different vehicles were examined and chemically analysed by CDC. The results obtained showed that there are a wide range of specifications in use, some for very similar discs, are very different. Carbon levels were found to vary between 3.14% and 3.80%. It was very clear that high carbon levels are becoming more acceptable, approximately 28% of the discs tested contained carbon levels above 3.50%, but only 10% of discs were below 3.30%.

The use of alloy elements was found to be very high, with some 76% of the discs tested containing significant elements, such as :-
- Chromium
- Molybdenum
- Copper
- Nickel
- Tin

either singularly or in combination i.e. Cr/Mo.

8.2 MATERIAL PERFORMANCE.

Four commonly available disc materials were evaluated at EFI, the materials chosen for this exercise were:-
- standard medium carbon non-alloyed cast iron equivalent to GG20/Gh190.
- a titanium bearing material containing 0.02 - 0.04% titanium with a medium carbon content
- a medium carbon molybdenum chrome bearing cast iron alloy containing 0.4% molybdenum and 0.2% chromium.
- a high carbon iron containing 0.5 - 0.8% copper. For the purpose of the exercise, one standard design of disc was chosen, namely a back vented disc of the type used on the Vectra.

The dynamometer testing established one basic fact, not surprisingly, wear due to friction increases as disc operating temperature increases. It was also established that of the standard materials under investigation, the high carbon material gave the best results in terms of resistance to thermal cracking.

9. ALTERNATIVE MATERIALS.

The drive to reduce vehicle weight, and improve efficiency, has led to the investigation of alternative materials for use as brake disc materials. To date, two alternative materials, carbon-carbon composites and aluminium metal matrix composites (MMC'S) have found limited use.

9.1 CARBON COMPOSITES

The carbon composite brake disc has found widespread use in the ultra high performance racing environment, typically on Formula 1 Grand Prix cars. Such discs are extremely expensive, but do offer the user a significant weight advantage. However recent tests with a Grand prix team have shown that there is no major brake performance advantage other than the weight saving, and the fact that discs will operate at temperatures approaching 1000 degrees centigrade. The inference from this is that higher braking loads can be applied and the massive amounts of heat generated can safely handled. The use of such a material on production road cars is most unlikely due to both the high cost factor, and its poor low temperature performance.

9.2 ALUMINIUM MMC MATERIALS.

It would appear that the main challenge to the traditional cast iron disc will come from an Aluminium MMC product. The properties of MMC materials have been widely examined and would appear to offer several major advantages, namely:-
- The thermal conductivity can be two or three times higher than cast iron
- An MMC disc could be 60% lighter than an equivalent cast iron component
- The thermal diffusivity, which is the rate of heat dissipation compared to that of storage, is four times that of cast iron.

Clearly an impressive material, the performance of which does depend on the nature of the composite dispersion. However unconfirmed reports would indicate that there have been problems with the limited application of this type of disc so far. It would appear that MMC discs are prone to problems relating to heat build-up caused by operational variation and noise. It has also been apparent that manufacturing problems, coupled with a high unit cost, have significantly contributed to the relatively rare use of these materials to date.

10. DISCUSSION - IS THERE A FUTURE FOR THE CAST IRON DISC ?

Obviously the writers have a vested interest in the subject under discussion, but the answer to this question must emphatically be yes. What is apparent however, is that a certain degree education and informative discussion is required. Unfortunately, cast iron through no fault of its own, has obtained a very old fashioned low tech image. It is the view of the author that this image has in part, not been helped by the National and International standards covering the material.

In twenty years of technical experience associated with automotive braking component industry, the writer has never seen a disc fracture at the junction of the friction ring and top hat, usually an area containing the thinnest section. Many discs have been examined that have failed due to heat induced cracking, usually resulting from a lack of thermal properties, but never failure due to outright lack of strength. Consideration of this fact would indicate that in the main, most discs have been over engineered from a material point of view. It is perhaps worth while to consider for a moment the facts surrounding the tensile strength of cast iron components.

It is generally accepted that the strength of cast iron decreases with increasing temperature, indeed Palmer reported (14) that above 300 degrees centigrade the strength falls sharply. Keiner & Werning (8) reinforce this view and further report that irrespective of the level of tensile strength at room temperature, the differences between the grey iron grades are minimal above 700 degrees centigrade. They strongly put the view that in general, discs produced in high strength materials in order to obtain high load bearing capacities and high hot/cold strengths, have been over engineered. Certainly the author's own practical experience would support this view.

Why should this be ? In part, the view the view must be put forward that grey irons have not been well served by the relevant national standards. Tensile strength is always specified in separately cast test bars, not strength in the critical section of the component. It is the authors view, again backed up by practical experience, that the properties of thin section components like brake discs should always relate to a critical section in the component. In the case of the brake disc this is usually specified in the friction ring, in the area of the top hat junction.

If the view is accepted that cast iron discs have, in the main, been over engineered there would appear to two avenues of development. Firstly, high carbon materials should in time become the accepted standard material for brake discs irrespective of vehicle size, weight and performance. There is plenty of work (8) & (15) to support this view. Secondly in order to compete with MMC materials, the weight of cast iron discs must be reduced. This will involve research into the development of production methods, alloy compositions and the microstructural integrity of the component. Novel casting methods may be required in order to produce near nett shape discs, thus improving the competitive edge of the component from a machining point of view. It may be that composite discs, having a cast iron friction rotor attached to light alloy bell housing may find a more wide spread application than the limited use in motor sport at present.

It is vitally important that industry from both ends of the production spectrum co-operate together to evolve a strategy for the development of cast iron. Academic institutions and research organisations must also be closely involved. The foundry industry must not be allowed to experience problems similar to those that occurred with the introduction of the titanium grades of disc materials. Unfortunately on that occasion, new specifications were introduced with seemingly little thought as to how the required specifications could be accurately controlled and little understanding of the consequences of adding titanium into the foundry system.

11. CONCLUSION.

Grey cast iron is the ideal material for the production of large quantities of high quality automotive components. The material is easily controlled to tight compositional requirements that satisfy the performance requirements of safety critical components such as brake discs. The authors have attempted to show in this paper that cast iron, far from being an old fashioned low tech material can, with the input of industry and research establishments, become a high tech material suitable for continued use well into the next millenium. The view is strongly put forward that efforts should be concentrated on high carbon cast iron disc materials.

However, until the day when passenger vehicles no longer require wheels, and one would therefore assume brake discs, it is fairly certain to say that the iron-carbon composite material known as grey cast iron will continue to be the first choice for the majority of braking applications and that Precision Disc Castings will be heavily involved at the forefront of the industry.

REFERENCES.

1. Rendall I, "The Chequered flag",
   Weidenfeld and Nicolson 1993

2. Godfrey J, "Ferrari Dino SP's Maranellos First Rear Engined Sports Prototypes".
   PSL Ltd. ISBN 1-85260-359-3

3. Engineering Data On Grey Cast Irons. BCIRA. Alvechurch 1977

4. Engineering Data On Nodular Cast Irons BCIRA. Alvechurch 1986

5. Hect R L, Dinwiddie R B, Porter W D, Wang H, "Thermal Properties of Grey Cast Irons. SAE Paper 962126.

6. Chapman B J, Mannion D, "Titanium Bearing Cast Iron For Automotive Braking Applications". Foundry Trade Journal, 1982 pp 232-246.

7. Chapman B J, Mannion D, "Cast Iron Brake Rotor Metallurgy"
   Institute of Mechanical Engineers. 1976 C35/76.

8. Keiner W, Werning H, "GG15HC High Carbon Grey Iron. An Ideal Material for Brake Discs and Drums" Konstrulieren & Glesten 15 (1990) No 4.

9. Wirth A, Whitaker R, "Developments in Cast Iron Rotor Technology"
   Advances in Automotive Braking Technology. Institute of Mechanical Engineers Leeds University April 3rd 1996.

10. Brecht J, Egner-Walter,A. "Influence of Material Selection on Stresses in Ventilated Brake Discs". SAE Paper 980595. Michigan 1998.

11. "Tool Material Optimises Brake Disc Machining"
    Foundry Trade Journal January 1983 pp102

12. Broskea T J, "High Speed Machining of Grey Cast Iron with Polycrystalline Cubic Boron Nitride", Carbide Tool Journal Vol 19 (5). Sept - Oct 1987 pp 17 - 20

13. Miller E, "Damping Capacity of Pearlitic Grey Iron and its Influence on Disc Brake Squeal Suppression". SAE Paper 690221

14. Palmer K B, "Mechanical and Physical Properties of Engineering Grades of cast Irons up to 500 centigrade". BCIRA Journal Report 1717. 1986.

15. Jimbo et al. "Development of High Thermal Conductivity Cast Iron for Brake Disc Rotors". SAE Paper 900002. International Congress and Exposition, Detroit 1990.

The effect of cast iron disc brake metallurgy on friction and wear characteristics

K ELLIS
European Friction Industries, Bristol, UK

ABSTRACT

After examination of a number of current vehicles, four disc brake metallurgy specifications were identified as being representative. These were tested in a common disc shape using the original equipment brake pads to determine the effect of disc metallurgy on the friction, wear and integrity characteristics. It was found that friction was only affected by the specification containing titanium as an alloying element. High carbon racing discs showed excellent integrity but average to poor wear whilst low/medium carbon molybdenum alloyed discs showed excellent wear but poor integrity. Further work will consider other pad materials and new disc metallurgies.

1. INTRODUCTION

The work detailed in this paper was carried out as part of the initial phase of a research programme sponsored by Precision Disc Castings. As such it was intended to provide a first view of the current state of automotive brake disc technology.

The first stage was to examine the original equipment (OE) discs from around one hundred of the best selling cars in Europe. The results of this examination, primarily for metallurgical composition but also for component design, showed that most discs could be grouped in to one of four grey cast iron alloy specifications. These specifications were:

A	PDC "Carballoy"	- a medium carbon grade equivalent to GG20
B	Ford spec SAM 1A 9100 C	- equivalent to GG25 and containing Titanium
C	PDC "Discalloy"	- a high carbon racing grade containing copper
D	GM/Opel spec QS 13M 000	- equivalent to GG35 and containing molybdenum and chromium

The next stage was to cast a common disc shape for testing purposes in each of these generic alloys. The shape chosen was the GM/Opel Vectra as it was typical of many current vented disc designs.

The aim was to then subject these discs to friction testing, wear testing and integrity testing to assess whether different disc metallurgy effected the performance characteristics of the disc brake. It was decided to use one common friction material for all of the testing. The material chosen was that supplied by GM/Opel as the OE material for the Vectra.

2. FRICTION TESTING

Testing was carried out was on an inertia dynamometer using the full brake assembly at an inertia of 57kgm^2. The test schedule used mirrored the testing required for ECE Regulation 90 aftermarket type approval. The test was in four sections:-
- a bedding run of 15 stops at 0.2g from 50km/h with a start temperature of 150°C (Fig 1)
- a Type 0 performance test. Three stops from 70km/h at 0.1g, 0.2g, 0.3g, 0.4g, 0.5g, 0.6g with a start temperature of 100 °C. The graph of deceleration against caliper line pressure was then produced using the average of the three stops as the pressures required to achieve 0.5g (P1) and 0.3g (P2) are required for the third and fourth parts of the schedule (Fig 2).
- a speed sensitivity test. Average of three stops from 65km/h, 100km/h and 130km/h at pressure P1 with a start temperature of 100 °C (Fig 3)
- a Type 1 fade test (Fig 4)

The results of the bedding run are presented as the average friction of the last five stops and the Type O tests as the values for P1 and P2 in the table below.

Disc material	Bedding friction (mu)	Type 0 (P1) (Bar)	Type 0 (P2) (Bar)
A	0.53	40	24
B	0.59	36	22
C	0.54	38	23
D	0.54	42	25

The results of the speed sensitivity are presented in figure 5, and the fade tests in figure 6.

The results show that there is no significant difference between specifications A, C and D for all sections of the test but noticeably different results for specification B. B exhibits a generally higher friction response and has a different speed sensitivity characteristic from the other materials. This phenomenon has been observed before (1,2) for grey cast irons containing titanium as an alloying ingredient.

3. WEAR TESTING

Testing was carried out using a full size disc and caliper on a "Krauss" type friction testing machine. The machine runs constantly with the brake being applied for short bursts and then released to allow the disc speed to recover. For the purposes of this research the brake pad was reduced in radial width so that a clear wear track could be produced in the disc. After preliminary tests established temperatures at which wear differences could be observed the final comparisons were carried out at 400 °C and 500 °C. The brakes were applied so that cycling occurred at +/- 15C around the target temperature for a total of 500 brake applications (See figure 7 for an example). The disc wear was then determined after the test by surface profiling the disc from a flat baseline outside the wear track (See figure 8 for a typical trace). Using this method an accurate measurement of the wear volume could be made relative to the total energy input during the brake applications.

The results of the wear testing (in mm^3/MJ) are presented in the table below.

Disc material	Nominal disc temperature	
	400 °C	500 °C
A	55.3	58.8
B	45.1	75.7
C	58.5	60.9
D	28.5	53.8

The results show that at the test temperatures specifications B and D exhibit the greatest sensitivity to temperature and that the lowest wearing disc specification is D.

4. INTEGRITY TESTING

An initial test was carried out on an inertia dynamometer which mirrored a disc integrity test currently carried out by the TUV in Germany to type approve aftermarket brake discs. Unfortunately the test showed no integrity problems or differences between any of the specifications. A more severe test was therefore undertaken using the "Krauss" type machine. The brakes were applied at a constant torque of 600Nm to raise the disc temperature from 100 °C (ambient on first cycle) to 600 °C in approximately 30 seconds. The cycle was repeated ten times and the disc then examined for signs of damage. Figure 9 shows a typical temperature and friction profile for the ten cycle test but it should be noted that the cool down times were actually in the region of 30 minutes. The test was run through five repeats of the 10 cycles and then ceased if no damage had been detected.

Even a test as severe as this only showed damage for one of the specifications. Failure was catastrophic with the disc cracking through both vanes and in to the top hat section. To further differentiate between specifications the test was repeated using a racing pad material which was much harder than the OE pad material and designed to run at higher temperatures. This material started to show catastrophic cracking of the disc much earlier in the cycles. Due to the earlier cracking only three temperature cycles were run initially, with inspections then being made at every cycle. The aim was to identify the onset of damage before a catastrophic failure occurred. In all cases no cracking was observed in the disc prior to the cycle in which the failure occurred. The table below shows the number of cycles required to cause catastrophic cracking for each specification tested against each pad material (50 cycles indicates that the test was ended with no cracking observed).

Disc material	Number of cycles to failure	
	OE pad material	Race pad material
A	50	20
B	41	3
C	50	50
D	50	3

The results suggest that the best disc by far for integrity was specification C which did not crack after 50 cycles against either material. This fits well to the fact that this specification is the material used in racing discs where the ability to withstand severe temperature fluctuations is critical.

5. CONCLUSIONS

The results showed that there are noticeable differences between the friction and wear characteristics of the four grey cast iron specifications. For friction testing the results suggest that titanium additions to the specification have a negative affect on friction causing higher than normal friction with greater instability. The titanium alloy discs also exhibited the highest high temperature wear and the worst integrity. The OE specification material (D) was the hardest and consistent with this showed the lowest wear at both temperatures and a poor integrity. The racing specification (C) conversely was the softest specification and showed a higher level of wear than the OE but had by far the best integrity of all the discs.

Further work is required in many areas.

(1) Friction testing using a number of different friction materials and at least three tests for each disc material to give good confidence of differences in friction.

(2) Wear testing for longer periods at lower temperatures to assess with greater accuracy any differences in wear at normal usage temperatures

(3) Friction, wear and integrity testing on alternative cast irons, cast steels and light alloys such as aluminium metal matrix composites.

6. REFERENCES

(1) Chapman B.J. and Hatch D., "Cast iron brake rotor metallurgy", I. Mech E. C35/76

(2) Chapman B.J. and Mannion G., "Titanium bearing cast iron for automotive brake applications", Foundry Trade Journal, 1982, pp 232-246.

7. ACKNOWLEDGMENTS

The author would like to thank Precision Disc Castings for supporting the research presented in the paper and for giving permission to publish it. Also to Mr. N.Keefe for the testing and support in producing the graphs for this paper.

Figure 1 - Typical result of bedding section

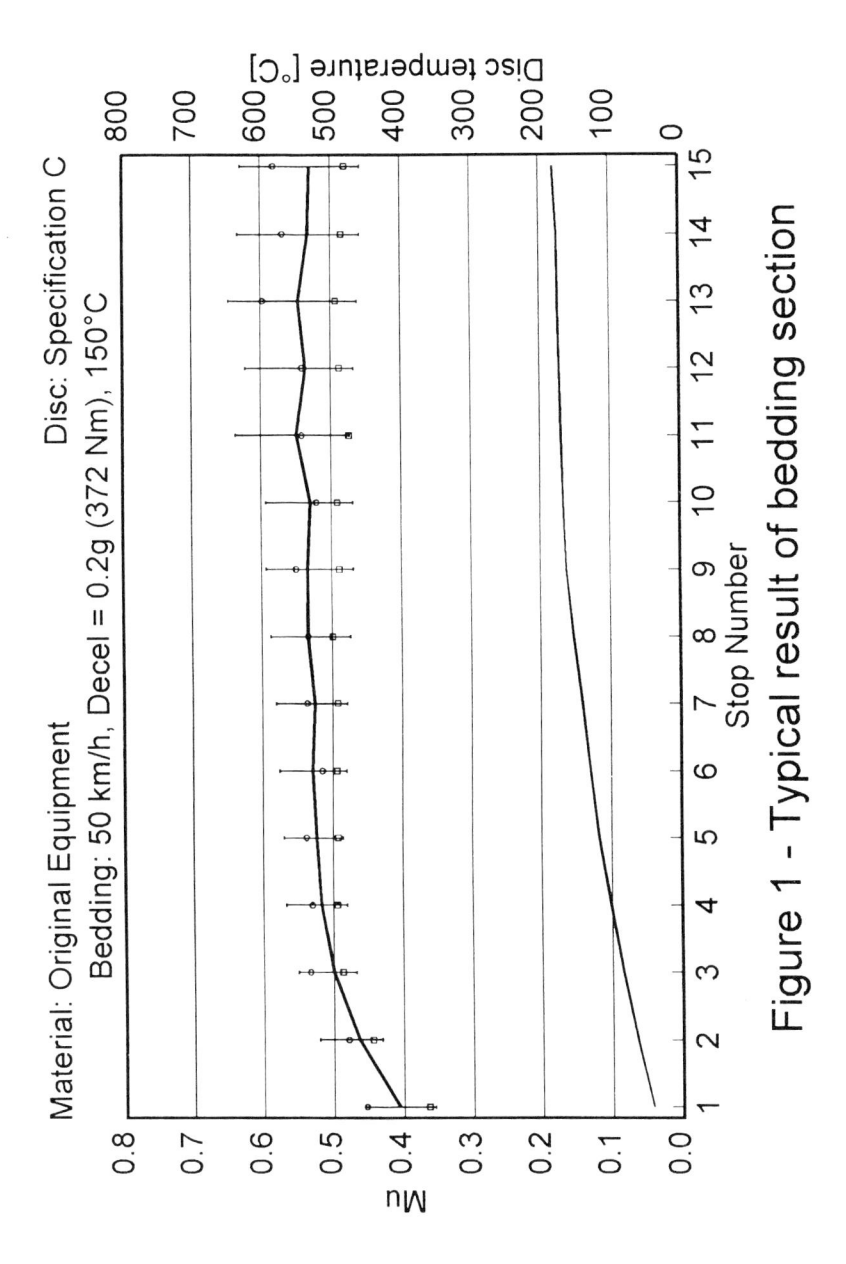

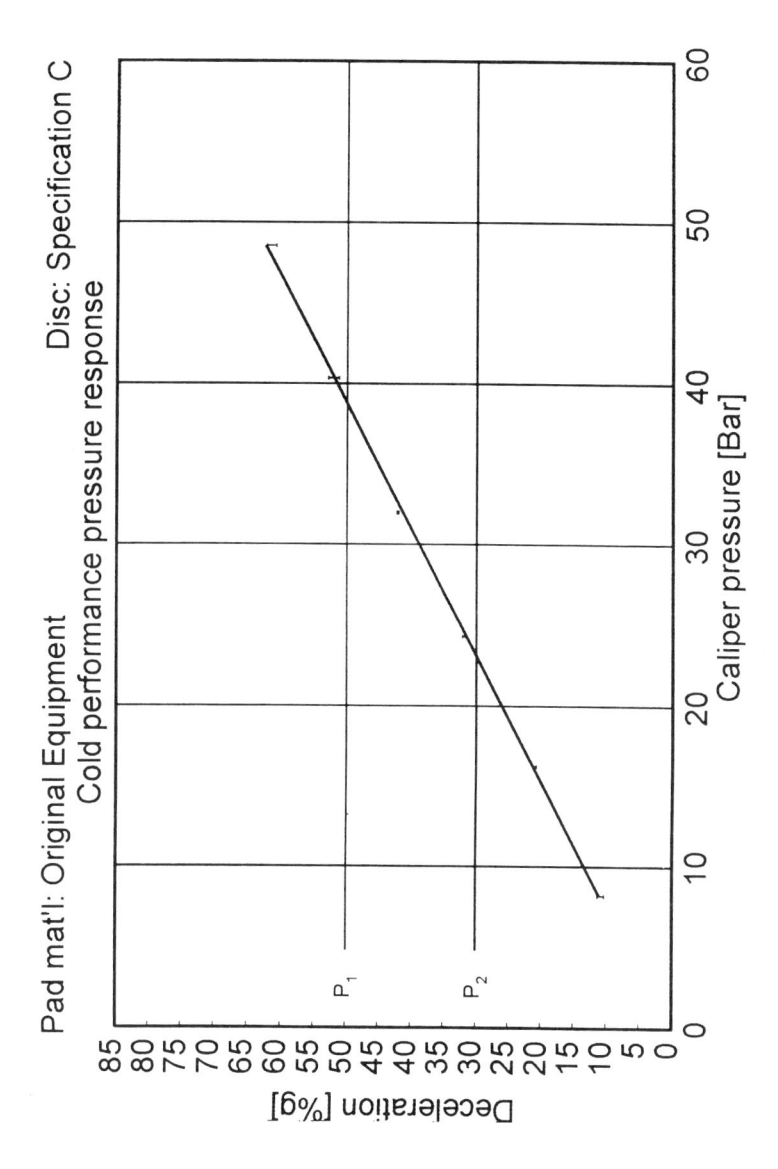

Figure 2 - Typical type 'O' friction test

Figure 3 - Typical speed sensitivity result

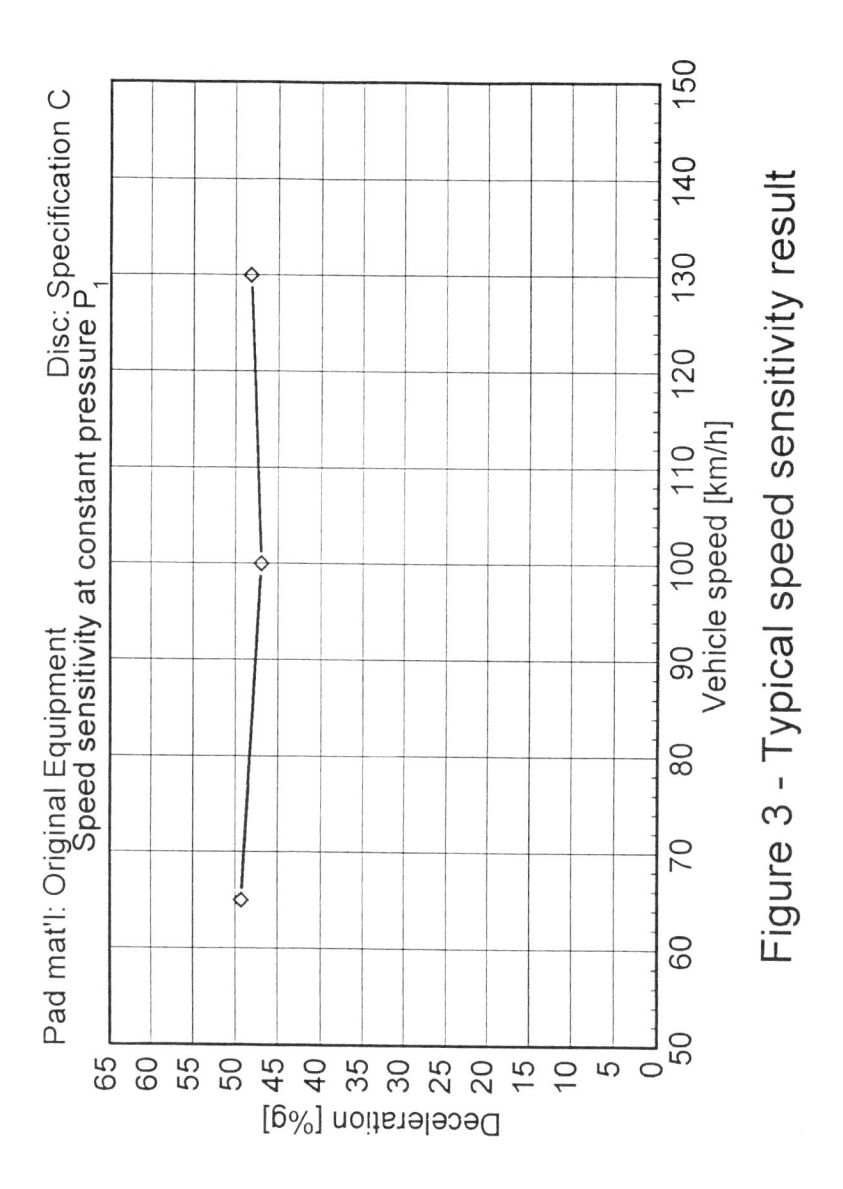

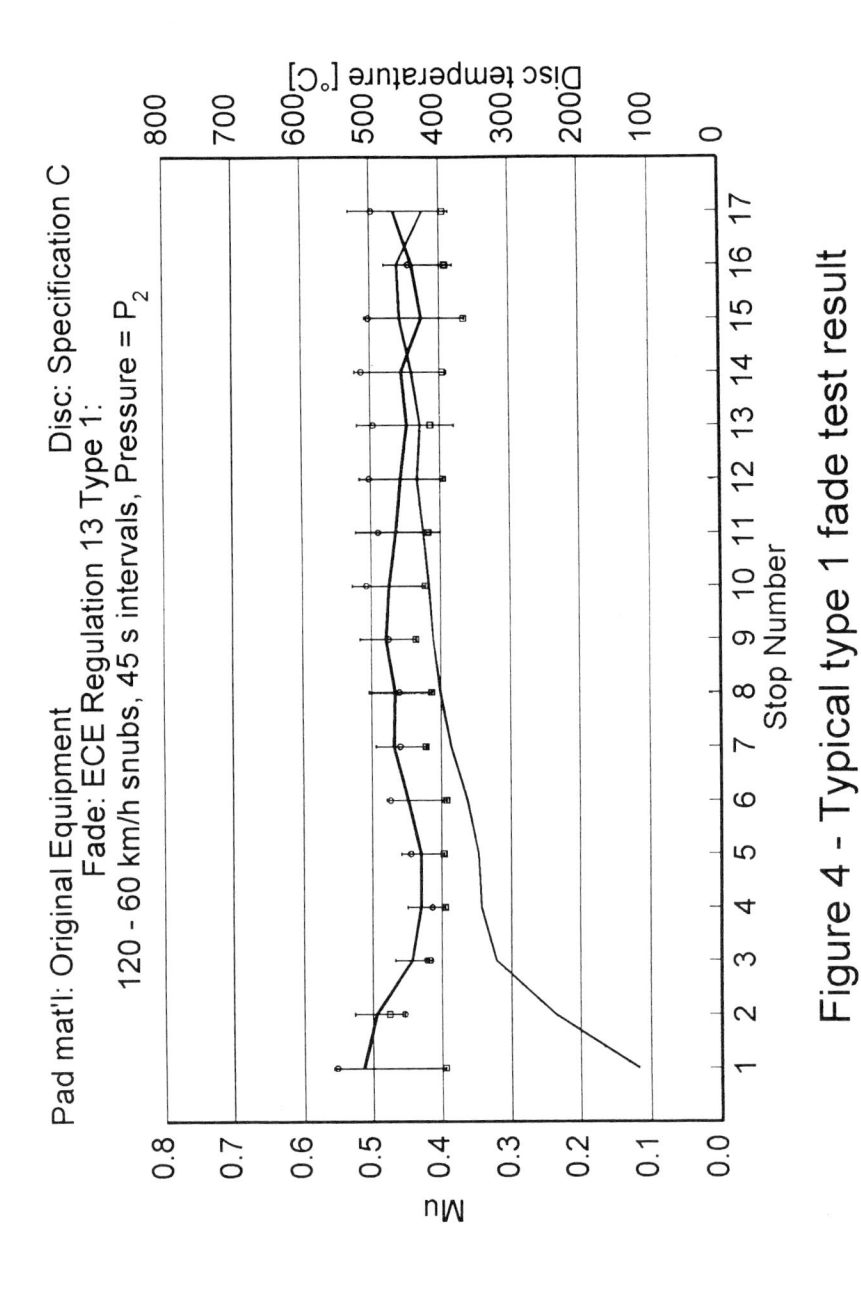

Figure 4 - Typical type 1 fade test result

Figure 5 - Speed sensitivity results

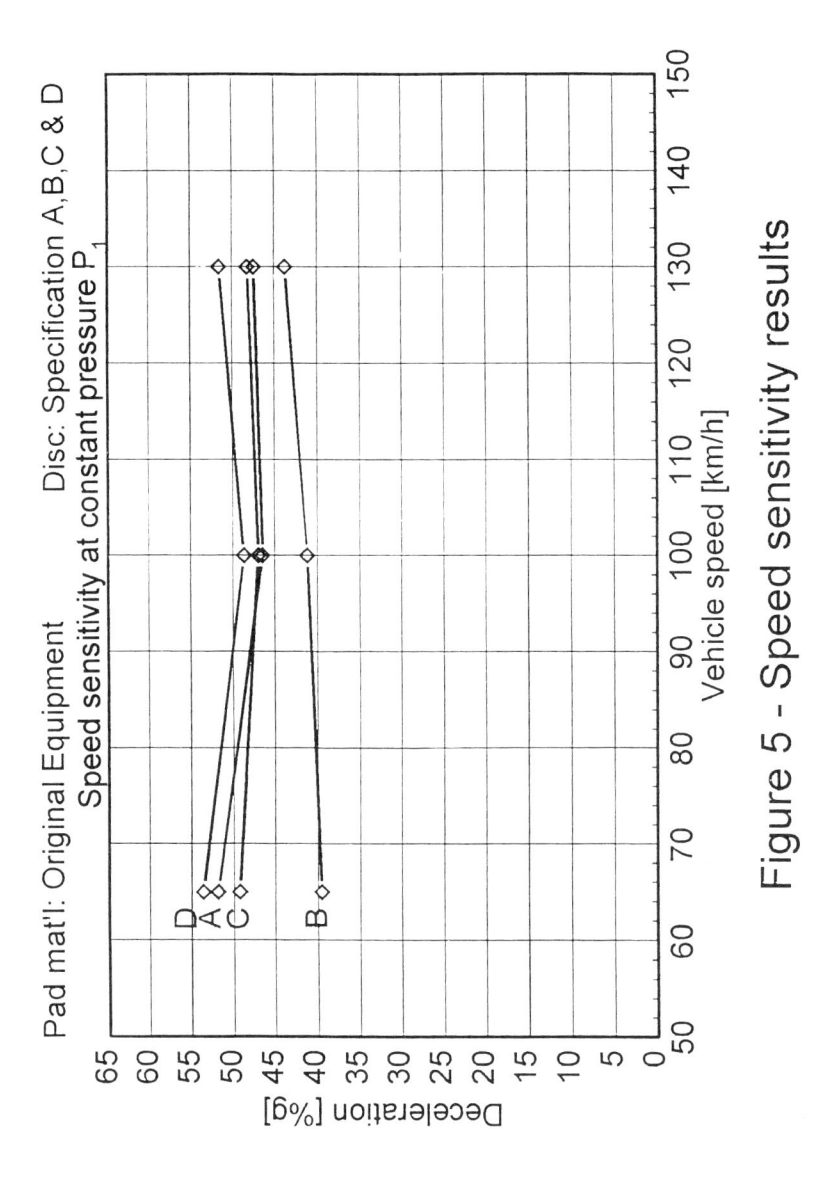

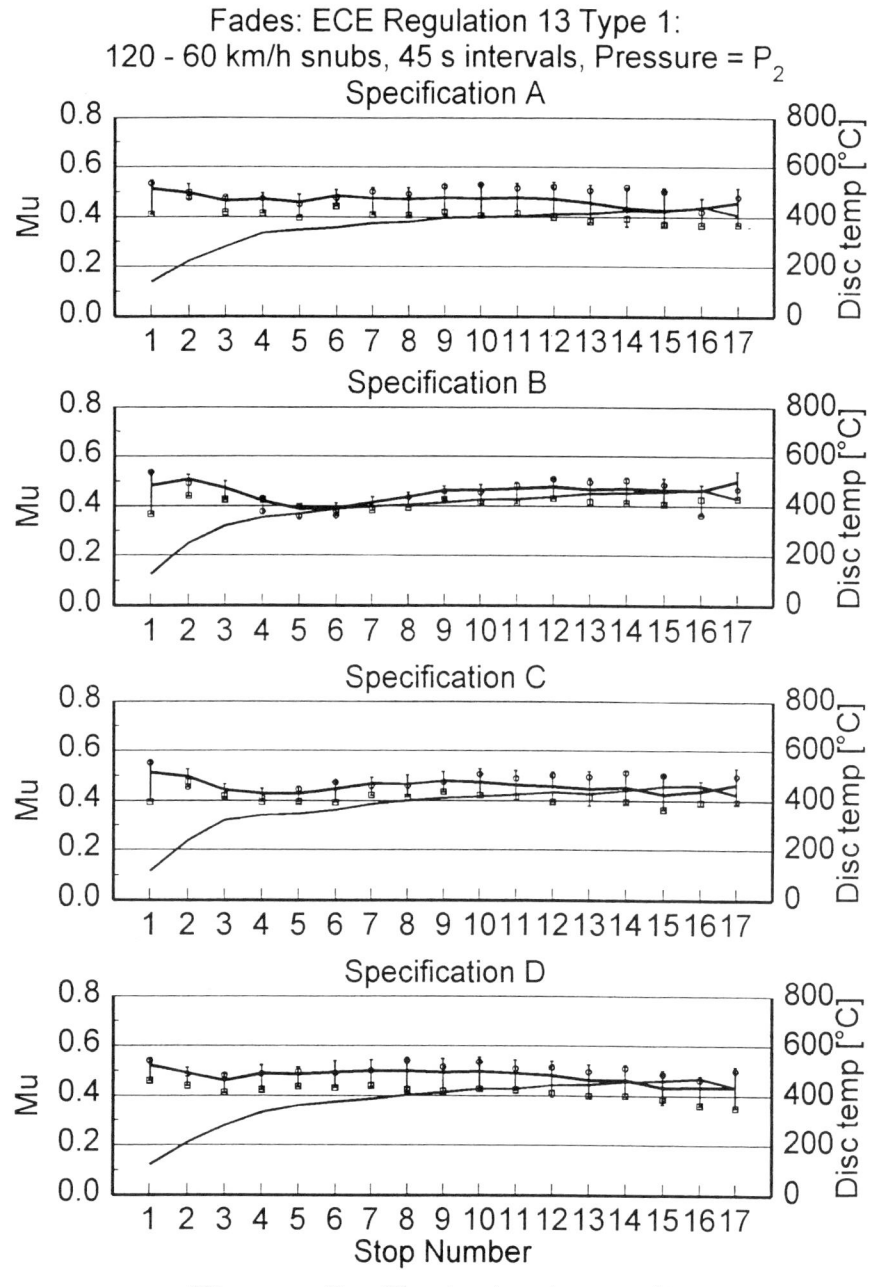

Figure 6 - Fade test results

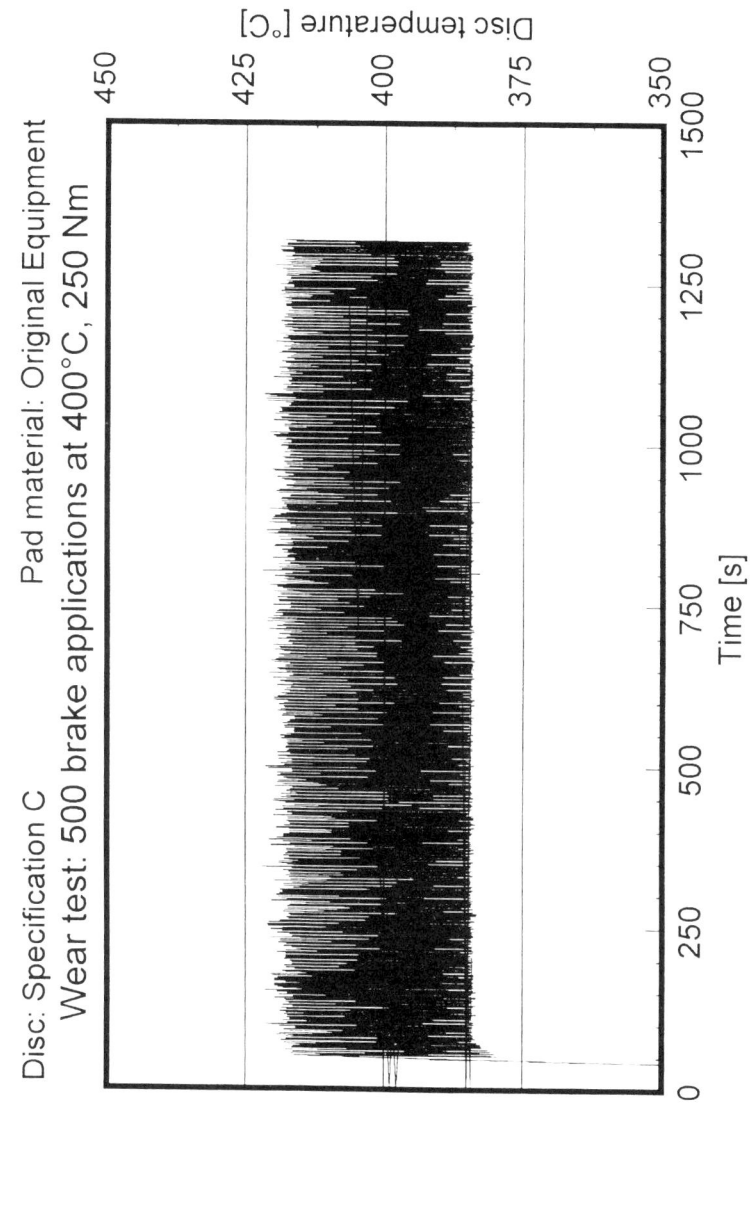

Figure 7 - Typical wear graph to illustrate temperature cycling

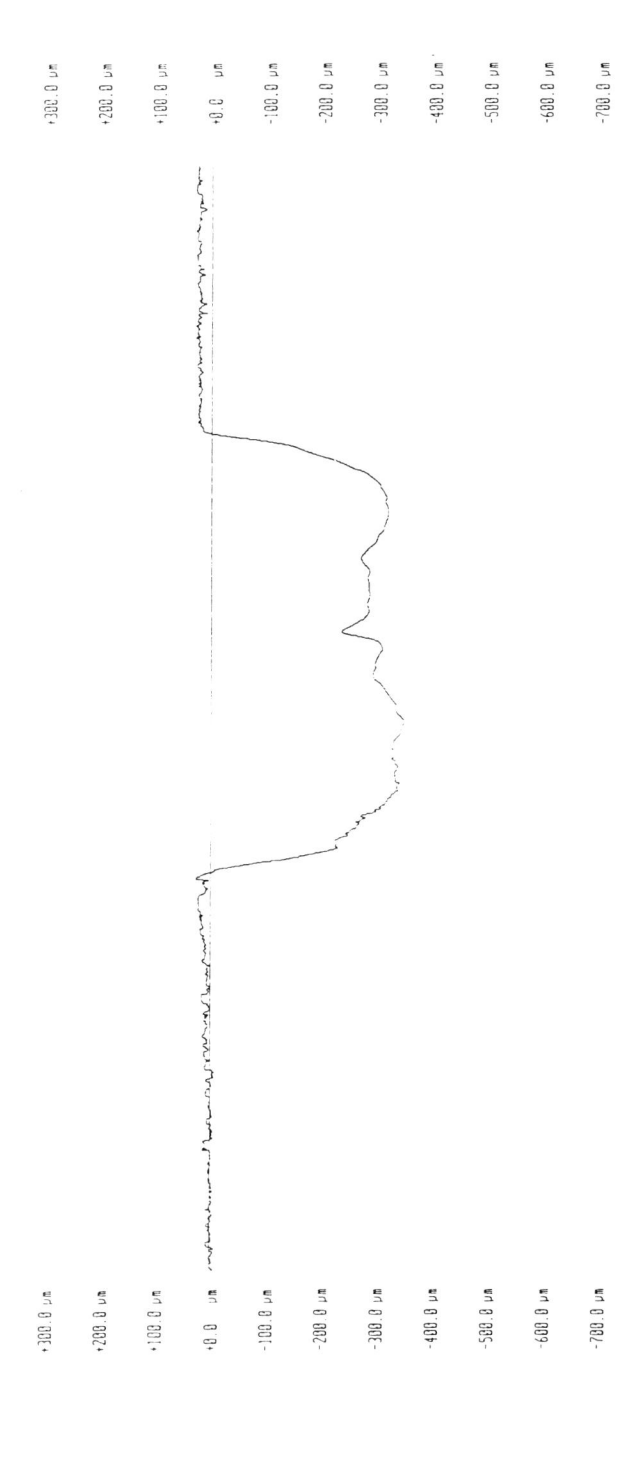

Figure 8 - Typical surface profile trace

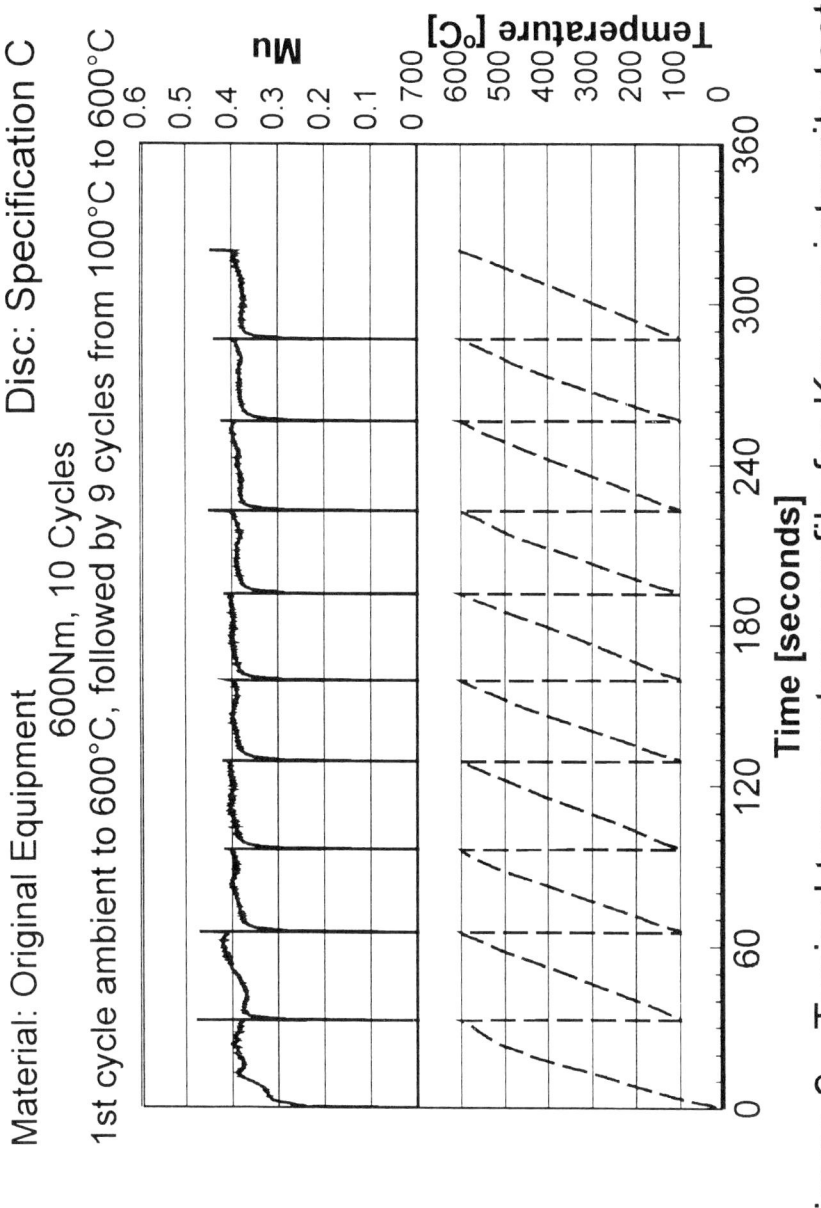

Figure 9 - Typical temperature profile for Krauss integrity testing

Aluminium metal matrix composite rotors and drums – a future trend

M J DENHOLM
Duralcan USA, Novi, Michigan, USA

1. ABSTRACT:

This paper discusses the use of AL MMC rotor and drum technology and how it might move light vehicle brake systems to new levels of customer satisfaction. Judder and roughness, noise, durability and drivability levels have all fallen short in recent years when compared with customer expectations and the improvements to other systems on the vehicle. AL MMC technology exhibits a remarkable fit, in addressing all of these areas simultaneously, while offering valuable unsprung, rotating mass reduction. While gains can be realized when AL MMC's are employed in direct substitution, this is not an ideal way to achieve successful performance and maximize AL MMC advantages. A more demanding design protocol is suggested, to provide low temperature braking and to maximize the benefits from the properties of these new materials.

2. INTRODUCTION:

Pure Aluminum has a solidus of 646°C. When reinforcing particles of any type are added to aluminum, the need to alloy the matrix to accommodate the special requirements of the particulate and the process, quickly emerges. For example, infiltration processes require the addition of Magnesium to accelerate the infiltration of matrix into the particulate preform. Mixed castable materials require the addition of Silicon to maintain fluidity without triggering undesirable side effects. In each case the solidus is depressed. In each case the solidus will be a little different. In every case the resultant MMC material will requires engineering a brake system with a lower operating temperature than typically designed for Gray Cast Iron (GCI). At first the requirement to design these materials may be seen as a commercial barrier. However when the outstanding diffusivity properties of these materials are considered, together with the inherent advantages of avoiding high brake operating temperatures, the opportunities to design for both performance and weight advantages can be seen as a big advantage.

By contrast, GCI systems are consistently required to operate at temperatures that cause local surface phase changes to the rotor, and cause damage to the lining material. While these conditions are not detrimental to safety or raw stopping performance. They are very detrimental to component life, and have a significant impact on driver comfort and customer satisfaction.

Most engineers begin their exposure to Aluminum Metal Matrix Composites (AL MMC) by testing a material substituted part. That is, taking an existing geometry, designed for GCI and substituting AL MMC in its place. While this approach is a reasonable first step it will not capture the full benefits of these remarkable materials. What follows is a discussion of the design approach more likely to result in a successful outcome, marked by improved performance, noise, comfort, life and appearance, all with a marked reduction in unsprung rotating mass.

3. MATERIAL PROPERTIES:

Before considering a design protocol it seems appropriate to review the relevant thermal, wear, and corrosion resistance properties of AL MMC's as they compare with the benchmark of GCI. An ever increasing choice of AL MMC materials is available and selection of the most effective alternative is important for each application. As a general rule we can say that SiC reinforced materials will provide the best thermal properties for brake applications unless the limiting factor is high operating temperature. In that case Alumina reinforcement might be considered as this allows the use of matrix alloys with higher solidus. This is however, a complex discussion worthy of treatment as a separate subject.

3.1 Thermal Diffusivity
Thermal diffusivity might best be thought of as the material's tendency to conduct heat away vs. the tendency to store heat. High diffusivity is desirable in a rotor material if it is the result of high conductivity.

3.2 Heat Capacity
Heat Capacity might be thought of as Specific Heat expressed per unit volume instead of per unit mass: a more appropriate measure where two materials of very different specific gravity are being compared.

3.3 Conductivity

A comparison of conductivity is the most direct indication of difference between various AL MMC's and GCI. It is important to note that this property assumes even greater importance in the operating temperature ranges of AL MMC's because radiation is not a factor at these low temperatures. By contrast, GCI relies heavily on radiation as an escape mechanism when very high temperatures are reached. Fortunately this mechanism offsets the rapidly diminishing diffusivity of GCI with rising temperature. GCI loves to store heat and this tendency increases with increasing temperature. In other words, as a heat-pumping medium it is inherently unstable.

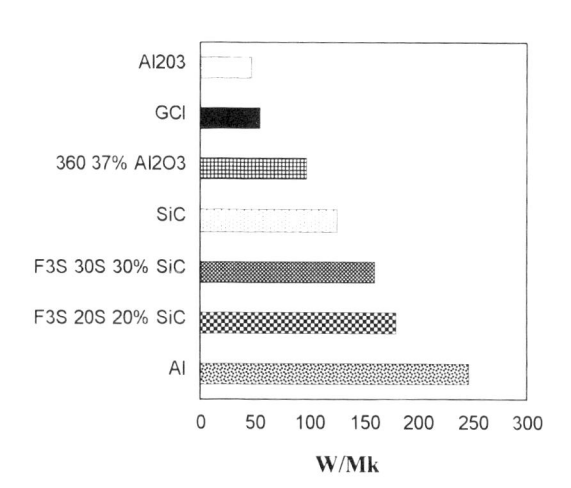

Table 1: Conductivity Comparison(1)

Material	W/Mk
Al2O3	~30
GCI	~50
360 37% Al2O3	~125
SiC	~150
F3S 30S 30% SiC	~160
F3S 20S 20% SiC	~180
Al	~250

3.4 Wear Resistance

The addition of ceramic particulate into aluminum was expected to improve many of the mechanical properties of the matrix material including wear resistance. The result was better than anticipated. At 20% by volume loading of SiC particulate, we see an order of magnitude increase over GCI (2). This level of wear resistance presents the possibility of lifetime rotors

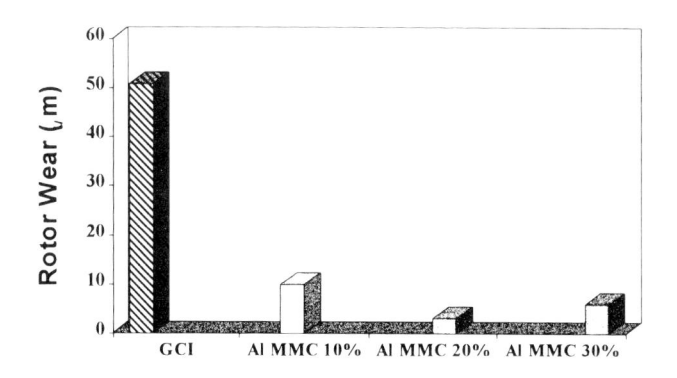

Table 2: Wear Rate Comparison

Material	Rotor Wear (μm)
GCI	~52
Al MMC 10%	~9
Al MMC 20%	~4
Al MMC 30%	~7

and drums, for the first time
4.0 A DESIGN PROTOCOL:
The thermal property advantages of AL MMC cannot be realized unless adequate airflow is available at the wheel end. Conduction and forced convection are the heat pumping means used by these materials. Every rule developed for GCI rotor design must now be re-examined. AL MMC rotors will respond to improvements in airflow both to and around the convection surfaces to a much higher level than GCI, making it worthwhile to persue. Small improvements in the ratio of volume to surface area will yield additional cost and weight savings without sacrificing performance. Reoptimizing diameters, braking path widths and caliper sizes all deserve attention for the system benefits in drivability, cost and weight.

4.1 Optimize Airflow to the Wheel End
Modeling by Messrs. Shen, Mukutmoni, Thorington and Whaite (3) has shown that airflow behavior around the wheel end is not intuitively obvious, but using computational fluid dynamics, provides accurate prediction of heat transfer coefficients. This tool offers the opportunity for significant optimization of cooling airflow in the early phase of vehicle design, when it can be accomplished most easily and economically. As we will see, adequate air supply to the rotor is the starting point for successful application of AL MMC to rotors, particularly for front brakes where energy input levels are at the highest level.

Sustained effort on the part of vehicle designers to reduce aerodynamic drag has been very successful, frequently at the expense of airflow to the wheel end. The challenge now is to find effective ways to create localized turbulence, sufficient to cool the brake without significantly changing the vehicle Cd.

Work being done by Dr. James Trainer and others on SAE recommended practice J1652 thru the Brake Effectiveness Subcommittee 4 of the SAE Brake Committee has revealed some valuable information on this issue. Front and rear brake (disc and drum) cooling curves were run on a General Motors B body car, at both 80 & 112 KPH vehicle speed. From These curves, heat coefficients were calculated and a +/- 10% acceptance band established. Corrections were made to normalize the data to $27°C(81°F)$ ambient. The curves for Disc and Drum brakes are shown in tables 2 and 3. When the same brakes were run on the

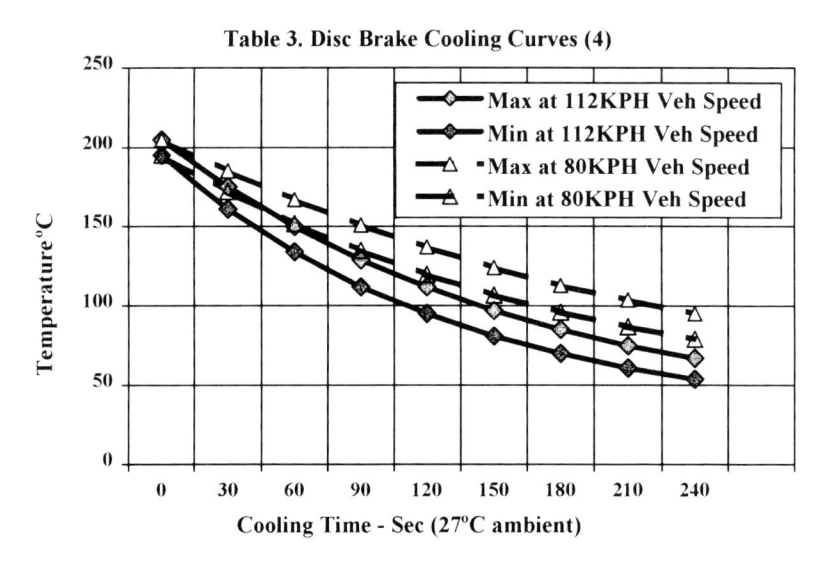

Table 3. Disc Brake Cooling Curves (4)

dynamometer, cooling air speeds had to be reduced to very low levels. In the case of the rear drum brake, a wheel had to be added and the ducted air speed reduced to Zero. This can only indicate that on this car there is very limited airflow to the wheel ends.

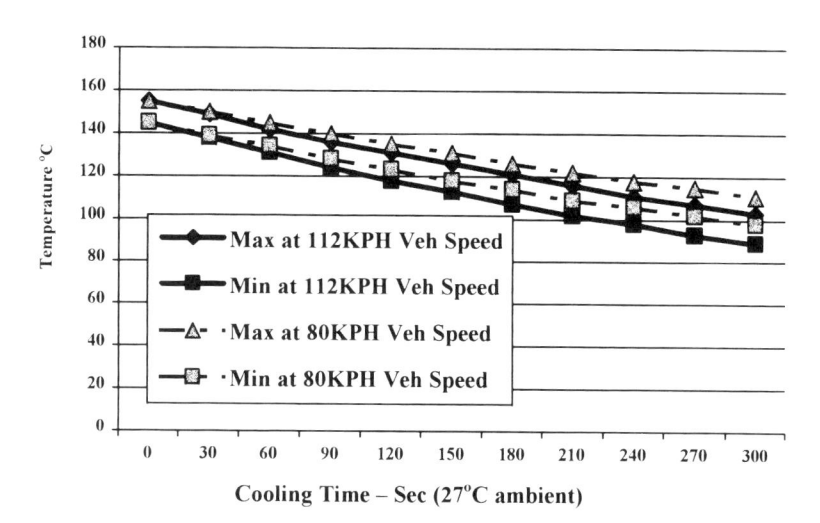

Table 4. Drum Brake Cooling Curves (4)

[Chart: Temperature °C vs Cooling Time – Sec (27°C ambient), showing four curves: Max at 112KPH Veh Speed, Min at 112KPH Veh Speed, Max at 80KPH Veh Speed, Min at 80KPH Veh Speed]

4.2 Optimize Rotor Design
Full optimization of an AL MMC rotor requires attention to 3 parameters.

4.2.1 Maximum Cooling Surface Airflow
Traditional GCI designs reflect the relative insensitivity of GCI to changes in airflow and low manufacturing cost. Large increases in cooling airflow result in almost unmeasurable reduction in operating temperatures. Significant GCI temperature reduction has historically been made using increased rotor mass, accepting the weight penalty.

Recent work done at the University of Illinois at Urbana-Champaign (5) illustrates the effectiveness of relatively small changes in design and the importance of attention to detail. Experiments performed on the Chrysler production LH front rotor resulted in a 70% increase in flow velocity at the rotor vane outlet. See fig 1.and table 1.

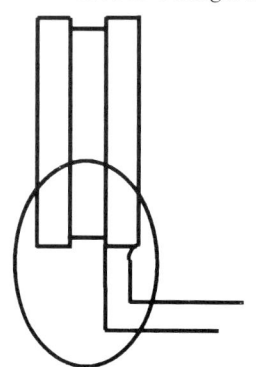

Fig 1a LH Production Configuration

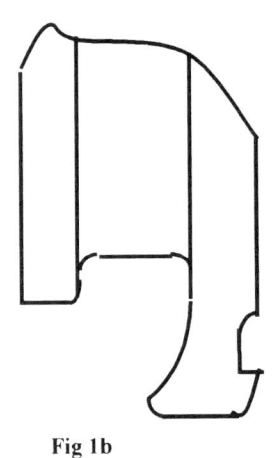

Fig 1b 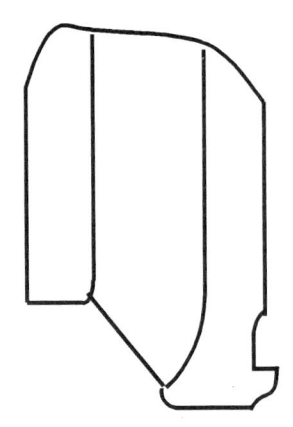   Fig 1c

Table 5.

Rotor RPM	Fig 1a M/s	Fig 1b M/s	Improved Flow	Fig 1c M/s	Improved Flow
600	0.957	1.264	32%	1.523	59%
800	01.159	1.688	46%	1.967	**70%**

The University study while of great interest and value, is only a detail improvement to traditional vented rotor design. New rotor concepts are needed which break from tradition, to fully capitalize on the AL MMC material characteristics and the more varied manufacturing options that these materials allow. Several creative alternatives currently in development, are showing great promise.

4.2.2 Optimize Volume to Surface Area Ratio
By re-adjusting the proportions of the rotor the ratio of volume to surface area can be changed to make maximum use of the AL MMC thermal properties. In making these changes, advantage can be taken of the significant improvement in rotor and lining wear rates. Lining area can be reduced in a way that will increase the effective radius and consequently reduce the caliper piston size, caliper weight and cost. Fig 2a shows a typical configuration that would be found in a 15ins-wheel package. Straight substitution of AL MMC will typically result in inadequate cooling to take advantage of the AL MMC thermal properties and will show symptoms of too little mass. This will be particularly evident with higher performing vehicles or those capable of high maximum speeds. It would be easy at this point to dismiss the technology as inadequate.

Fig 2a Production Configuration

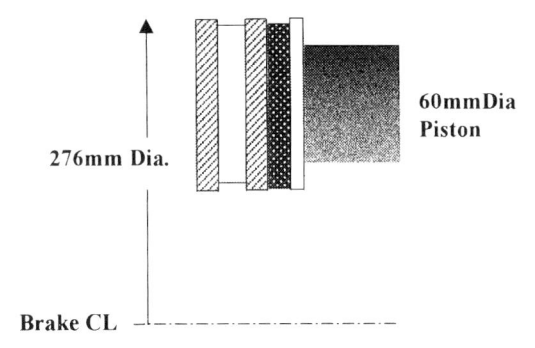

[Diagram showing 276mm Dia. rotor with 60mm Dia. Piston, Brake CL indicated]

Fig 2b illustrates a revision in the ratio of volume to area that might be made while preserving the caliper geometry and overall wheel-end package. This approach is a minimal attempt to optimize for the AL MMC properties by reacting to the kind of results achieved with the approach reflected in Fig 2a. If this is done in combination with the improvements to airflow shown in Fig 1c, good thermal performance can be expected.

Fig 2b Revised Volume to Area Ratio

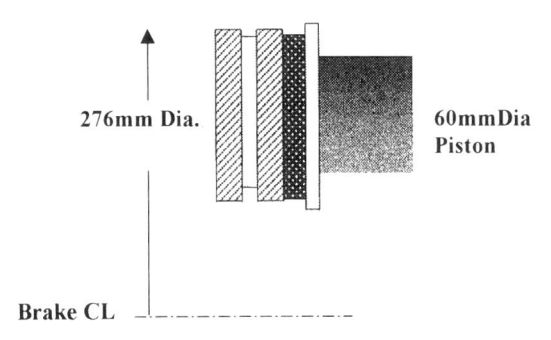

[Diagram showing 276mm Dia. rotor with 60mm Dia. Piston, Brake CL indicated]

By taking advantage of the much lower lining wear rates possible with AL MMC, the rotor diameter can be increased without concern about large increases in weight. The swept width of the rotor can be reduced giving a double increase in effective radius. The increased radius allows the use of a smaller, lighter, less expensive caliper. Fig.2c. shows an actual example of the magnitude of the change that may be possible. The 60mm caliper has been reduced to 54mm; a 19% reduction in piston area.

4.2.3 Optimize Volume for Heat Capacity

The final step in the optimization process requires that the heat capacity of the rotor be adequate to accept the worst case energy input condition without exceeding the maximum operating temperature of the rotor material being used. The test procedure used to accomplish this is at the discretion of the vehicle manufacturer or the 1st tier brake system

Table 6 : Effect of Designing for Mass

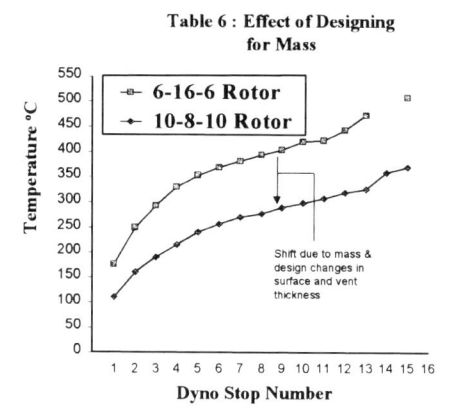

[Graph showing Temperature °C vs Dyno Stop Number for 6-16-6 Rotor and 10-8-10 Rotor, with annotation "Shift due to mass & design changes in surface and vent thickness"]

supplier. The AMS test can be used in combination with one or more maximum speed stops. It is important to remember that traditional test procedures for burnish, fade and recovery

Fig. 2c. Reoptimized Brake

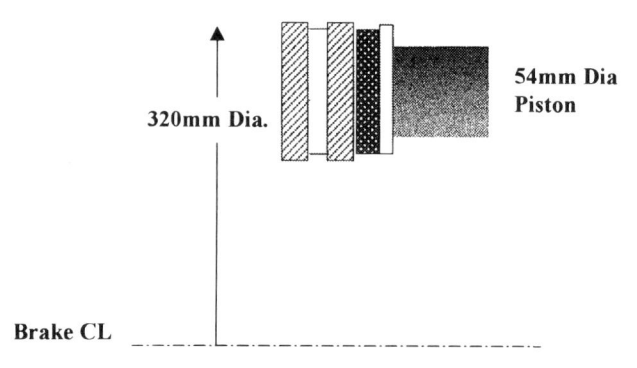

[Diagram showing 320mm Dia., 54mm Dia Piston, and Brake CL]

have all been developed around GCI characteristics and need to be revisited as we develop better understanding of the AL MMC technology. As an example, drag testing of AL MMC rotors has typically shown very low temperatures and has never demonstrated a design limiting condition or a concern about fluid temperatures as has been seen with GCI. Drag testing is probably unnecessary for AL MMC light vehicle brakes.

4.3 Design for Manufacturing
The combination of lower density, greater casting process flexibility, and the need to use polycrystalline diamond tools for cutting Al MMC materials, changes the design rules considerably, if minimum manufactured cost is to be achieved. Recognizing that AL MMC materials are more expensive that GCI; makes it particularly important to employ every possible method to offset this cost.

4.3.1 Material Selection
Material choices continue to increase. As a rule, increased loading increases material and machining cost. Castable materials have the lowest overall cost and SiC reinforcement provides better thermal properties than Al_2O_3. Al_2O_3 reinforced alloys however, provide about 50°C higher operating temperatures.

4.3.2 Net Balance
The low density of the Al MMC materials plus the availability of several near net shape casting processes, makes the elimination of a balancing operation possible. For this to be achieved however, selection of an accurate casting process and careful attention to the setup in the machining processes are needed.

4.3.3 Casting.
The ideal casting process is the one that eliminates the need for machining. The accuracy level now required for both discs and drums ensures that some machining of key surfaces will always be required. Castings should be designed with the absolute minimum machining stock. Where possible the use of cores should be avoided as they always represent an incremental cost. High Pressure Die-Casting and Squeeze Casting processes run well with AL MMC's at loading levels of < 30%.

4.3.4 Machining.
Cutting of AL MMC works best when polycrystalline diamond tools are used. Again simple substitution does not provide an optimum machining process. Minimize metal removal, Avoid radii less than 3.0mm for roughing cuts and 1.5mm for finish cuts. This will reduce the cost of tools and increase their life. Tapping operations should be avoided, and wherever possible holes should be cast in. Finish reaming for final accuracy is less expensive than drilling.

4.3.5 Corrosion resistance
The inherent corrosion resistance of these materials eliminates the need for any corrosion prevention requirements. Thermal properties of parts subjected to extended exposure to corrosive conditions can be expected to retain thermal performance close to new part levels. Table 7. Illustrates the level of corrosion resistance that can be used for design purposes.

Table 7. Corrosion Resistance Comparison (Neutral Salt Spray Test: ASTM B-117)

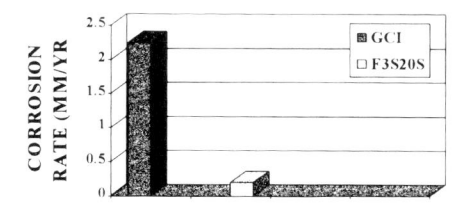

5. SUMMARY

To take the full advantage of AL MMC technology, the following design protocol is strongly recommended.
- Optimize airflow to the wheel end.
- Maximize rotor effective radius and minimize caliper size
- Maximize rotor convection surface area
- Maximize rotor air flow capacity
- Adjust rotor volume for desired maximum operating temperature.

6. REFERENCES:

(1) Kingston Research & Development Center, Alcan International Limited.
(2) SAE Paper # 970787. Development of Disk Brake Rotor Utilizing Aluminum Metal Matrix Composite: Nobuyuki Oda, Yukuhiro Sugimoto, Takahiro Higuchi, and Kouji Minesita, Mazda Motor Corporation.
(3) SAE Paper # 971039. Computational Flow Analysis of Brake Cooling: Fred Z. Shen, Devadatta Mukutmoni, Kurt Thorington, and John Whaite, General Motors Corporation.
(4) SAE Brake Effectiveness Subcommittee 4 of the SAE Brake Committee
(5) SAE Paper # 971033. Ventilated Brake Rotor Air Flow Investigation: Michael D. Hudson and Roland L. Ruhl. University of Illinois at Urbana-Champaign

Production and selection of iron powder for the friction brake industry

C E EVANS
Pyron, Niagara Falls, USA

ABSTRACT

There are three (3) main methods to produce iron powder – Atomization, CO Reduction, and Hydrogen Reduction. Each process produces a different iron particle which effects brake performance.

These production methods will be examined as well as the representative iron particles produced. Particle shape, surface irregularities, apparent density, and internal porosity will be examined of the iron particles represented.

Formulators, compounders, engineers should be aware that all iron powders are not alike. These different iron powder particles can seriously effect final friction brake characteristics.

Iron powder is an important additive in brake formulations for semi-metallic, low metallic, and even non-metallic brakes. Iron particles effect the brakes critical characteristics such as: coefficient of friction, noise, weight, rotor and brake wear, and cost – just to name a few. Iron powder, however, is not a commodity – all iron powders are not alike. Different production methods produce different iron particles.

This presentation is to familiarize the friction brake manufacturers as well as the friction industry with the 3 basic iron powder manufacturing processes – describe each process – and compare the typical iron powder particles of each process relative to the friction industry requirements.

Atomization, CO Reduction and Hydrogen Reduction are the three (3) most common processes used to manufacture iron powder. Each iron powder particle representing these 3 processes will show different properties and characteristics due to the process. Exactly what are these different properties? They are particle shape, surface area, and apparent density. From this information, brake manufacturers and compounders can determine exactly what iron powder properties would be best suited for their friction applications.

The first process we will examine is Atomization. Atomization is by far the most popular process for producing iron powder in not only North America, but the world. The bulk of conventional atomized production is by the water atomization process.

In this process, molten iron is subjected to a high pressure water spray producing fine iron droplets. These droplets are quenched and collected in a water cooled tank. After magnetic separation, dewatwering, and drying, the droplets are then heated in a belt furnace at approximately 1000C in a reducing gas atmosphere of either nitrogen / hydrogen or pure hydrogen gas. Milling, screening, and blending follow.

This description is an over-simplification of a highly engineered process. (Fig.1). What we are interested in, however, is the resultant shape and characteristics of the atomized iron particle.

Viewing a cross-section (Fig.2) an atomized iron particle has little or no internal porosity; a very smooth surface; an extremely dense particle. The typical physical properties for an atomized iron powder are as shown. (Fig.3). Because of these properties, coupled with its physical shape, atomized iron powder is not the product of choice for friction brake formulations.

The second process to be examined is the CO/Carbon Monoxide reduction process. This process is unique to the Hoeganaes Corporation in North America, located in Riverton, New Jersey, and also the Hoganas Corporation located in Hoganas, Sweden.

Magnetite, Fe_3O_4, is processed and prepared for blending with a mixture of 85% coke and 15% limestone. This mixture is dried, crushed, and charged in a tunnel kiln and heated to approximately 1200C. During the furnacing many reactions occur. After the reduction process above, the iron is ground and screened and then reduced and annealed in a protective atmosphere at 700 – 1000C. This CO process is completely different from the atomized process, resulting in a completely different iron particle. (Fig.4).

This CO produced iron particle has a more irregular surface area; more surface gaps than the atomized particle. The CO process produces a larger specific surface area which indicates more internal porosity and a lighter weight iron particle than atomized. (Fig.5). Typical physical properties for the CO reduced iron powders are as shown. (Fig.6).

The third and last process to examine is the Pyron Hydrogen Reduction Process, unique to Pyron Corporation located in Niagara Falls, N.Y.

The Pyron Hydrogen Reduction Process begins with mill scale which is scalped, magnetically separated, cleaned, and then ground in a continuous ball mill to -100 mesh size. After screening, the mill scale is oxidized at 980C in a gas fired multiple hearth furnace. This process converts the ferrous oxide (magnetite) in the mill scale to ferric oxide (hematite).

After the oxidation process, the ferric oxide is hydrogen reduced at 980C. The hydrogen gas in the furnace reduces the ferric oxide to metallic iron. A simple milling operation converts this cake to iron powder. A typical hydrogen reduced iron particle looks like this. (Fig.7).

This large amount of surface and internal porosity is a result of this unique process. A highly porous and irregular surface results in high internal porosity as shown. (Fig.8).

The internal and external configurations enable these iron particles to completely bond with any brake manufacturing process, be it resin or sintered bonded. (Fig.9). The range of physical characteristics for hydrogen reduced iron is as shown.

Comparing the cross-sectional photomicrographs shown before, (Fig.10) you are able to see for yourselves the difference in iron particles all due to different manufacturing processing.

Pyron produces a specifically designed friction iron powder that combines all of the desired frictional attributes (Fig.11) – low weight, high surface area, highly irregular surface, high green strength, and a precise screen size. Pyron Hydrogen Reduced R-12.

R-12 iron powder has the highest surface area and internal porosity, as well as the lowest apparent density range of any friction iron powder produced today. (Fig.12). R-12 iron powder, produced specifically for the friction brake industry, has been accepted and enjoys an excellent reputation for worldwide use due to its specifically designed frictional properties. What makes this product so attractive is its unique particle shape and low apparent density. (Fig.13).

With an A.D. in the range of 1.0 g/cc to 1.5 g/cc, friction compounders understand that it takes less R-12 by weight to fill a volume than other conventional products.

Because the R-12 has the lowest A.D. of all iron powders, the brake weighs less. Next, because the surface is so irregular, (Fig.14). the iron particle bonds well to the bonding agent so there is virtually no chance of particle vibration, or particle separation which causes drum scoring. The hydrogen reduction process also reduces scoring possibilities by producing soft iron particles with no carbon-rich areas. (Fig.15). And finally, R-12s porous internal characteristic results in a more uniform surface with no "iron rich" areas to distort friction values. The following are SEM slides of R-12 illustrating the particles unique surface and internal characteristics. (Fig's.16–18).

The next group of slides compare test results of Pyron Hydrogen Reduced Iron Powder versus competitor product values. (Fig.19). These slides were provided by some of our brake customers who prepared the tests and shared their data with us. (Fig.20). These slides compare the advantages of using a hydrogen reduced sponge iron powder in friction formulations comparing Friction vs. Temperature as shown, and Lining Wear vs. Temperature. (Fig.21).

Finally, it is important to remember the subject of this presentation – there is a difference in iron powders. All iron powders are not alike. When selecting an iron powder for a friction compound, be aware of the types and grades available. Each iron particle has its own separate and unique characteristics due to its manufacturing process.

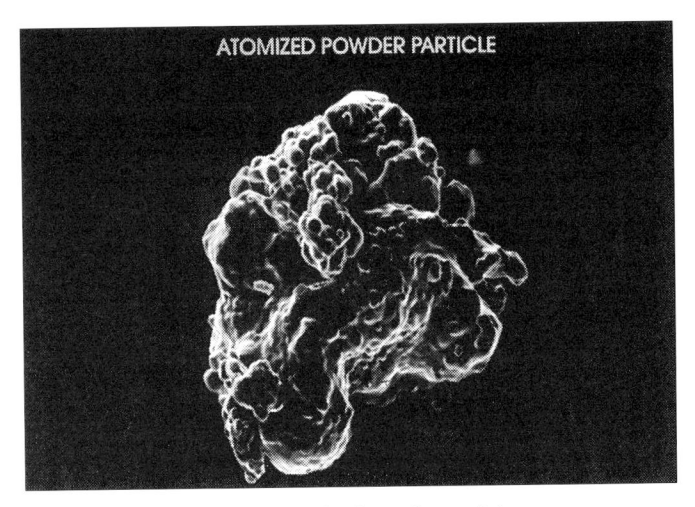

Figure 1: Atomized powder particle

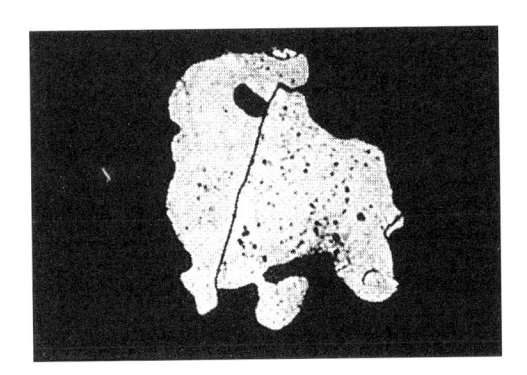

Figure 2: Cross-section, atomized

ATOMIZED POWDER PROPERTIES
Apparent Density: 2.50 – 3.00 g/cm^3
Green Strength: 1000 – 1500 psi
Specific Surface Area – 1.00

Figure 3: Atomized powder properties

Figure 4: CO reduced powder particle

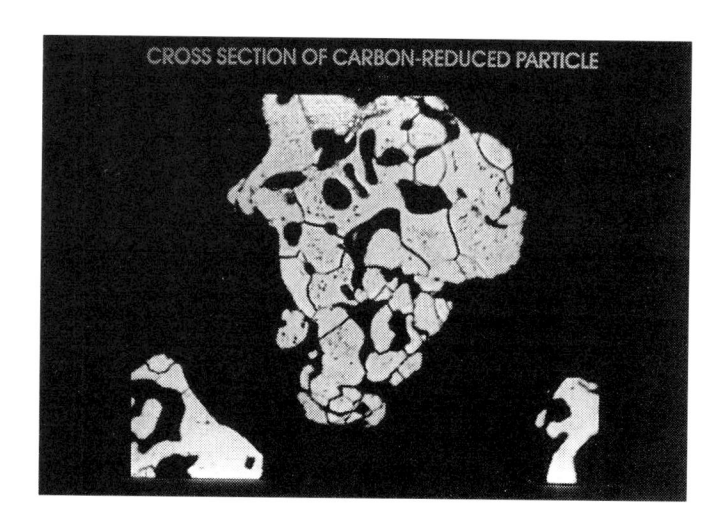

Figure 5: Cross-section, CO reduced

CARBON-REDUCED POWDER PROPERTIES
Apparent Density: 1.80 – 2.65 g/cm^3
Green Strength: 1600 – 5000 psi
Specific Surface Area – 1.16

Figure 6: CO reduced powder properties

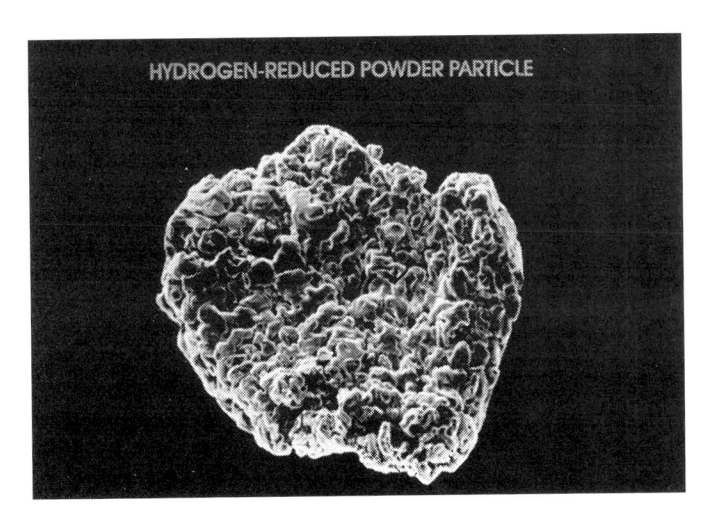

Figure 7: Hydrogen reduced powder particle

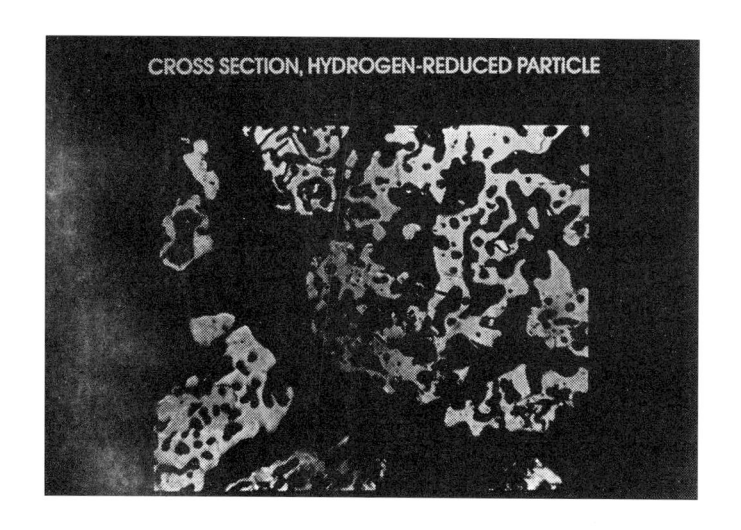

Figure 8: Cross-section, hydrogen reduced

HYDROGEN-REDUCED POWDER PROPERTIES
Apparent Density: 1.00 – 2.50 g/cm3
Green Strength: 1800 – 8500 psi
Specific Surface Area – 1.34

Figure 9: Hydrogen reduced powder properties

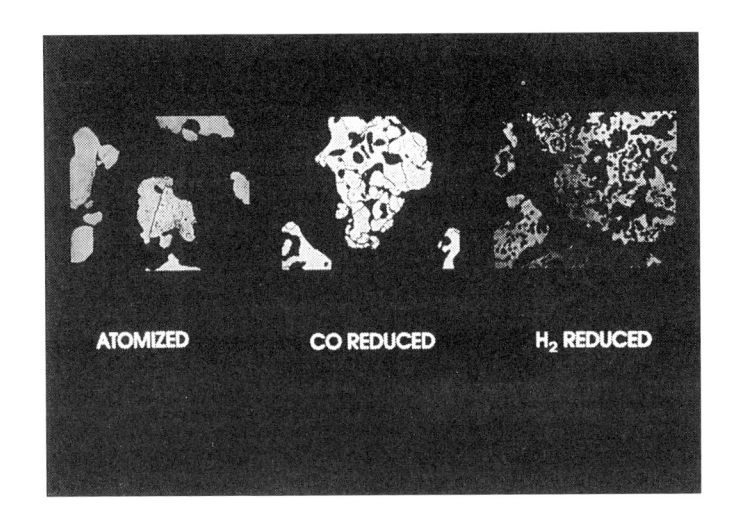

Figure 10: Comparison cross-sections

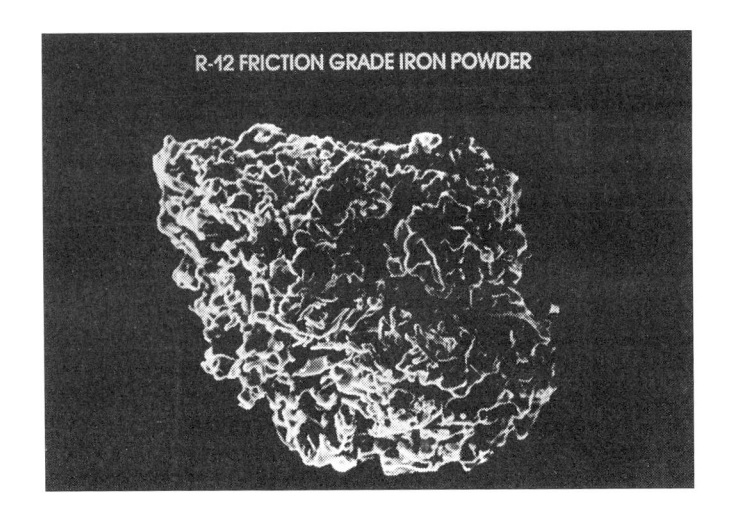

Figure 11: R-12 friction grade iron powder

Figure 12: R-12 @ 200X

	R-12 Properties				
Flow, s/50 g. (Hall) poor					
Appartent Density 1.0 to 1.5 g/cc					
Screen Analysis					
USA No.	16	1000 mm	00.0	-	
	20	850	00.0	-	02.0%
	60	250	30.0	-	70%
	100	150	10.0	-	35.0%
	325	45	11.5	-	29.5%
	-325	-45	00.0	-	12.0%
Chemistry					
	Carbon		0.000	-	0.050%
	Manganese		0.250	-	0.550%
	Phosphorus		0.000	-	0.20%
	Sulfur		0.000	-	0.006%
	H2 Loss		0.000	-	2.500%
	Acid Insol.		0.000	-	0.400%

Figure 13: R-12 properties

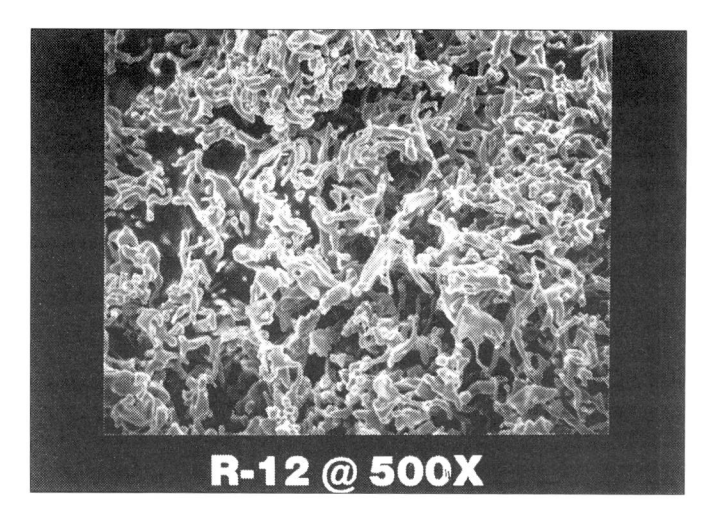

Figure 14: R-12 @ 500X

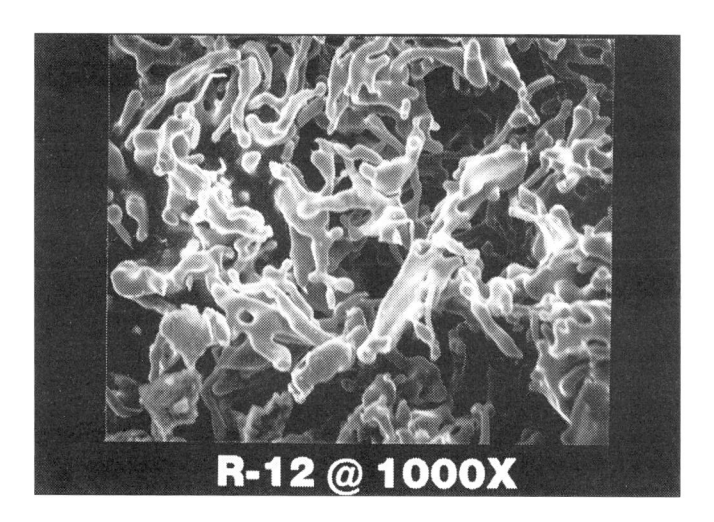

Figure 15: R-12 @ 1000X

Figure 16: R-12 @ 200X/1000X

Figure 17: R-12 @ 1250X

Figure 18: R-12 @ 125X/800X

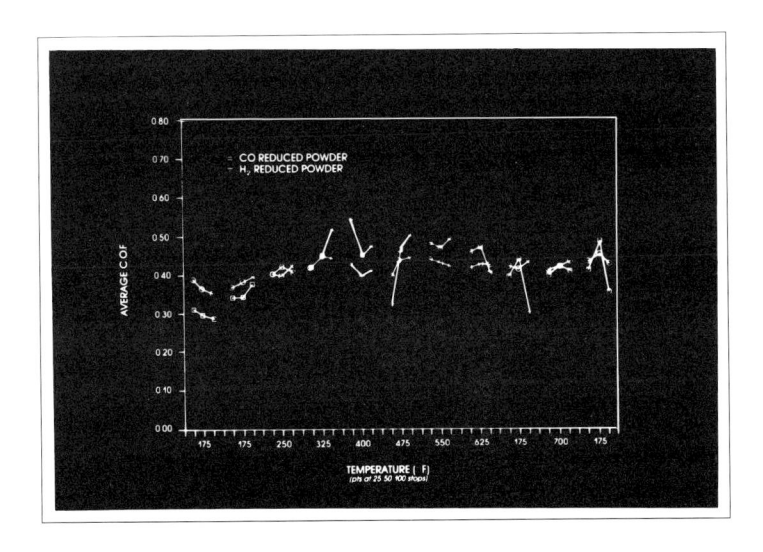

Figure 19: Friction vs temperture

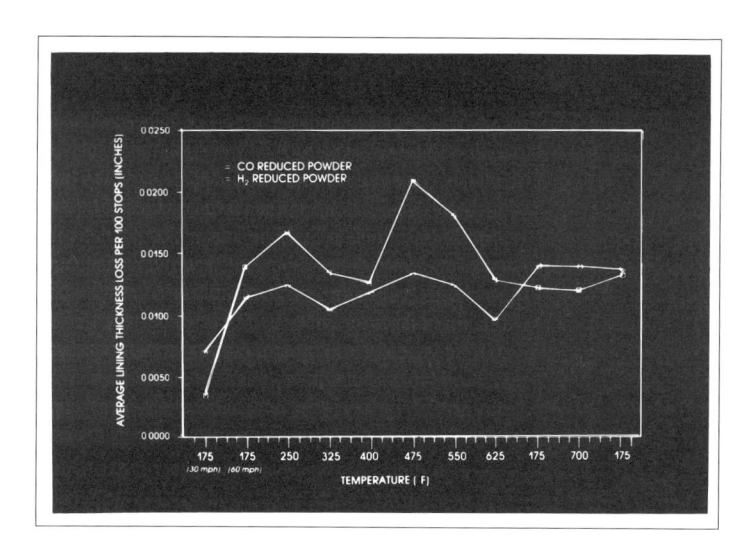

Figure 20: Lining wear vs Temperature

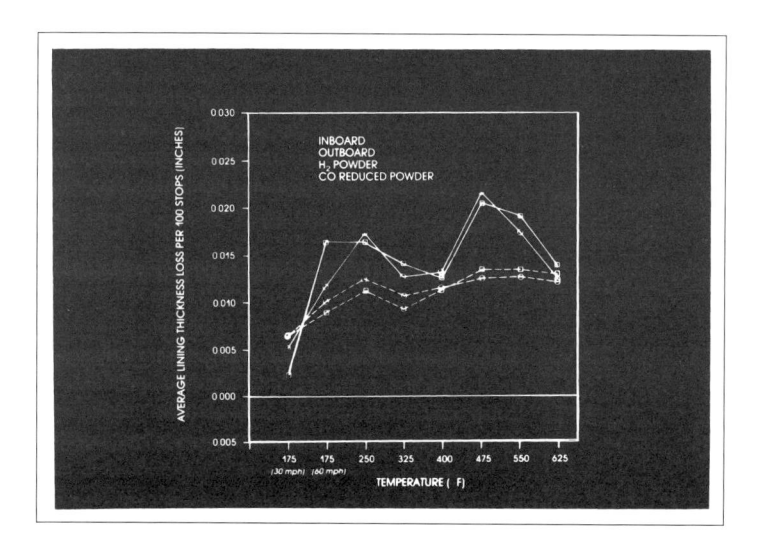

Figure 21 Lining wear vs temperature

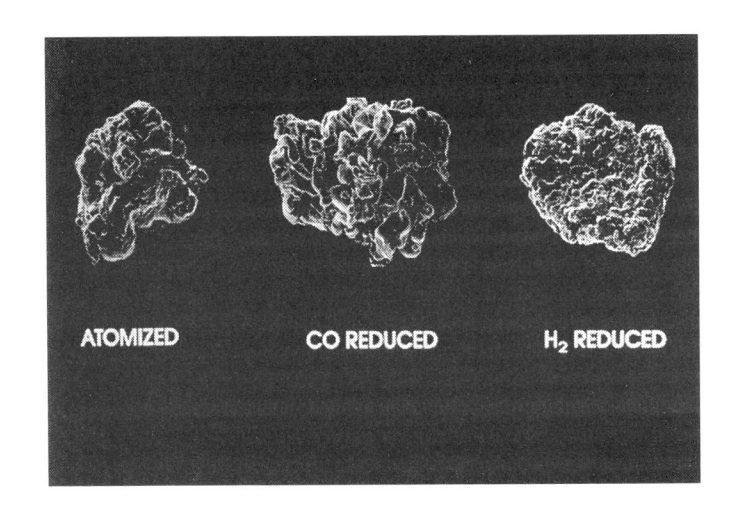

Figure 22

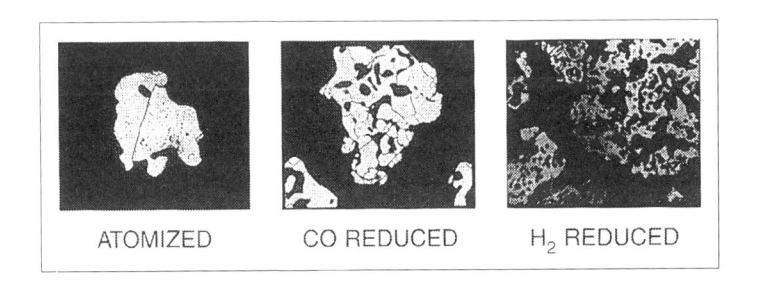

Figure 23

Thermo-mechanical instability in braking and brake disc thermal judder: an experimental and finite element study

T K KAO and **J W RICHMOND**
T&N Technology Limited, Warwickshire, UK
A DOUARRE
Ferodo-Abex, Noyon, France

SYNOPSIS

The paper covers the development and application of finite element techniques to thermo-mechanical instability in braking and brake disc hot judder.

It begins with a brief review of the phenomena of thermo-mechanical instability in braking and disc thermal judder. Then the paper describes and analyses experimental results of hot judder testing carried out on a dynamometer.

Illustrations are then given as to how the finite element models can be developed to help understand the effect of the complex phenomena described above. This is firstly demonstrated by the development an axisymmetric thermo-mechanical model which is capable of dealing with localised heat generation and wear. Then, recent development of a more sophisticated 3D thermo-mechanical model is described, with examples to illustrate its application to hot judder simulation.

Through use of the 3D thermomechanical simulation together with the detailed analysis of experimental results, it is found that the out-of-phase thermoelastic "buckling instability" is a significant factor causing hot spotting and hot judder. This new finding contributes to the fundamental understanding of the phenomena and widens the scope to improve the brake system performance in the relevant aspects.

1. INTRODUCTION: A BRIEF REVIEW OF THE PHENOMENA AND ITS MODELLING TECHNIQUES

1.1. DISC BRAKE HOT SPOTTING AND EXCITATION OF TORQUE VARIATION

Progress in the development of engines and vehicle chassis has led to brake systems of ever increasing energy absorbing density. To provide adequate braking force the disc and pad as a friction pair operate at temperatures up to 800 °C, interface pressures up to 10 MPa and rubbing speeds up to 22 m/s. Under some conditions an automotive brake may develop thermal distortion, which in the extreme promotes visual evidence of hot spotting, leading to problems of rotor cracking and judder.

Various types of hot spot found in automotive brake system were described in detail in [1]. One remarkable characteristic of hot spots is that they are often regularly spaced around the rotor.

Evidence of thermal distortion has also been observed by Woodward et al. at T&N Technology and Ferodo **[2]**. The temperature time histories across the brake disc surface have been recorded by a thermal imaging technique. From these measurements it can be seen that heat generation is restricted to one or two narrow bands, indicating the presence of a localised pressure distribution. The temperatures seen at the hot spots were found to exceed 800 °C. The presence of such high, localised temperatures has two effects. Firstly, when cast iron is elevated to temperatures exceeding 700 °C and subsequently cooled rapidly, a permanent phase change from pearlite to martensite can occur. Because of the higher volume of martensite compared to pearlite the disc can locally grow several microns , which leads to disc thickness variation. In addition, the coefficient of friction is not constant, but is a function of temperature, friction speed and contact pressure. Martensite will show a different coefficient of friction with the pad material than pearlite. Secondly, high compressive hoop stresses are generated in the surface layers of the cast iron rotor in the regions around the hot bands or hot spots. These can lead to plastic flow and can ultimately cause radial cracking of the rotor to occur after relatively few cycles. Commercial vehicle disc brakes are particularly prone to this type of failure.

The changes in shape of the original components and the disc thickness variations (DTV) can cause variations in normal force. Both the variations in normal force and the variation of local friction coefficient can result in variations in the brake force and consequently brake torque variations (BTV), which, as a source of excitation, is considered a major component of the forces and moments that cause judder.

1.2. DYNAMIC ASPECT OF JUDDER

Judder is a forced vibration driven by such excitation generated at the interface. Thickness variation of the rotor causes physical interference between the rotor and pads while passing through/over the brake lining, and internal forces and moments vary accordingly. The frequency is consequently directly proportional to the wheel speed and therefore also to the forward velocity of the vehicle. It is thus usual to relate judder frequencies to wheel speed; for example frequencies at twice the number of wheel rotations per second are called second order.

Thermally excited judder has frequencies of typically 6 to 20 cycles per revolution of the wheel. Consequently, low frequency vibrations (approximately 10 - 100 Hz according to disc rotation speed) are transmitted from the disc brake to the body as body shake and steering shake, and are audible as a drone, causing discomfort to both driver and passengers. Because of the relatively low frequencies involved, they are usually felt rather than heard.

At the vehicle design stage it is important to approach a situation where the brake and the other adjacent components do not have a critical natural frequency or harmonic corresponding with the operational frequencies generated by DTV or unstable thermal deformation. Analysis of the front suspension of a typical small family saloon car has revealed a natural frequency of 15 Hz. This equates to a linear velocity of 25 m/s and is coincident with typical judder frequencies, giving rise to resonance as the vehicle decelerates. Whilst efforts should be made to reduce or eliminate the cyclic torque, steps should also be taken to de-sensitise the system and vehicle.

In disc brakes, variations of normal force can also be the cause of movement of the piston if the pads can move. This can induce pressure fluctuation in the hydraulic system, which in turn can influence the normal forces in other brakes in the car.

1.3. HOT JUDDER AND COLD JUDDER

Two broad classifications exist for judder vibrations which derive from the source of the non-uniformity of the braking force. These are "cold" and "hot" judder. The former is associated with disc thickness variations induced by "off brake" wear [5,6,7]. Disc rotors are able to develop thickness variation because the brake pads lightly touch the rotor in some sectors and not at all in other sectors. The reason for this is limited piston rollback, which induces a slight drag on the disc rotor. The disc runout, pad retraction and pad compressibility are among the important factors affecting cold judder. The important role of material transfer from brake pad to disc in creating disc thickness variation has also been demonstrated [8]. The specific characteristics of the brake pads that will counteract disc thickness variation are minimum rotor wear when the brakes are not applied and some degree of rotor wear when the brakes are applied. "Hot" judder, on the other hand, is associated with a thermal cause. It is seen that for a variety of friction materials there is an increase in torque variation at elevated temperatures [9]. It is hot judder which is addressed in this paper.

1.4. CORRELATION BETWEEN HOT SPOTTING AND HOT JUDDER

As strong evidence of the correlation between hot spoting and thermal judder, reference is made to an investigation made of a brake disc after juddering [10], in which a number of bluish, oval discolorations were observed on the braking surface, pointing to local overheating. When a comparison was made of the order number of the dominant judder vibration and the number of hot spots (which were divided relatively evenly across the friction ring), it was demonstrated that these coincided.

In addition, thermal deformation of the disc rotor was twofold in nature [11]: and composed of "waviness" (varying lateral run-out throughout the rotational dimension) and uneven rotational expansion (i.e. the thermal thickness variation throughout the rotational dimension). A good agreement between measured torque fluctuation and simulated thermal deformation has been demonstrated.

1.5. APPROACH TO DISC BRAKE JUDDER USING THERMOMECHANICAL MODEL

The generation of friction heat at the friction material surface is proportional to the work done as defined by local interface pressure, sliding velocity and coefficient of friction, and therefore corresponds to the form of the interface pressure distribution. In a conventional disc brake analysis, the interface pressure is assumed either constant (constant pressure assumption) or inversely proportional to the radius (even wear assumption). However, under dynamic braking conditions the interface pressure is seldom uniform and varies with time, being continuously modified by a combination of (a) thermal distortions arising from friction heat generation, (b) mechanical distortion due to the applied actuation force, (c) wear of the friction pair materials and formation of transfer film, and (d) initial disc thickness variations and runout.

It is apparent that an accurate analysis of transient temperatures in high-energy sliding systems must account for the above-mentioned factors. The finite element method has been extensively used for thermal analysis applied to brake components. However, most of the analyses are still based upon the uniform contact assumption.

The first step forward in the specific use of a technique for the study of frictional effects under high energy sliding contact conditions was made for an aircraft-type annular disc brake [12]. This approach has been used as a basis and extended for the study of frictional energy generation and dissipation incorporating the thermophysical, friction and wear properties of resin bonded composite friction materials by Day and Newcomb [13] and for the study of railway brake systems [14]. A similar thermomechanical finite element model was used by Brooks et.al. for the study of disc brake judder [15]. All these models were limited to an axisymmetrical approach to the system. Recently, Floquet et al presented a layered FFT-FEM method to deal with geometrically periodic but nonaxisymmetric moving solids such as ventilated disc brakes [16]. The method is based on a partition of the system between moving axisymmetric parts (hub caps) and repetitive non axisymmetric parts (vents). The former part is analyzed using the FFT-FEM method and the latter one with the 3-D FEM technique.

More recently, Du et.al. has numerically implemented a perturbation finite element analysis for thermoelastic stability [17], where instability was associated with real eigenvalues of exponential growth rate.

An integrated axisymmetric model has been reported previously [19,20], in which the most important factors affecting hot spotting due to the localised contact pressure and temperature distribution are included. The present work is aimed towards further development of this integrated model, extended for 3D modelling of hot spotting and hot judder. Parallel to the development of the theoretical model, it was important to carry out experimental testing to provide a platform upon which the modelling work is based, and then subsequently validated.

2. EXPERIMENTAL HOT JUDDER TESTING AND ANALYSIS OF THE RESULTS

2.1. DYNAMOMETER HOT JUDDER TEST AND DTV MEASUREMENT SYSTEM

On a dynamometer initial disc thickness variation and runout can be installed with a high degree of precision. Initial (manufactured) DTV values lie between 0.006 mm and 0.010 mm for most passenger car rotors (commercial vehicle discs can have 100 microns of runout/DTV as manufactured). Disc rotors can warp (superimposed waviness) due to the clamping of the wheel or uneven tightening of the wheel nuts or bolts, or inadequate stress relieving of the casting. Most manufacturers have installed disc rotor runouts better than 0.080 mm but would like to improve this value to lower than 0.040 mm. DTV of 0.015 mm can show up as brake judder to an experienced driver in a sensitive car.

Taking these factors into account, a series of torque variation tests were carried out. Initial disc thickness variation and runout were installed between 0.020 and 0.040 mm, and between 0.0 and

0.060 mm respectively. Pad and disc were bedded and burnished accordingly. Then seven constant duty stops were carried out from 70 mph (966 rpm) to 40 mph (552 rpm) with the pressure required to achieve a constant deceleration of 0.19 g. A period of 35 second cool down at 70 mph was allowed between stops.

The disc thickness was measured around the disc at three radii using three pairs of capacitive transducers [21]. Three pairs of transducers were used in total, so that measurements were made at the outer, inner and centre radius of the rubbing path (**Fig.1**). Disc run-out was estimated by taking the output from the transducers on one side only (usually fist side). Disc thickness is calculated by subtracting from the absolute separation of a pair of transducers the sum of their individual distances from the disc surface; DTV is then the circumferential variation of this quantity.

2.2. TYPICAL TEST RESULTS

Fig.2 demonstrates the measured air gap during the 6th stop by using three pairs of transducers shown in **Fig.1**. Three curves (1,3,5) of measured air gaps grow upwards, while the other three sensors on the opposite side go downwards. Thus the growth of coning during this stop is clearly illustrated.

Fig.2 can be analysed in greater detail for individual revolutions. This leads to **Fig.3** which demonstrates the measured air gap for three revolutions. It can be seen from **Fig.3** that the disc rotor "warps" twice every revolution during this braking period.

As mentioned before, by subtracting from the absolute separation of a pair of transducers the sum of their individual distances from the disc surface, disc thickness can be calculated. **Fig.4** was thus obtained from **Fig.3**. The circumferential variations of the obtained disc thickness at the outer, middle and inner radii are clearly shown. It is noticed that along the middle radii the measured DTV has two peaks whereas at the outer and inner radii there is only one dominant peak.

Extending this technique to the whole braking period, we obtain **Fig.5**, which demonstrates the growth of DTV along the outer, middle and inner radii during the 6th stop.

For the same three revolution period, there is a corresponding pattern of torque variation shown in **Fig.6**. It is noted that there are two peaks in torque variation during each revolution, one dominant and the other minor. This is considered to correlate the two DTV peaks of the middle radius measurements shown in **Fig.4**.

Extending the torque variation data to cover the whole braking period, we obtain **Fig.7**, which demonstrates the growth of torque variation during the 6th stop. This corresponds to the growth of DTV shown in **Fig.5**.

2.3. MORE DETAILED EXPLANATION OF THE EXPERIMENTAL RESULTS

More detailed explanation of the experimental results were derived in conjunction with the theoretical results obtained by 3D simulation. This will be given in section 4.3.

2.4. MATERIAL RESPONSE TO DTV

For any material the amount of torque variation is strongly linked to the magnitude of the DTV. For a given DTV, different materials may lead to different torque variations. **Table 1** shows the effect of pad compressibility on disc rotor coning, rotor thermal distortion, disc thickness variation as well as on the torque variation. The strong dependence of DTV and torque variation on pad material is clearly illustrated.

3. AXISYMMETRIC FINITE ELEMENT THERMOMECHANICAL MODEL

3.1. FEATURES OF THE AXISYMMETRIC THERMOMECHANICAL MODEL

A brief account was given previously of an analysis technique developed at T&N Technology and Ferodo [19,20]. In this, a special contact element was formulated, and used to derive the frictional force and heat generated at the friction interface. Predictions of the friction pair performance changes with time can be assessed as a wear model is embodied in the contact element formulation.

Fig.8 illustrates schematically the special elements which are placed at the interface between the friction material and disc, and which are used to inject heat into an axisymmetric model of the braking system. The features of the element are:

- heat generation at each contact position is a function of the local sliding speed, friction coefficient and contact pressure;

- the thermal and stress analyses are fully coupled, as is required to follow properly the rapid change of pad and disc geometry. Thus any coning or local growth of thickness will alter the contact pattern and vice versa;

- the thermal resistance between the pad and disc caused by the transfer film is represented;

- the temperature/pressure history at each point is monitored and the local wear is calculated. In this way the effect of wear upon the pressure distribution and heat generation can be taken into account;

- the friction coefficient can be made a function of local pressure, temperature and sliding velocity;

- there is no need for the artificial constraints that have been applied by previous workers; for example it is unnecessary to constrain the centre line of the disc for coning. This makes it possible to consider realistic problems and examine design issues for brake system.

The method and element developed represent a significant improvement over any previously used techniques. This approach has been successfully used for practical applications including simulating multi-stop fully transient braking phenomena over times exceeding 300 seconds. The theory, development and validation of the model is described in greater detail in [20].

3.2. EXAMPLE ANALYSIS

Fig.9 illustrates two examples of the development of localisation of the pressure on the finger side of a disc during a brake application as predicted by this model. **Fig.10** illustrates the corresponding temperature distributions in the disc and pad for these two examples at the end of the brake application. The explanation of these results will be given in the following sections.

3.3. PREDICTED MATERIAL RESPONSE TO LOCALISATION

For a given brake condition, different materials may lead to different levels of localisation. Such trends are well predicted as shown in **Fig.9** and **Fig.10**.

With reference, for example, to **Fig.9 (b)**, it is assumed that the friction pad and rotating disc are both initially relatively smooth with minimal thickness variation and that, upon initial application of the brake, the contact pressure varies only slightly. The sliding velocity, however, increases with the radius, and the pressure at positions where the disc is thicker will be higher than normal. In consequence there is a tendency for the highest rate of heat generation to occur between the centre and the outer radius of the disc. This will very quickly lead to a concentration of contact area to a region depending on its thermal deformation. Because of the concentrated contact area, the temperature distribution on the friction surface becomes very non-uniform, with temperatures being quite high in contact zones, and lower elsewhere. The corresponding thermal deformation then leads to more concentrated contact, and hence, more and more non-uniform temperature. Thus narrow contact bands are formed on the disc surface as shown in **Fig.9 (b)** and corresponding narrow hot bands on the disc surface as shown in **Fig.10 (b)**.

For pad materials of different compressibilities, it is observed that softer pads improve the interface contact pressure distribution, and hence result in a more even temperature distribution. Such a trend is well predicted by the model as demonstrated by **Fig.9 (a)** and **Fig.10 (a)** for a soft pad compared with **Fig.9 (b)** and **10 (b)** for a hard pad under identical brake conditions.

3.4. THERMOELASTIC INSTABILITY MOVEMENT OF HOT SPOTS AND DTV SIMULATION BY AXISYMMETRIC MODEL

The example shown in **Fig.9 (b)** also demonstrates the thermoelastic instability movement of hot spots. In this example, the concentrated contact pressure at the hot spot band leads to concentrated wear in that region. The consequence of this concentrated wear is a shift of the high pressure contact to another area. Similar thermoelastic instabilities to those predicted are often observed experimentally [19].

Under certain braking conditions especially due to the varying history of wear pattern, the repeated moving of the hot band can lead to the formation of the consequent multi-hot bands, as illustrated in **Fig.11**. The pattern of formation of hot band (or hot bands) depends on the history of wear, which in turn depends on the braking pattern and hence on the history of coning and localisation.

Fig.12 illustrates the global displacement of a disc rotor at the end of a brake application. With initially defined local thickness variation, the model can be used to simulate the growth of the local thickness variation and the corresponding effect on localisation of pressure and temperature.

4. A 3D FINITE ELEMENT THERMOMECHANICAL MODEL

4.1. FEATURES OF 3D FRICTIONAL SLIDING THERMOMECHANICAL MODEL

Although the formation of the hot band and its associated growth of surface thickness can be predicted by the axisymmetric thermomechanical model, the judder and the associated torque variation are due to the global geometric variation around the circumference. Consequently, a 3D model is necessary if the global geometric variation around the circumference is to be simulated.

The basic concept used in the development of the axisymmetric friction contact interface element sketched in **Fig.8** was extended for the 3D model. However, due to the feature of the frictional sliding contact between pads and disc rotor required in the 3D model, the notion of node-to-segment contact was used instead of the classical "gap" elements. A 2D node-to-segment contact is illustrated in **Fig.13**, which has been further extended to 3D to simulate the frictional sliding contact between pads and disc rotors. In order to have accurate results for the area of the contact surfaces and the normal reactions, contact surfaces are used.

The actual contact surface is often very different from the pad surface and the thermal distortions may give variations of surface contact during braking. To simulate the varying contact condition, the penalty method is used **[22]**. The normal reaction on the contact surface thus becomes:

$$N = k * g \quad (\text{if } g <= 0)$$

or

$$N = 0 \quad (\text{if } g > 0)$$

where g is the gap between I and the segment J-K (**Fig.13**), and k is the contact stiffness. The experimental measurements done on real pads at T&N Technology **[23]** has demonstrated that the contact stiffness of the surface layer of brake pads is lower than the stiffness of the bulk material of the pads.

The features of the interface element in the axisymmetric thermomechanical model as mentioned in the previous section are also maintained in the 3D thermomechanical model. In addition, special care has been taken to speed up the nonlinear contact convergence. This is particularly important since 3D thermomechanical simulation involves massive CPU computational effort as well as massive data storage.

4.2. EXAMPLE OF ANALYSIS

Fig.14 shows the finite element discritization of a conventional solid disc brake assembly and the distribution of contact pressure between two opposite pads and the rotor just before braking. The solid disc was chosen because it demands less computational capacity and time than a ventilated disc. In **Fig.14** the interface elements between the pads and the rotor on both piston and the finger sides are also shown.

Fig.15 illustrates the temperature contour after 4 seconds braking. The installed initial DTV and runout (overall 40 microns) have one peak and one trough per revolution. The hot spot formed due to localised contact is clearly shown.

Regarding the thermal distortion, **Fig.15** shows that the rotor has not only thermal coning (which is axisymmetrical), but also" warping". This is due to the uneven distribution of temperature on the opposite sides of the rotor. Such deformation cannot be predicted by 2D axisymmetric models.

To visualise the development of such thermal "warping", **Fig.16** illustrates the global geometric displacement along the circumference of the rotor both on the piston side and finger side. Each curve represents the distribution of the displacement in the piston loading direction in a period of time. The development of "warping" is clearly demonstrated.

The difference between the displacements of opposite sides (**Fig.16.(a)** vs. **Fig.16.(b)**) is the growth of disc thickness variation, which is shown in **Fig.17**. In **Fig.17** there are two peaks, which correspond to two hot zones at the rotor, one at the piston side, and the other at the finger side, each out of phase about 180 degree.

The torque variation as recorded by variation of the normal force on the pad is shown in **Fig.18**. covering three revolutions. There is a dominant peak accompanied by a minor one. The growth of torque variation shown in **Fig.18** corresponds not only to the growth of thickness variation shown in **Fig.17** but also the development of thermal "warping" shown in **Fig.16**. Extending this diagram, we obtain **Fig.19**, which illustrates the growth of torque variation within a 5 second braking period due to the development of disc "warping" (**Fig.16**) as well as the growth of DTV (**Fig.17**).

The development of this 3D model is still in the early stage. However, the results obtained to date are qualitatively in good agreement with those obtained experimentally.

4.3. "WARPING" OR "THERMOELASTIC UNSTABLE BUCKLING" AS AN IMPORTANT FACTOR CAUSING HOT SPOTTING AND HOT JUDDER

The experimental and simulation results have lead us to consider the fundamental mechanism of brake hot spotting and hot judder further.

In **Fig.15**, the peak temperature on the piston side surface of the rotor is out-of-phase with the peak temperature on the finger side surface of the rotor. Where there is a hot zone or hot spot on the piston side surface, there is a cold zone on the opposite surface, and vice versa. Corresponding to **Fig.15**, one experimental example shown in **Fig.20** also demonstrates such a case that the hot zone on one side of the rotor is out of phase of the hot zone on the opposite side [27].

It is due to such out-of-phase non-uniform temperature distribution that the rotor "warps" or "buckles" as shown in **Fig.15**. When the rotor "warps" or "buckles", it causes more localised out-of-phase contact, and thus induces more pronounced out-of-phase hot spotting. When such localisation becomes severe, disc cracking may occur. This is also a significant factor causing torque variation and hence hot judder, in addition to the causes usually attributed to DTV excitation. We called it "thermoelastic unstable buckling", which is different from the usual buckling and different from the concept of classic "thermoelastic instability" **[24,25,26]**.

As already mentioned, the development of such an out-of-phase "warping" can be seen by comparing **Fig.16.(a)** and **Fig.16.(b)**, where the material on the opposite sides of the rotor deforms in parallel with each other.

This finding is also supported by the experimental results. If we revisit **Fig.3**, we can note that when the air gap curves 1,3,5 (measured at one side,see **Fig.1**) go up-wards, curves 2,4,6 (measured on the opposite side) go down-wards, and vice versa. This means that the two opposite surfaces of the rotor move in parallel during the rotation. Thus, the two peaks for every rotation shown in **Fig.3**, are the indication of the two-peak mode of disc "warping".

Under certain brake conditions, those two peak "warps" may develop further at the late stage of braking, appeared as multi-peak "buckling", for example, the seven peak mode "buckling" indicated by air gap measurement (**Fig.21**, [28]). Those multi- peak mode "buckling" corresponds well with the out-of-phased multi-hot spots, for example, the seven-peak mode "buckling" events correspond well with the measured hot spots shown in **Fig.22**, where seven hot spots are distributed evenly, with the opposite sides out-of-phase. This experimental example again significantly supports the present concept of "thermoelastic unstable buckling".

5. CONCLUSION

5.1. SIMULATION OF 3D THERMOELASTIC INSTABILITY: A FIRST STEP FORWARDS

Various thermal analyses have been made previously of disc brake judder, but owing to the difficulty of taking circumferential temperature variation using conventional techniques, transient temperature distribution and thermal deformation has seldom been analyzed, and it has not been possible to determine the exact mechanism leading to the deformation causing the judder.

In this respect, the technique described in this paper represents the first such attempt.

Given the complexity of the mechanisms involved, the work to date on the 3D modelling of hot judder is still at an early stage. With continuous advancement of computer power, the model will be further enhanced to include more details of physio-tribological aspects within the friction sliding contact element. Naturally, this must be approached in parallel with an experimental exploration to establish the relevant correlations quantitatively.

5.2. "THERMOELASTC UNSTABLE BUCKLING": A SIGNIFICANT FACTOR TO HOT SPOTTING AND HOT JUDDER

In this paper it is demonstrated experimentally and theoretically that in addition to the thermal growth DTV the "thermoelastic unstable warping or buckling" induces excitation to hot spotting and disc cracking, as well as to torque variation and hot judder.

This new understanding widens our scope to improve brake system performance with reference to hot spotting, disc cracking, torque variation and thermal judder.

6. ACKNOWLEDGEMENT

The authors would like to acknowledge the considerable contribution made to this work by many at both T&N Technology and Ferodo. Particular thanks are due to Mr. A.Moore and Mr.E.Little for the dynamometer tests and Dr.P.Shenton for the experimental testing of materials. Thanks are also due to Mr.D.Holme for many inspiring communications and discussions concerning disc "warping" and "hot spotting".

7. REFERENCES

[1] A.E.Anderson, R.A.Knapp, Hot spotting in automotive friction systems, Wear, 135 (1990), 319-337.

[2] A.Woodward, H.Hodges, M.W.Moore, Advanced measurement Systems for Routine Assessment of Brake Performance, Proceedings of IMechE, Braking of Road Vehicles, C444/028/93, 1993

[3] K.H.Oehl, H.G.Paul, Bremsbelage fuer Strassenfahrzeuge: Entwicklung und Erprobung, Verlag Modern Industrie, 1991, Seite 57-60, Rubbeln durch Dickendifferenzen.

[4] J.-P. Pompon, Guide du Freinage des Poids Lourds, Tome 1, Ferodo Abex, 1995, p.127-128, Le Bruit de Broutement.

[5] A. de Vries, M. Wagner, The Brake Judder Phenomenon, SAE 920554, 1992.

[6] M.J.Haigh, H.Smales, M.Abe, Vehicle Judder under Dynamic Braking Caused by Disc Thickness Variation, Proceedings of IMechE, Braking of Road Vehicles, C444/022/93, 1993

[7] P.Ferdani, R.S.Williams, E.J. Little, Current Research in Friction Materials, I.Mech.E., C524/038/97, 1997

[8] M.Boerjesson et.al. The Role of Friction Films in Automotive Brakes Subjected to Low Contact Forces, Proceedings of IMechE, Braking of Road Vehicles, C444/026/93, 1993

[9] W.Stringham, et.all., Brake Roughness- Disc Brake Torque Variation, Rotor Distortion and Vehicle Response, SAE930803, 1993.

[10] W.Kreitlow, et.al., Vibration and 'Hum' of Disc Brakes under Load, SAE 850079

[11] H.Inoue, Analysis of Brake Judder Caused by Thermal Deformation of Brake Disc Rotors, FISITA, 865131, 1986.

[12] F.E.Kennedy, F.F.Ling, A Thermal, Thermoelastic, and Wear Simulation of a High-Energy Sliding Contact Problem, Trans.ASME, J.Lubric.Tech., 1974, 96, Part 3, 497-507.

[13] A.Day, T.P.Newcomb, The Dissipation of Friction Energy from the Interface of Annular Disc Brake, Proc.I.Mech.E., 69/84, 1984.

[14] P.Dufrenoy, D.Weichert, Prediction of Railway Disc Temperatures Taking the Bearing Surface Variations into Account, Proc.I.Mech.E., Part F: J.Rail & Rapid Transit, Vol.209, 1995, 67-76.

[15] P.C.Brooks, D.Barton,D., D.A.Crolla, A.M.Lang, D.R.Schafer, A New Approach to Disc Brake Judder using a Thermomechanical Finite Element Model, AUTOTECH'93, I.Mech.E., C462/31/064, Birmingham, U.K., 1993

[16] A.Floquet, et.al., Realistic Braking Operation Simulation of Ventilated Disc Brakes, Trans. ASME, Vol.118, July 1996

[17] S.Du, et.al., Finite Element Analysis of Frictionally Excited Thermoelastic Instability, Journal of Thermal Stresses, 20, 185-201, 1997.

[18] M.K.Abdelhamid, Brake Judder Analysis Using Transfer Functions, SAE973018, 1997.

[19] T.K.Kao, J.W.Richmond, M.W.Moore, The Application of Predictive Techniques to Study Thermo-elastic Instability in Braking', SAE942087, 1994.

[20] J.W.Richmond, T.K.Kao, M.W.Moore, The Development of Computational Analysis Techniques for Disc Brake Pad Design, in <Advances in Automotive Braking Technology, Design Analysis and Materials Developments>, ed.by D.C.Barton, MEP, 1996.

[21] T.R.Holton, T.Hodges, A.J.Woodward, The Use of Advanced Measurement Systems in Friction Material Development, T&N Technical Syposium, 1995, Paper 11.

[22] Z-H.Zhong, Finite Element Procedures for Contact-Impact Problems, Oxford University Press, 1993.

[23] P.Shenton, Internal Communication, T&N Technology, 1997.

[24] J.R.Barber, Thermoelastic Instabilities in the Sliding of Conforming Solids, Proc.Roy.Soc., Series A, Vol.312, 1969, PP.381-394

[25] K.L.Johnson, Contact Mechanics, Cambridge University Press, 1985.

[26] K.Lee, J.R.Barber, Frictionally-Excited Thermoelastic Instability in Automotive Disk Brakes, ASME J.Tribology, Vol.115, pp.607-614, 1993

[27] E.Little, SAE Presentation and Private Communication, 1997.

[28] J.D.Holme, Private Communications, T&N Technology, 1998.

Table 1: Hot DTV & BTV: Effect of Pad Compressibility

Pad	Hard	Soft
Initial DTV (mm)	0.020	0.020
Temperature (after 90 revolutions) (°C)	232	208
Coning (ref. to Fig.2) (mm)	0.250	0.200
Waviness (two peaks per rev: ref.to Fig.3) (mm)	0.050	0.040
DTV growth (90 revolutions: ref.to Fig.5) (mm)	0.011	0.008
Thickness growth (90 revolutions) (mm)		0.015
Growth of BTV (90 revolutions: ref.to Fig.7)	18 N.m	10 N.m
Torque variation (global: ref.to Fig.7)	60-78 N.m	50-60 N.m
Torque variation (single revolution)	81 N.m	62 N.m
DTV (ref. to Fig.4) (mm)	0.038	0.040

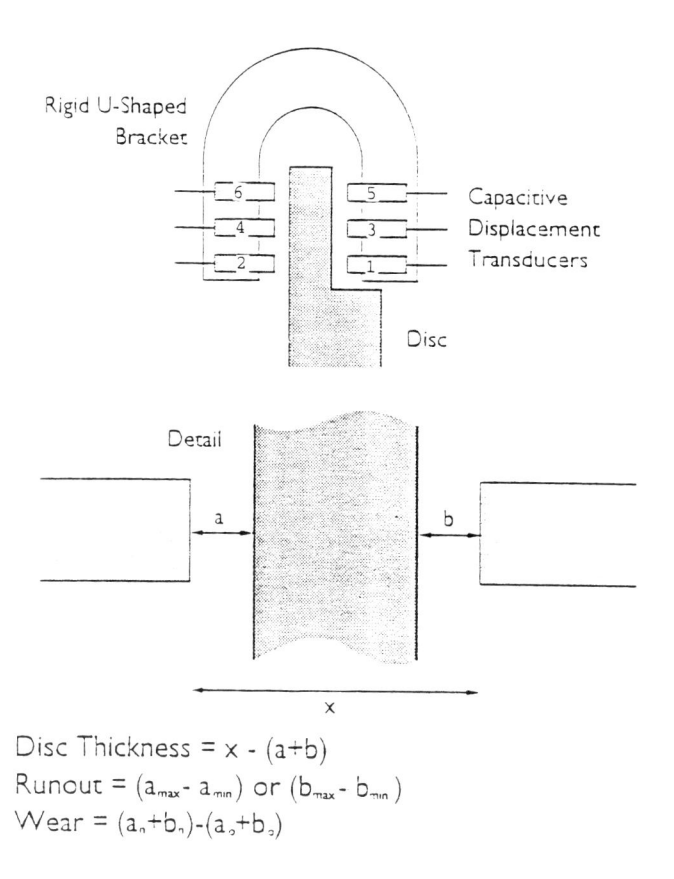

Disc Thickness = x - (a+b)
Runout = $(a_{max} - a_{min})$ or $(b_{max} - b_{min})$
Wear = $(a_n + b_n) - (a_o + b_o)$

Fig.1. Configuration of DTV, Run-out and Wear Measurement System

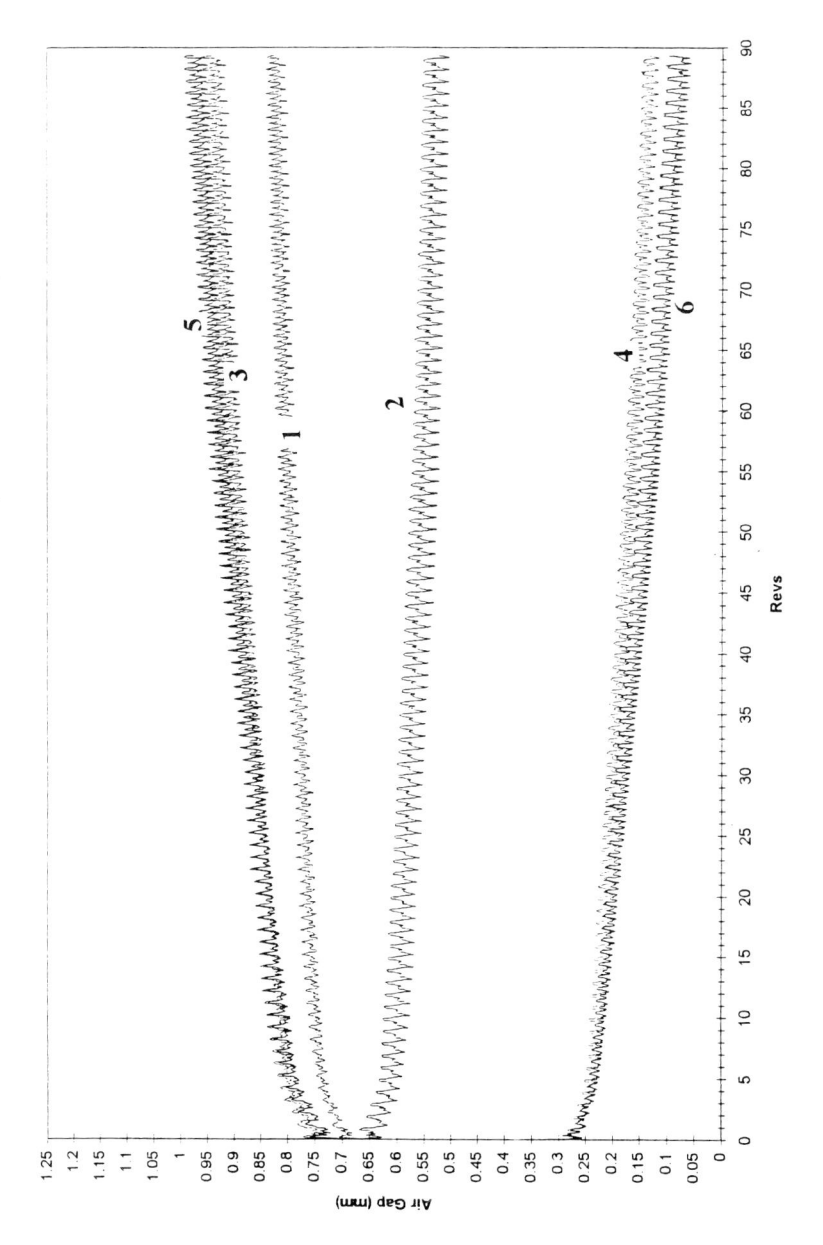

Fig.2. Measured Air Gap showing the Development of Disc Coning During the 6th Stop (ref.to Fig.1)

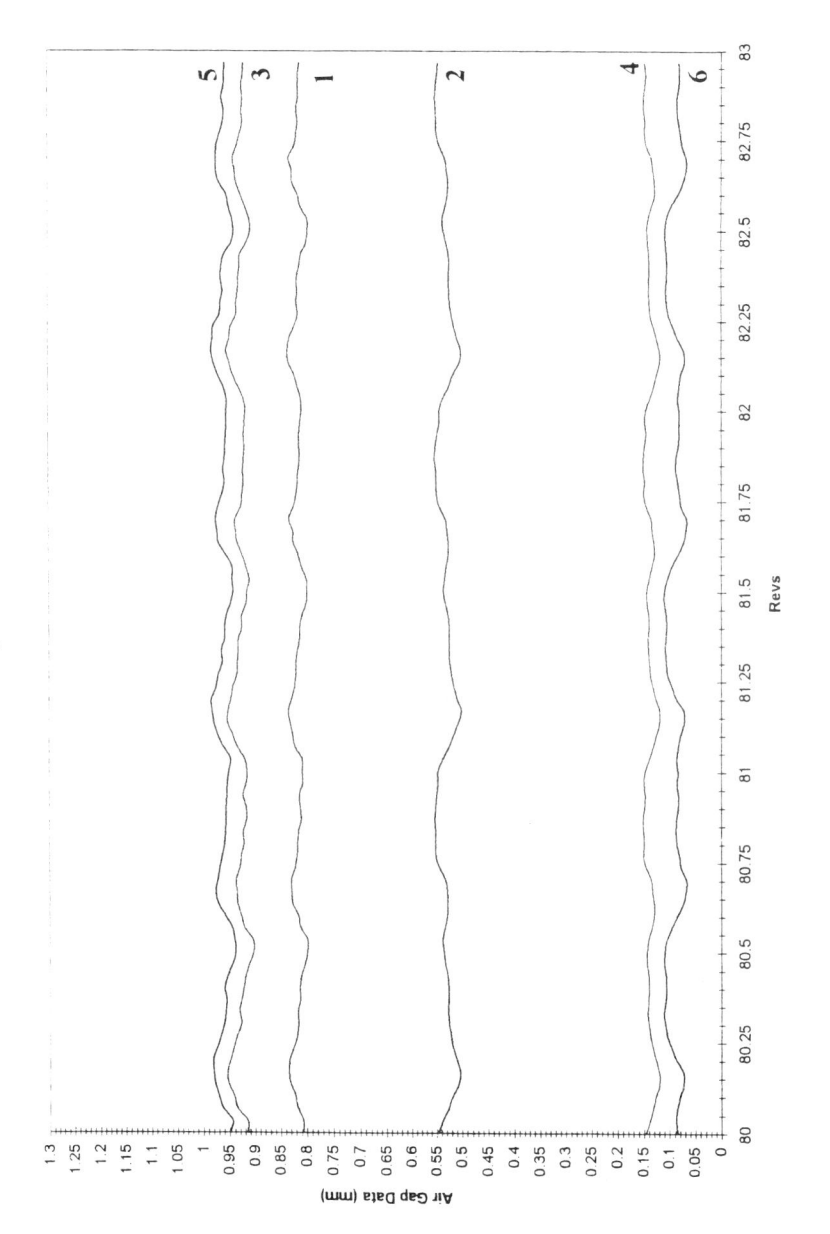

Fig.3. Details from Fig.2, showing Disc Waviness during Three Revolutions (ref. to Fig.1)

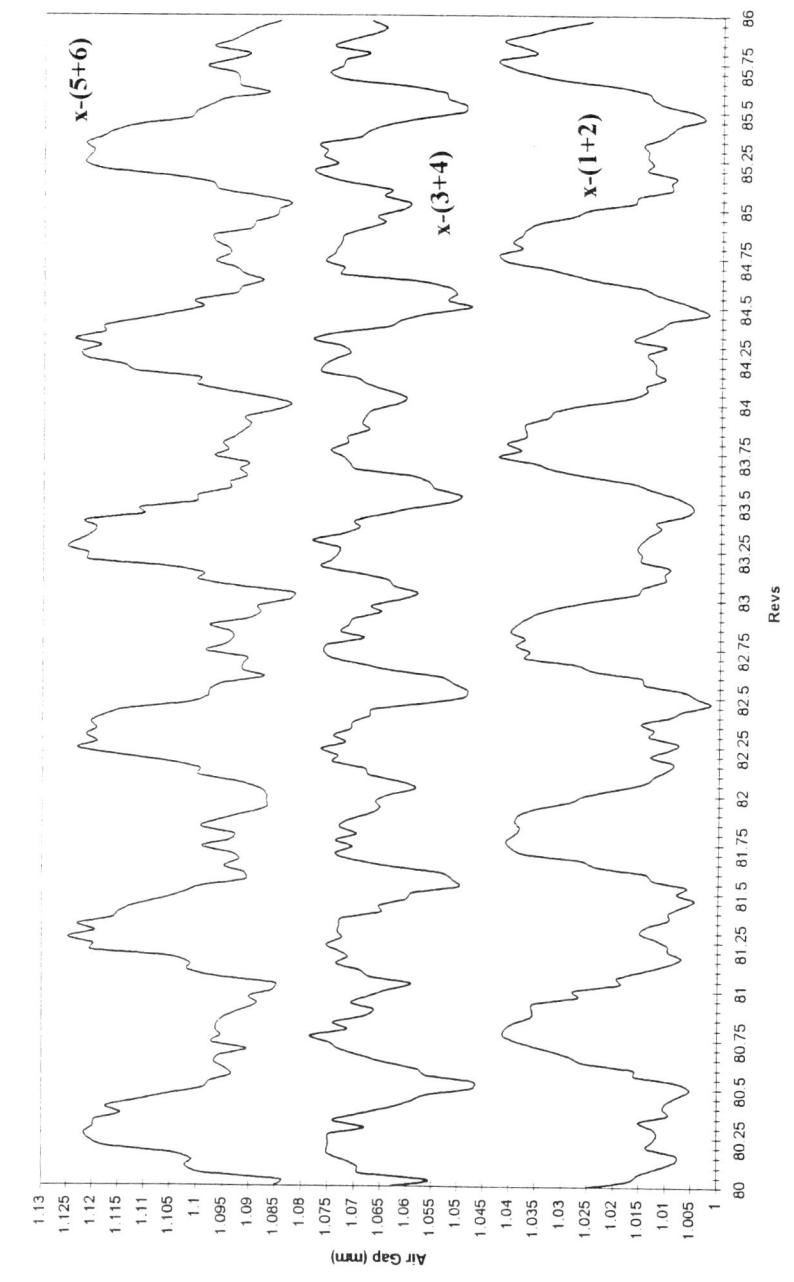

Fig.4. Combined Data (x-(1+2), x-(3+4), x-(5+6)) from Fig.3, showing DTV History (ref.to Fig.1)

Calibrated Channel Air Gap Data Test 100010 Stop 6

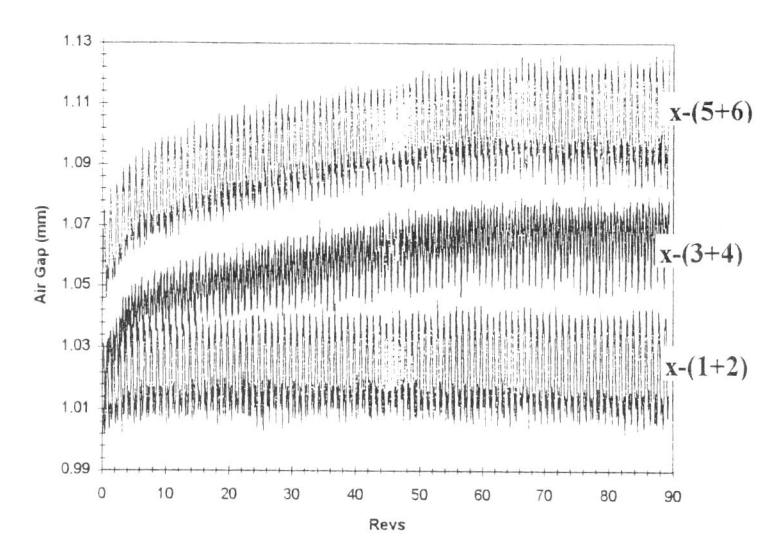

Fig.5. Combined Data (x-(1+2), x-(3+4), x-(5+6)) from Fig.2, showing DTV History (ref.to Fig.1)

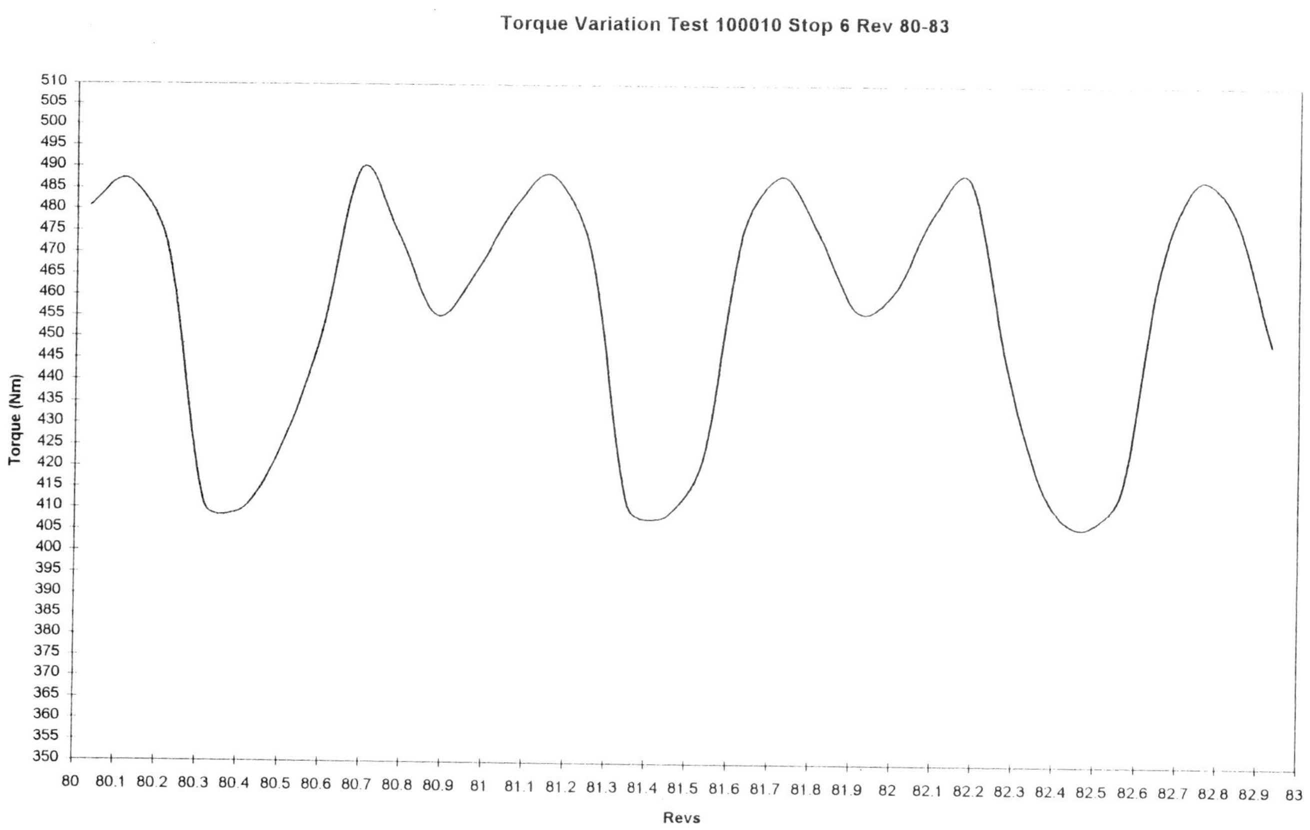

Fig.6. Measured Torque Variation (covering the Period corresponding to DTV shown in Fig.4)

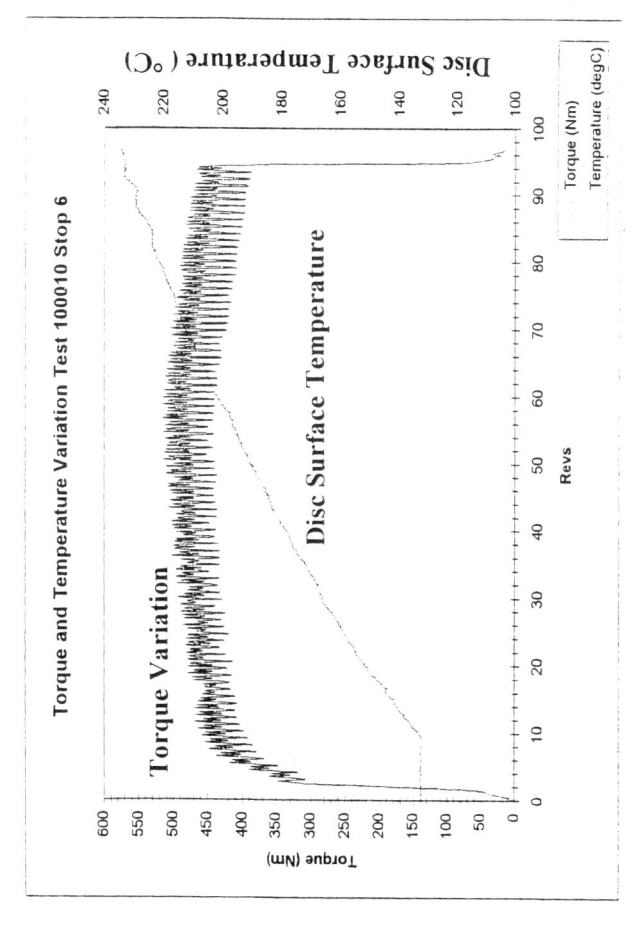

Fig.7. Measured Torque Variation during the 6th Stop (corresponding to DTV shown in Fig.5)

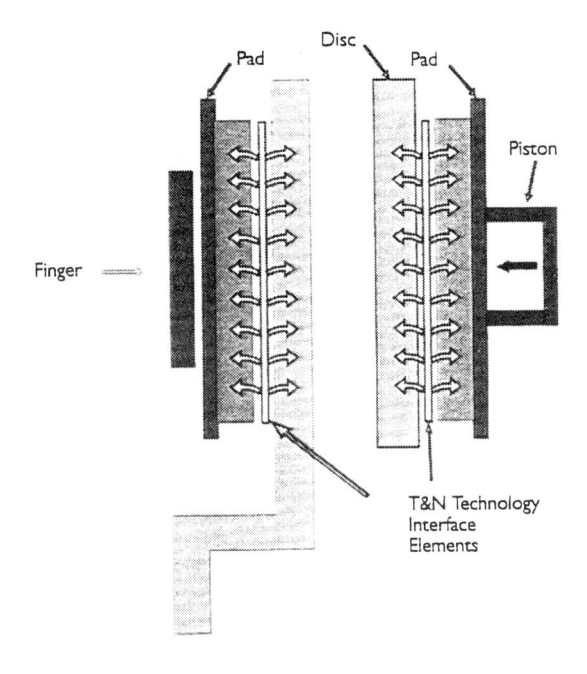

Fig.8. Schematic Illustration of the Axisymmetric Thermomechanical Model

(a) soft pad: E=137 N/mm² braking time 45 sec at 1 N/mm²

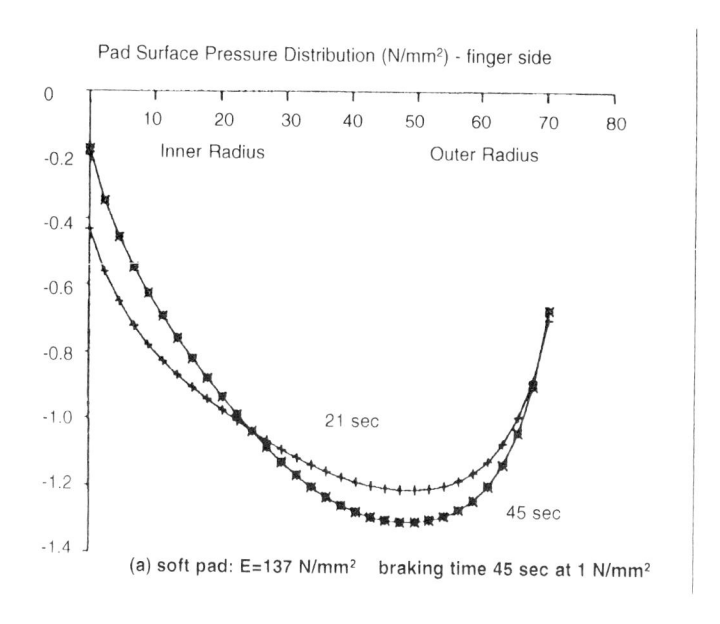

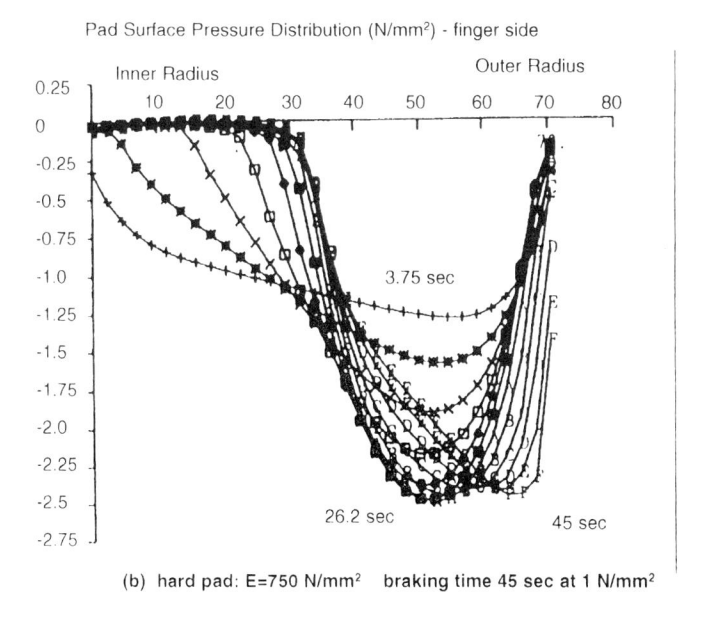

(b) hard pad: E=750 N/mm² braking time 45 sec at 1 N/mm²

Fig. 9. Development of localisation during a brake application

(a)

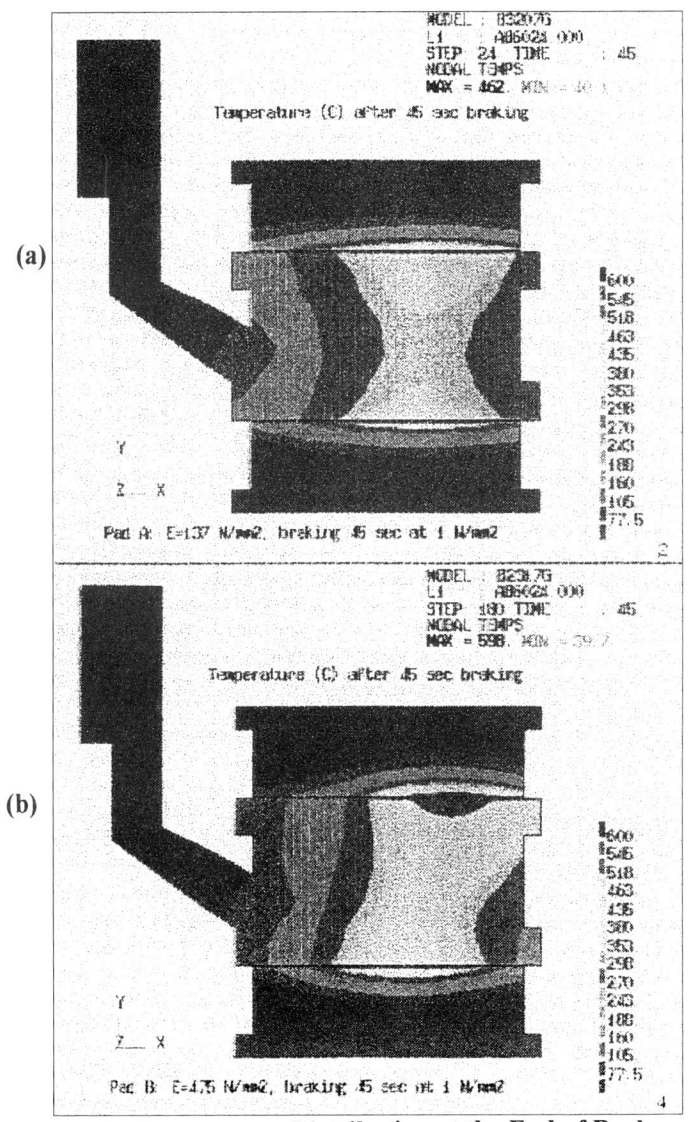

(b)

Fig.10. Temperature Distribution at the End of Brake Application of Fig.9: (a) with soft pads; (b) with hard pads.

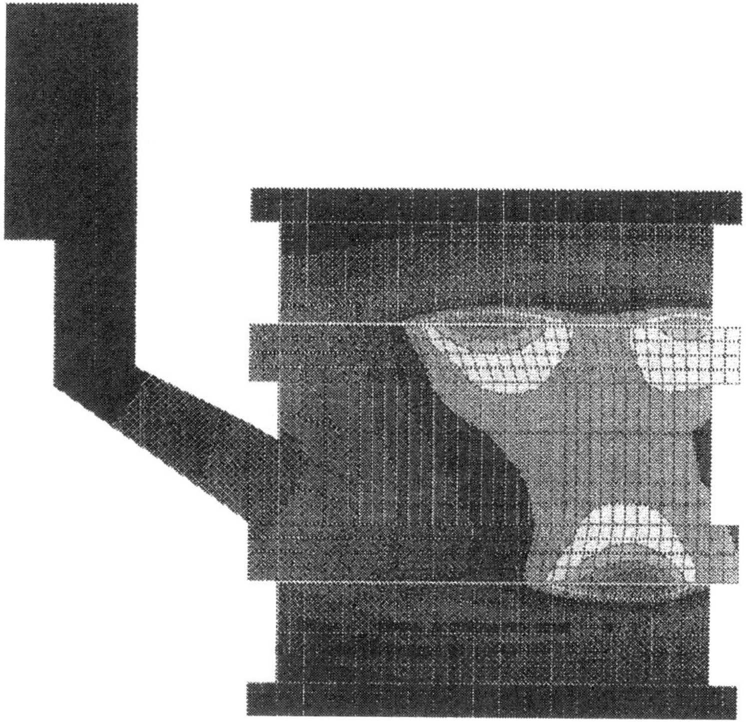

Fig.11. Formation of Multi Hot Bands.

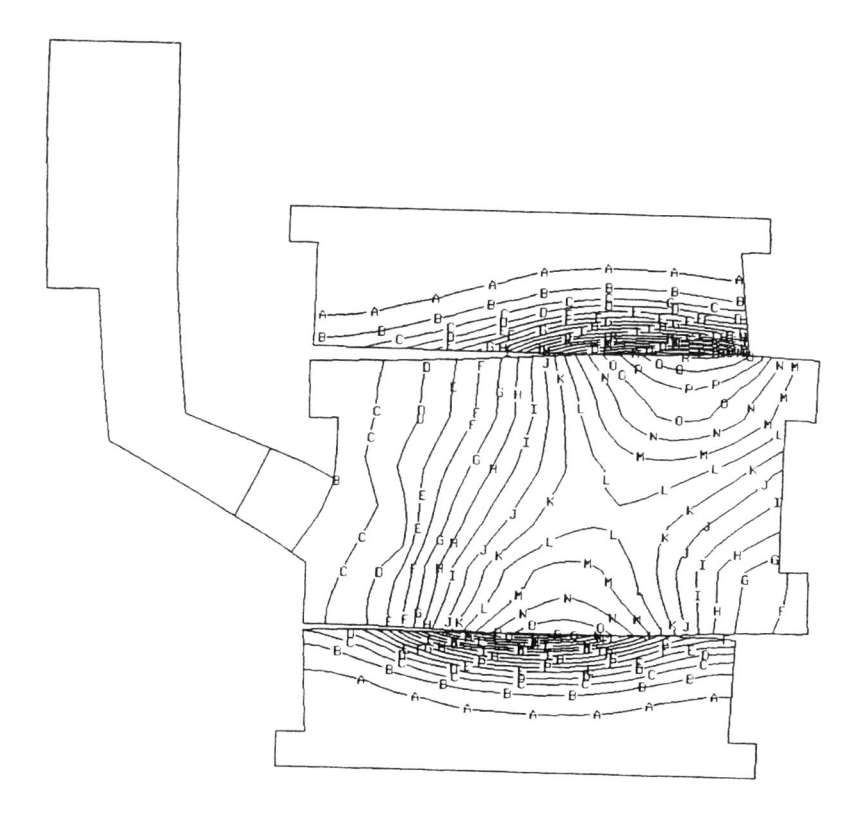

Fig.12. Global Displacement of Disc Rotor at the End of Brake Application.

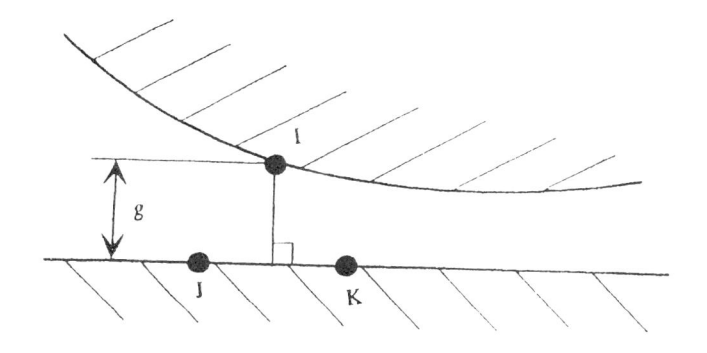

Fig.13. Node-to-Segment Contact (Conceptual Illustration in 2D)

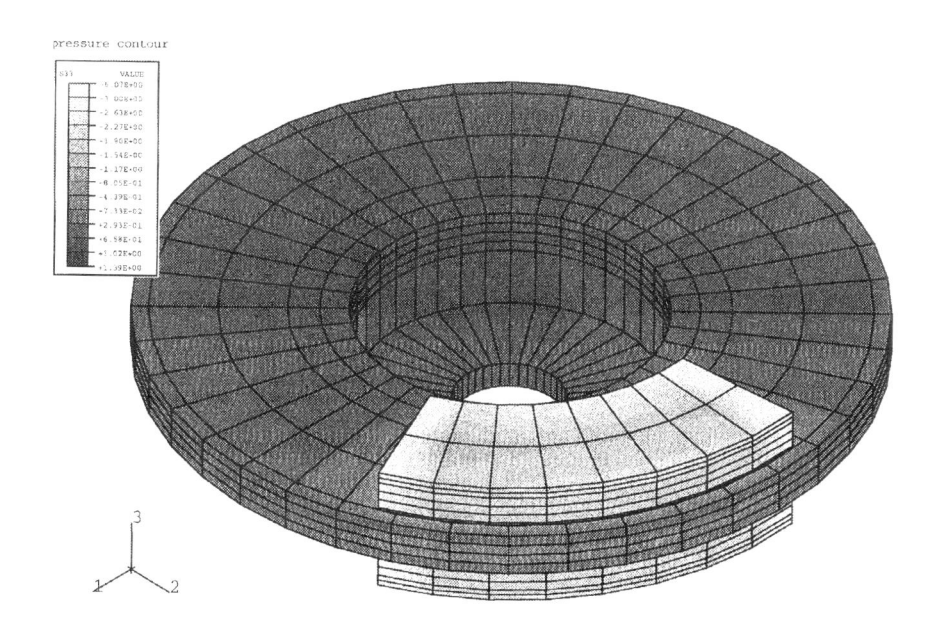

Fig.14. 3D Thermomechanical Model of Brake Operation with Contact Pressure Contour (before Braking)

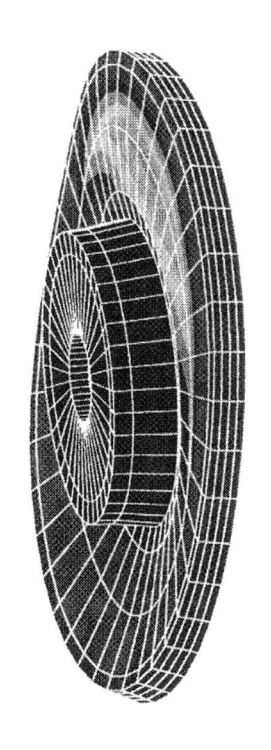

Fig.15. Simulated Temperature and Waviness of the Disc Rotor (after 4 Second Braking)

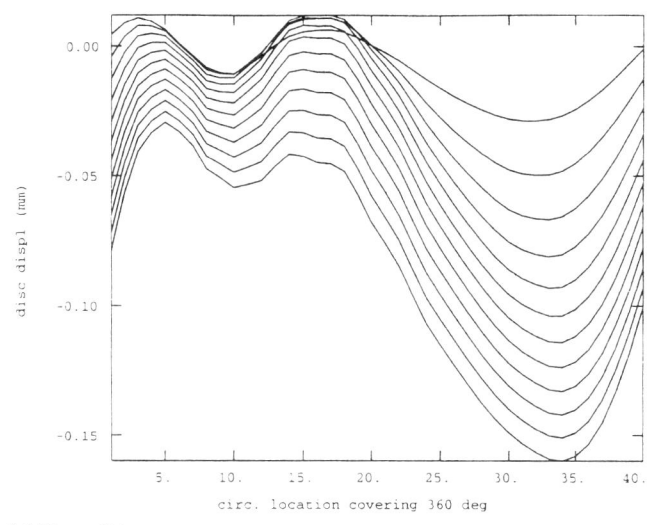

(a) Piston Side

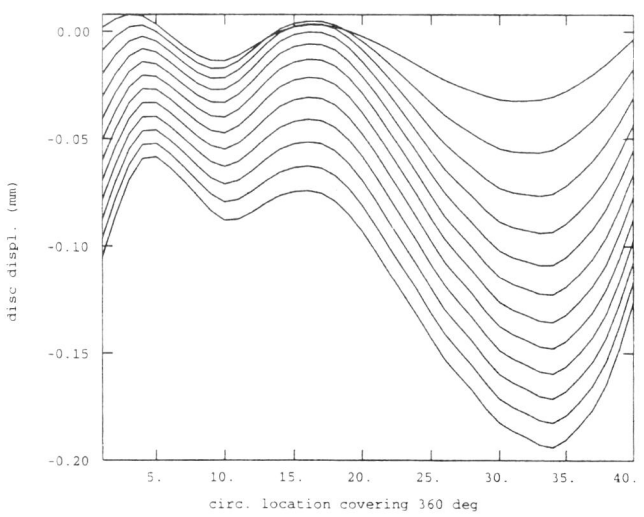

(b) Finger Side

Fig.16. Simulated Global Displacement of Disc Rotor

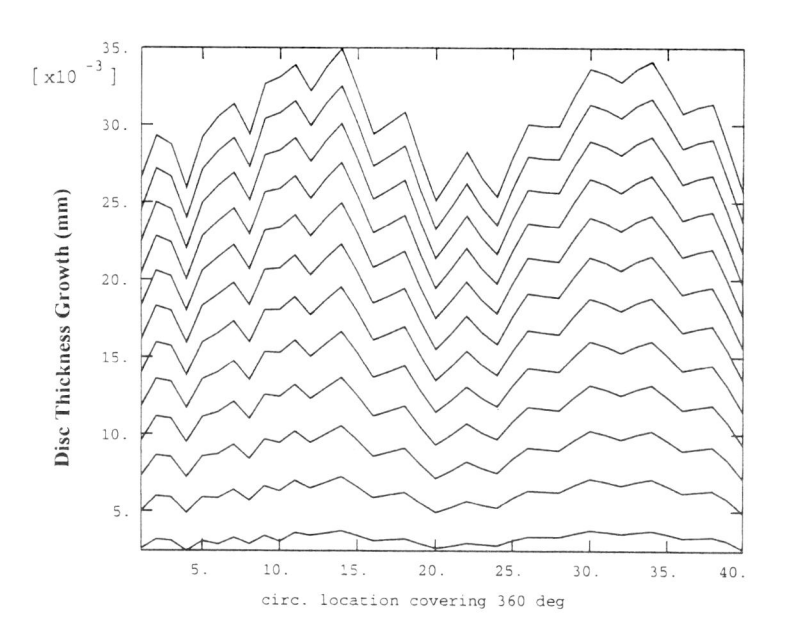

Fig.17. Simulated Disc Thickness Growth (from Fig.16.(a) and (b))

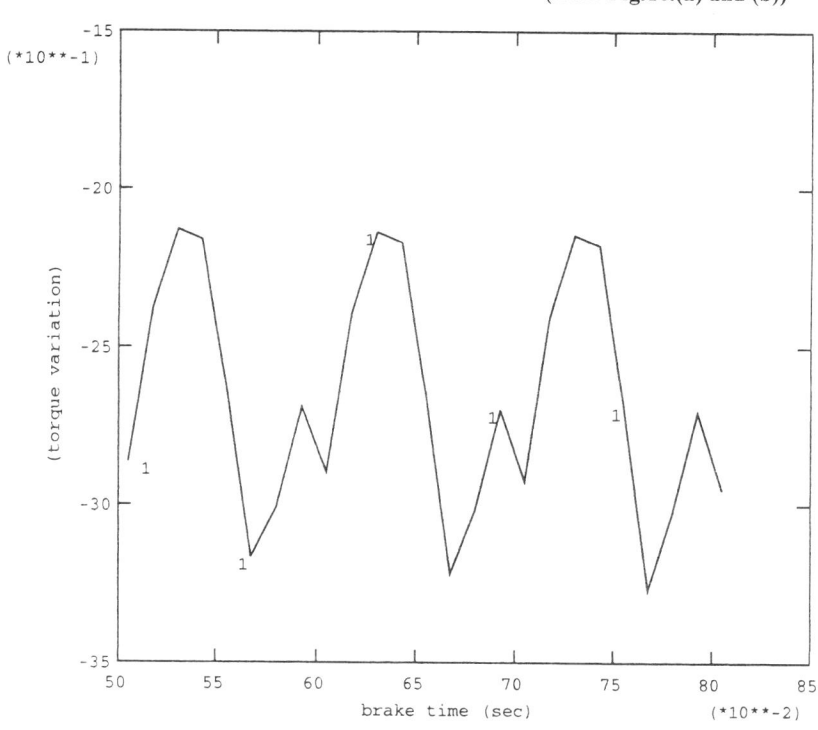

Fig.18. Simulated Torque Variation covering three revolutions.

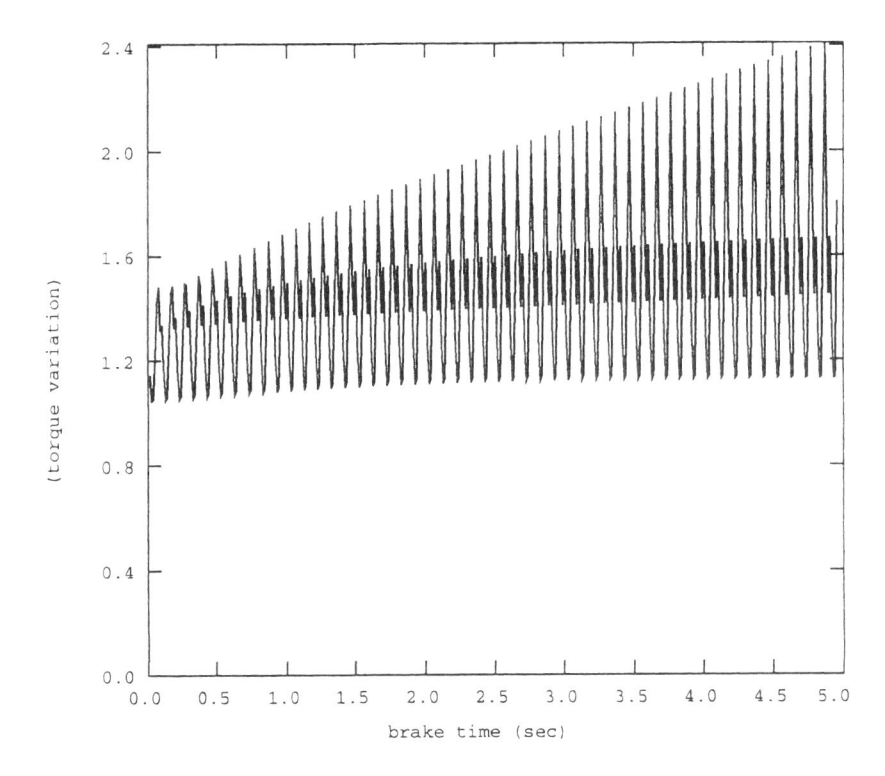

Fig.19. Simulated Torque Variation covering 5 seconds.

Fig.20. Experimental Example showing that the Hot Zone on one Side is Out-of-Phase with the Hot Zone on the Opposite Side. (Note: Piston side in foregraound, fist side shown in reflection in background)

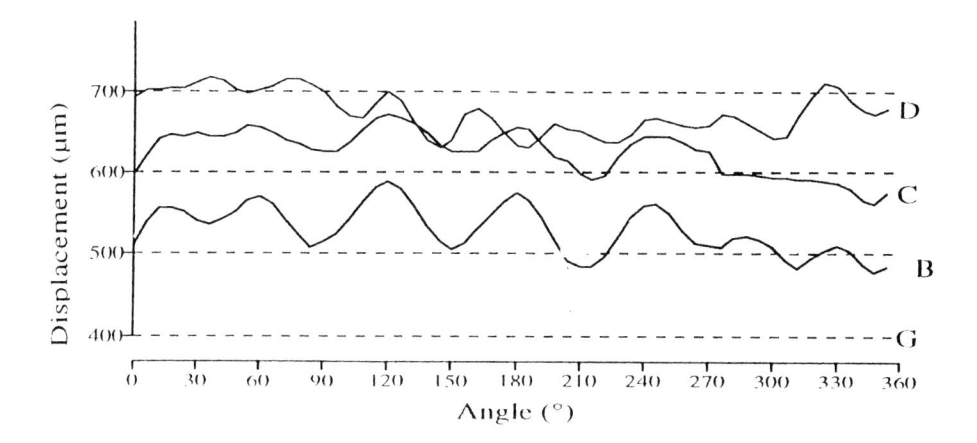

Fig.21. Measured Air Gap illustrating Seven-Peak "Thermoelastic Unstable Buckling"

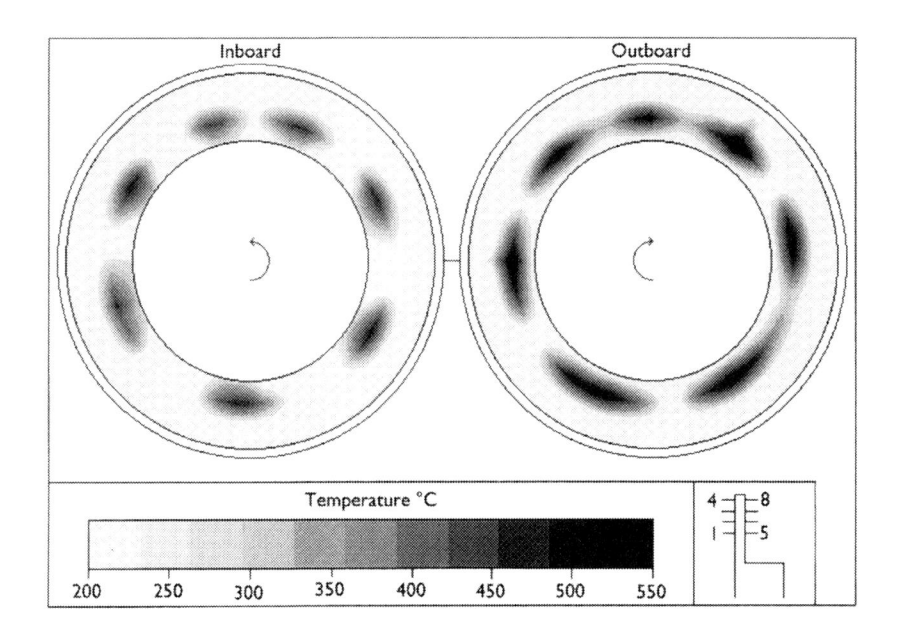

Figure 22. Measured rotor temperatures illustrating seven hot spots distributed out-of-phase on the opposite sides (corresponding to Fig.21)

The brake disc – development challenge and factors for success in the pneumatic disc brake system for heavy trucks

V KÜHNE
Knorr-Bremse Systeme für Nutzfahrzeuge GmbH, Munich, Germany

ABSTRACT:
The performance of the brake disc is of great importance for the success of the pneumatic disc brake in heavy commercial vehicles. In this case performance means especially life cycle and robustness, i.e. robustness regarding disc cracking, but also weight, cost and ease of servicing.

The development steps in the fields of design and material optimisation, especially on the basis of conventional brake disc designs with the corresponding functional improvements are described in detail. Furthermore a second generation brake disc is presented, based on a completely new connection method between brake disc and hub, which will set new standards in the field of the disc brake relative to costs, robustness, weight and ease of servicing.

THE BRAKE DISC

1. INTRODUCTION

In the disc brake system (fig. 1) the brake disc has, apart from the caliper which produces the application force, an outstanding influence with regard to the efficiency of this kind of wheel brake.

The demands on an efficient brake disc are extremely varied. In principle this component has to transmit reliably the brake torque, which is produced by means of friction forces from the applied brake pads. With heavy commercial vehicles up to 30.000 Nm are demanded, a fact, which imposes high standards on the design of the brake disc. At the same time the brake disc must show an optimised thermal economy and an excellent robustness. Optimised thermal economy means, to keep the temperature level of the brake disc during the brake application as low as possible. This is of great importance, since the pad wear increases exponentially with raising brake disc temperatures. A low temperature level is reached - especially on the disc surface - by means of very good heat conductivity and sufficient heat capacity of the brake disc, whereas the restriction concerning weight sets certain limits, as well as by means of a relatively good ventilation of the brake disc.

Brake disc robustness means wear resistance and indifference to cracking. This matter will be treated very intensively in the following. We are here in a classical conflict of aims, since on the one hand hard materials favour wear but on the other hand they lead to negative effects on the cracking because of the lack of ductility.

Last but not least the ease of servicing is of importance. That means, the brake disc should be economical and the exchange of it on the vehicle should be simple.

2. PROBLEM AREA - DISC CRACKING

The experience made in using disc brakes in heavy commercial vehicles shows, that the complete crack is the most common cause for an unplanned exchange of brake discs.The initial point of the cracking are small hair line cracks on the surface of the brake disc, caused by extreme temperature strain on the surface. They are on principal unavoidable but noncritical, since they are eliminated by natural wear. A problem only arises, when these hair line cracks - especially caused by deformation of the brake disc - begin to grow uncontrolled. The influencing factors for disc cracking are shown in fig. 2.

The optimisation of brake discs is carried out by means of special test benches, where the number of cycles up to the complete crack is a measure for the robustness of the brake disc. With this procedure it is possible to carry out comparable tests.

Fig. 3 shows optimising steps with regard to improved materials and design with a conventional brake disc. Concerning the design, the stiffness of the disc was successfully increased, which has a favourable effect on the deformation behaviour.

Fig. 4 shows how other factors have an influence on the robustness of the brake disc with regard to cracking. This chart also shows the brake disc of the second generation of Knorr, which represents a quantum leap concerning cracking.

3. THE SECOND GENERATION BRAKE DISC

3.1. The new brake disc is a flat disc with teeth, which is fixed by means of intermediate elements on the respective cam of the hub (fig. 5). The intermediate elements assure the torque transfer and the axial fixing with free mobility of the disc teeth at the same time in radial direction.

This mode of fixing assures, that the brake disc can expand freely when heated up during the brake application and thus the deformation of the brake disc is minimised in each operating status. With this method one of the main causes for crack growth is eliminated permanently. Fig. 6 shows the Wheel End of a planetary hub reduction axle with the second generation brake disc.

3.2. Technical features

Fig. 7 shows the advantages of a second generation brake disc in comparison with a conventional brake disc with regard to cost, weight, crack insensitivity and the guaranteed transfer of the torque.

3.3. Standardisation

On introducing the brake disc of the second generation there is the possibility of using identical components on the front and rear axles of a commercial vehicle. The same is true for trailer axles. Fig. 8 shows typical Wheel Ends with conventional brake discs, in comparison with Wheel Ends with identical brake discs.

3.4. Ease of servicing

The flat brake disc is a requirement for showing divided discs (Fig. 9). For the service case this has clear advantages, for now there is no longer a need to dismantle the caliper and/or the planetary hub reduction axle of the rear axle (Fig. 10).

4. SUMMARY

With the brake disc of the second generation Knorr-Bremse has put a milestone in the further development of the system "disc brakes for heavy commercial vehicles", since this innovation is superior to the conventional brake disc in all aspects - technically and economically. Thus a further condition is created which continues to intensify the success and the acceptance of disc brakes in heavy vehicles.

KNORR-BREMSE Systeme für Nutzfahrzeuge SfN

Disc Brake System

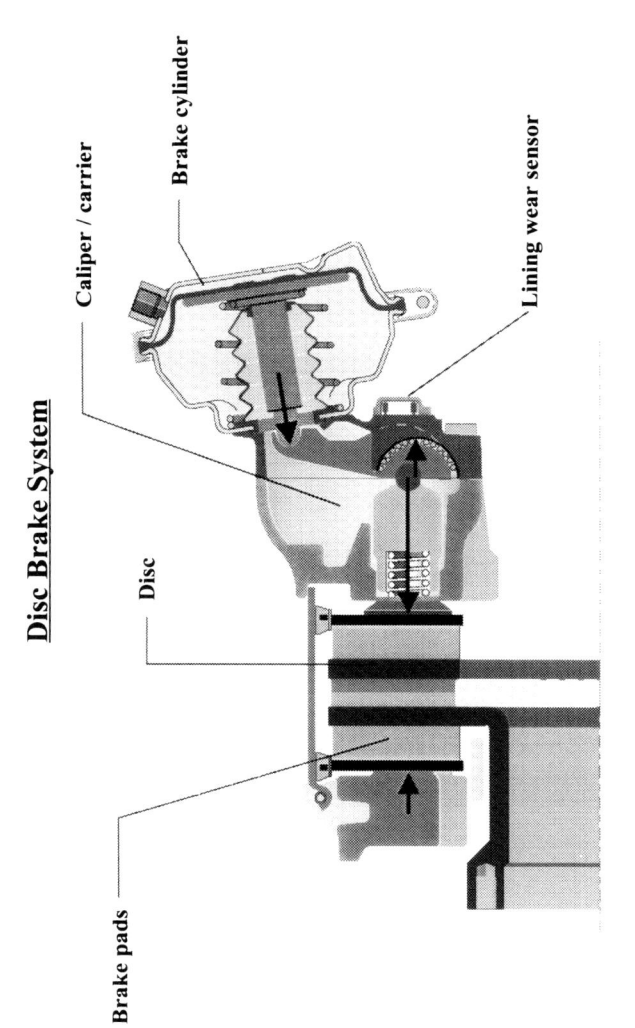

Figure 1

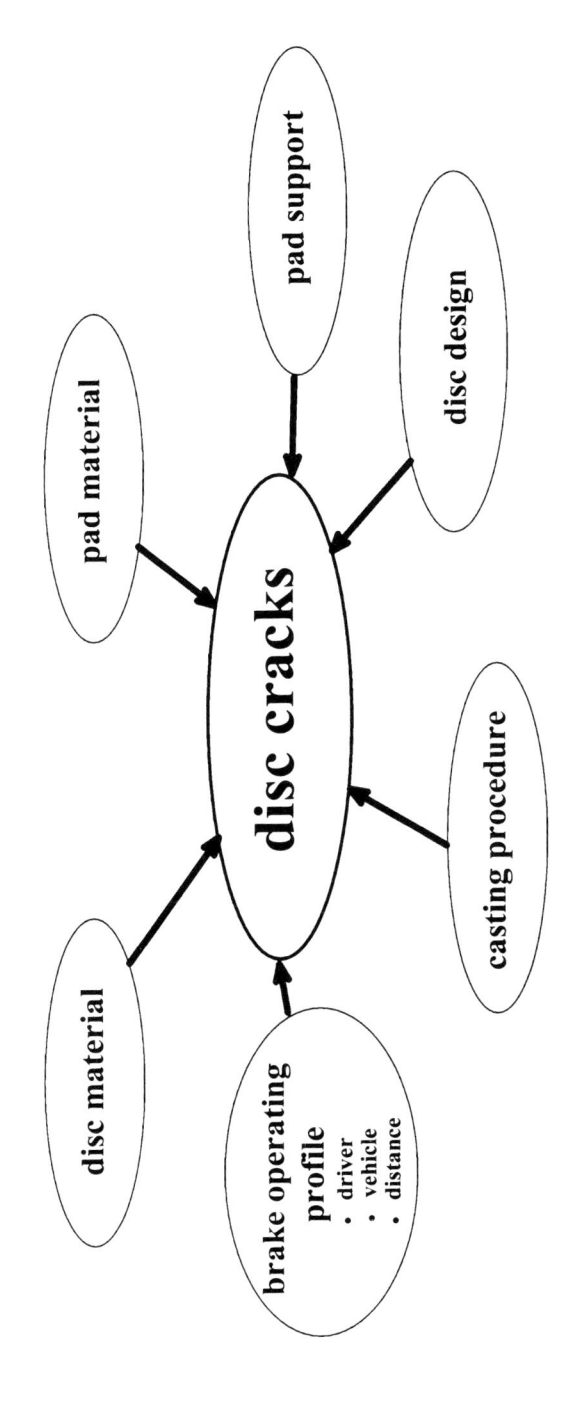

Figure 2

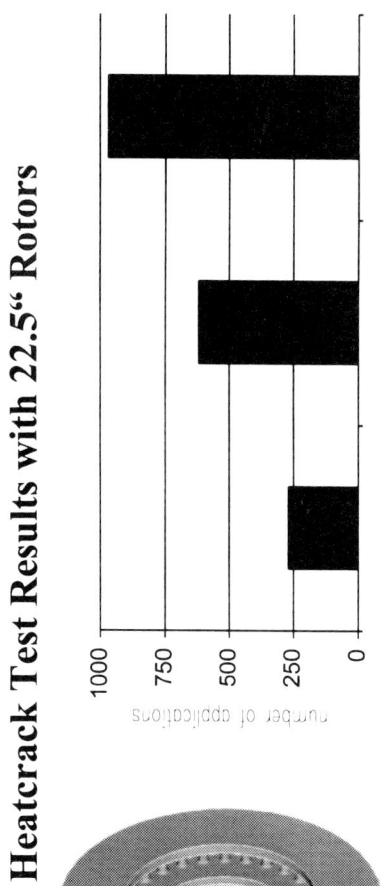

KNORR-BREMSE Systeme für Nutzfahrzeuge   SfN

Heatcrack Test Results with 22.5" Rotors

cooling channel width	17 mm	12 mm	12 mm
rotor material	casted iron high carbon	casted iron high carbon	casted iron high carbon + molybdenum
weight [kg]	31	33	33

Figure 3

KNORR-BREMSE Systeme für Nutzfahrzeuge SfN
Development Steps for Minimising the Rotor Problem

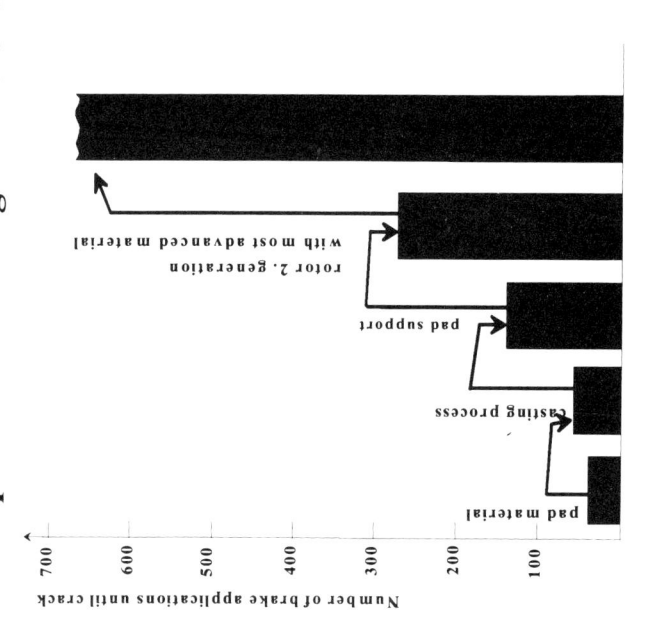

Figure 4

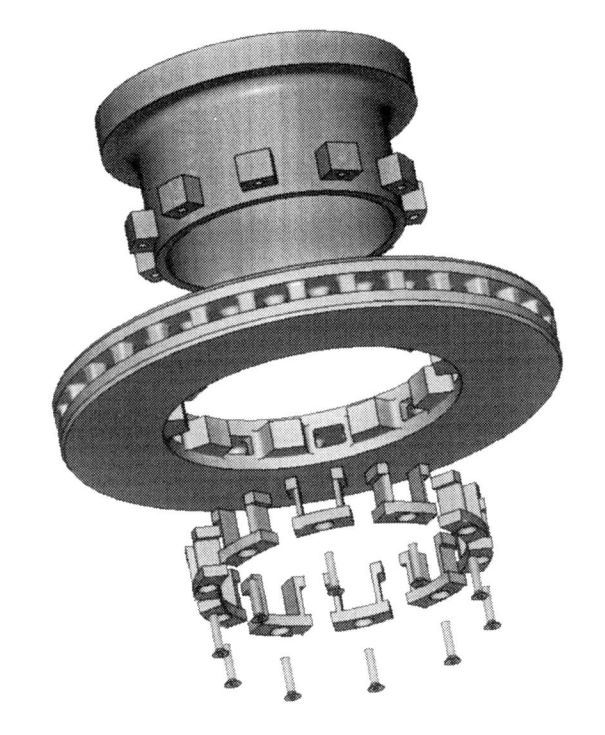

Figure 5

KNORR-BREMSE Systeme für Nutzfahrzeuge SfN

Axle Module with Rotor 2nd Generation

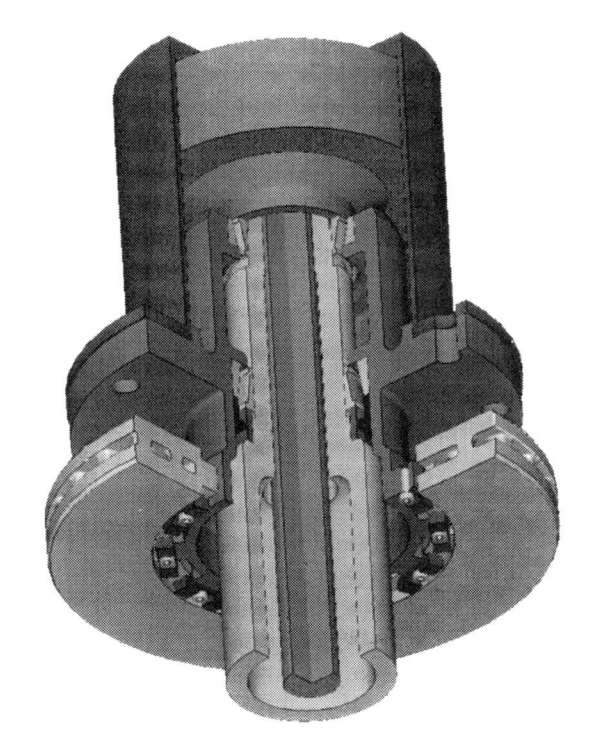

Figure 6

KNORR-BREMSE Systeme für Nutzfahrzeuge **SfN**

Features of the 2nd Rotor Generation

*System Advantages in Comparison
with a Conventional Rotor*

Costs of disc:

minus 20 %

Weight reduction:

15 kg/Axle

Robustness:

▶ Cracks — Factor 2

▶ Torque — Factor 3

Figure 7

KNORR-BREMSE Systeme für Nutzfahrzeuge   SfN

Standardisation with Rotor 2nd Generation

Rear axle

Front axle

Conventional rotor

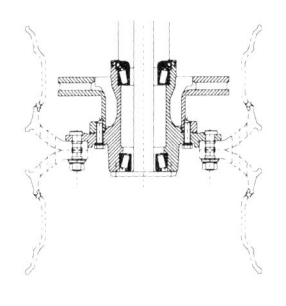

Rear axle

Front axle

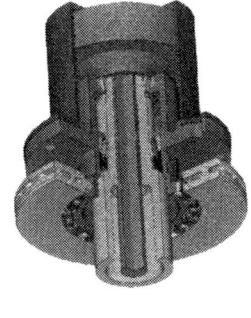

2nd Generation rotor

Figure 8

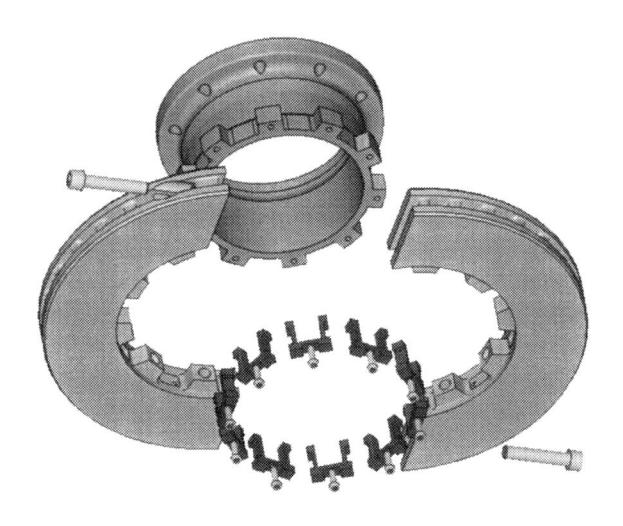

KNORR-BREMSE Systeme für Nutzfahrzeuge SfN

Two Part Splined Rotor

Figure 9

KNORR-BREMSE Systeme für Nutzfahrzeuge SfN

Easy Rotor Service

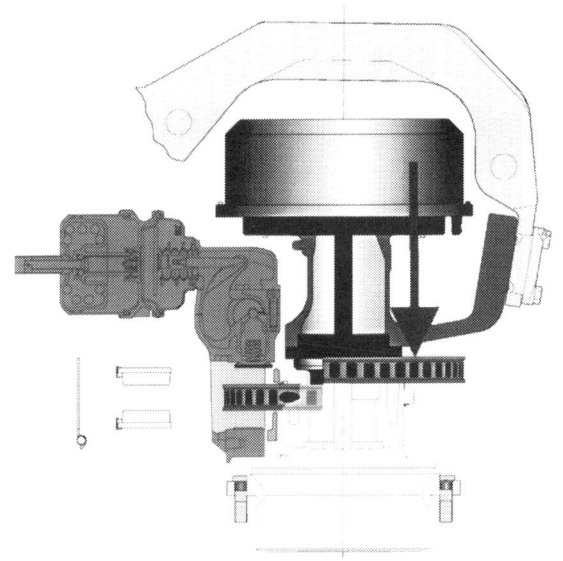

Figure 10

The influence of heavy vehicle brake drum geometry on brake torque stability

T VARMA and **R McLELLAN**
Carlisle, Motion Control Industries Inc., Charlottesville, Virginia, USA

SYNOPSIS

This paper presents and discusses preliminary results of testing carried out on full-scale inertia brake dynamometers for heavy vehicles at the Carlisle Technical Center, Charlottesville, Virginia. The results establish a relationship between brake drum geometry and torque fluctuation. The torque fluctuation is analyzed during one revolution of a brake drag application of a heavy vehicle S-Cam drum brake. The results also show that friction material formulation can significantly affect the overall sensitivity of a heavy vehicle foundation brake to irregularities in brake drum geometry.

1 INTRODUCTION

The Simplex S-Cam drum brake design remains the mainstay of heavy vehicle foundation brakes in the US. While European truck and brake manufacturers have concentrated development on air actuated disc brakes for category N3 and N4 heavy vehicles over the past decade, the US industry has, until recently, continued its emphasis on air actuated drum brakes for class 7 and 8 heavy vehicles. Combining the US market demands for a low purchase price, and the recent FMVSS 121 braking requirements for reduced stopping distances, foundation brake component suppliers have been pressured into re-emphasizing the role of brake design on overall heavy vehicle braking performance.

Most performance tests for vehicle brake applications evaluate the brake forces (torque) developed by the foundation brakes[1] as opposed to the friction coefficients of materials which are of interest from a materials development standpoint. It is well recognized, and documented [2,3,4], that the torque is dependent on various mechanical and geometric aspects of the foundation brake components as well as friction material characteristics and lining dimensions. Day and Harding[5] studied the effects of lining geometry on friction pairs in contact relative to torque output. Their investigations were from a lining geometry perspective involving heel-toe and crown contact in curved friction pairs. However from a brake drum standpoint, there is a dearth of published literature on the effects of geometrical variations in brake drums on the torque output of friction pairs.

According to studies by Lang and Smales[6], the μ/velocity characteristics of a friction pair is fundamental to the generation of groan and torque instability and is known to occur when the following equation is satisfied:

$$Fd\mu/dv > c$$

where c is the damping in the system and F is the limiting static friction. Hum, squeal and squeak (varying amplitude torque fluctuations) are all geometrically induced instabilities. Their work included point masses added to rotors after studying the vibration characteristics of discs and drums. They observed that an inertial damper at the anti-nodal point of the vibrating drum yields positive damping.

From a materials standpoint, Aronov et. al.[7] have studied the effect of the normal load and system stiffness on the interaction between friction, wear and vibration using two dissimilar surfaces, steel rubbing on cast iron. They concluded that different regimes of friction are encountered in non-lubricated friction pairs: (1) A steady state regime characterized by a friction coefficient that is independent of the normal load, and which had a small amplitude of slider oscillations. (2) A nonlinear friction regime characterized by a friction coefficient that increases with normal load and an increase in amplitude of the slider oscillations. (3) A self-excited regime characterized by periodic oscillations of the slider. According to studies done by Ishihara, et. al.[8] higher linear stiffness of the friction material causes an increase in friction force fluctuation (squeal) with pressure. They reported the elimination of the squeal by decreasing the stiffness. Blau[9] in studying the effect of transfer layer on the torque trace refers to momentary seizures causing fluctuations in the trace. Various traces are depicted with differing characteristics attributable to transfer films produced by material wear debris. Blau also refers to a variable "event time" which he uses to explain the periodic fluctuations in frictional forces. Accordingly, if the event times are longer than the duration of each discrete event, then the nominal friction will be restored. If not, there will be an apparent change in the nominal value of the friction coefficient.

It should be noted that previous studies have concentrated on fluctuations in torque output which occur at much higher frequency than those in the current study. The primary concern of this study is to identify and reduce sources of axle-end component fatigue and wear (air chamber mountings, camshaft bushings, cam lobe flatting, tire flat-spotting) and premature antilock brake system kick-in. The frequencies associated with these concerns are of the same order as wheel rotation while the frequencies targeted by previous studies have been orders of magnitude higher, taking them into the audible range.

2 TEST CONDITIONS

2.1 Friction Materials

Two friction materials with significantly different compressibility (High compressibility – HC and Low compressibility – LC) were each installed on a commercially available S-cam heavy vehicle drum brake. A drag application of the brake was then made against drums with controlled geometry for 40 revolutions at three speeds (32, 64 and 96 km/h) and two pressures (34.5 and 69 kN/m^2). The HC material had a compressive modulus of 17.1 X 10^3 MN/m^2 and the LC material had a compressive modulus of 52.8 X 10^3 MN/m^2. Figure 1 below shows a graph of the temperature dependence of the compressive modulus of the lining materials studied.

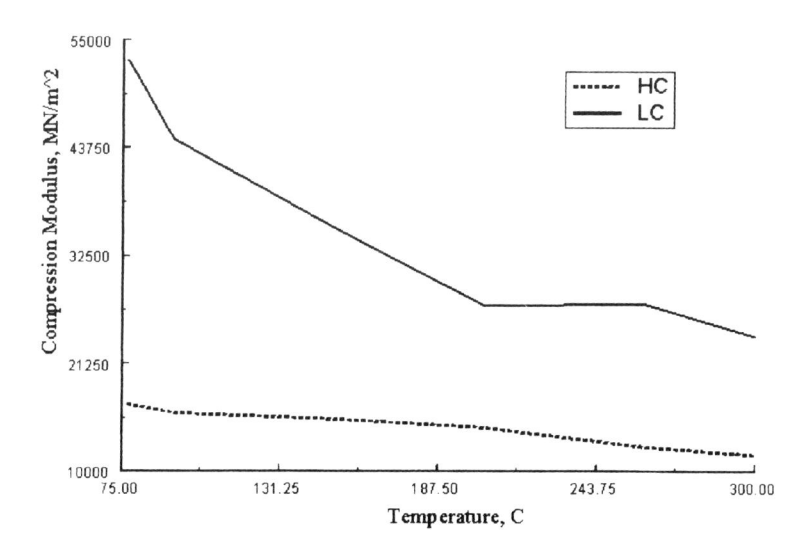

Fig. 1 Temperature dependence of compression modulus of the friction materials

2.2 Rotors

The grey cast iron brake drums used were standard, commercially available with identical chemical makeup. These drums were re-machined to give controlled levels of geometrical variation. A 420 mm X 180 mm cast iron drum was used with a nominal wall thickness of 12.7 mm. It had a squealer band with 30.5 mm nominal thickness and 45 mm nominal width, 15 mm from the mouth of the drum. When required, flat spots were introduced after the re-machining operation. The various drum geometries used were as follows (Since the squealer band is machined both inside and out at the factory, its thickness is excluded below):

(1) Control Drum (BL) - Thickness of 12.7 +/- 0.63 mm with a maximum Inner Diameter (ID) runout of 0.13 mm.
(2) Drum with a Thickness Variation (TV) – Same as (1) for ID runout but with thickness Varying from 11.4 to 16.5 mm. The thickest and thinnest sections were 180^0 apart.
(3) Drum with Flat Spot (FS) – Same as (1) but with two flat spots introduced 180^0 apart. The maximum ID runouts at the flat spots were 0.63 and 0.96 mm for LC and HC pairs respectively.
(4) Drum with high ID Runout (RO) – Same as (1) for thickness but with an ID runout of 0.5 mm.

2.3 Procedure

A constant pressure, constant speed drag application (as opposed to a stop) was chosen to give consistent energy input into the drum eliminating any output torque variability due to speed variation. An initial drum temperature of 93^0C was chosen to minimize any output torque variation resulting from drum expansion. Drum temperature was used for control during the test cycles. Short duration drags (40 revolutions) were selected to minimize heat build-up during the brake application while allowing sufficient time for brake application induced torque fluctuations to minimize. At the start of each test condition, a light burnish procedure of 50 stops from 64 km/h at 3 m/s^2 was done to eliminate the effect of surface finish. Linings were machined due to the mean drum radius to maximize initial pre-burnish contact.

2.4 Experimental Parameters

- 50 Burnish stops with a 3 m/s^2 deceleration from 64 Km/h.
- Materials: High compressibility & low compressibility.
- Pressure: 34.5 kN/m^2 and 69 kN/m^2
- Speed: 32, 64 and 96 Km/h (32 Km/h = 162 rpm).
- Time: 14.8, 7.4 and 4.9 s. (Equivalent to 40 revolutions at each speed).
- Temperature: Initial drum temperature of 93^0C.
- Drums: As described above.
- 5 drags were done at each condition (Speed, pressure, drum configuration and material).
- Rolling Radius = 528.3 mm, Drum size = 420 mm X 180 mm, Air chamber = Type 30, Slack Adjuster Length = 139.7 mm.

3 TEST DATA

Figures in appendix 1 detail the raw data generated during testing. The torque traces of the various friction pairs at different pressures and speeds are presented.

4 RESULTS

Table 1 below summarizes the results of the analysis. The sensitivity factor S, is a statistical measure of dispersion of maxima and minima of the torque from the mean value and is considered representative of torque fluctuation. The summary presents sensitivity values, mean torque values, peak amplitudes and degrees of drum rotation ($\alpha1$ and $\alpha2$) between peaks as a function of pressure and speed for various friction pairs. $\alpha1$ refers to the degrees of rotation going from the small peak to the large peak and $\alpha2$ from the large peak to the following small peak.

Table 1 Summary of data analysis

Friction Pair	Speed, km/h	Pressure, kN/m^2	Sensitivity, S	Mean, Nm	Amplitude1, Nm	Amplitude2, Nm	$\alpha1$, Deg.	$\alpha2$, Deg.
LC-BL	32	34.5	27	181.4	32.5	0.0	360	N/A
LC-BL	32	69.0	38	875.5	42.0	0.0	360	N/A
LC-BL	64	34.5	25	137.9	30.2	0.0	360	N/A
LC-BL	64	69.0	39	916.8	50.9	0.0	360	N/A
LC-BL	96	34.5	27	166.7	28.0	0.0	360	N/A
LC-BL	96	69.0	28	768.6	24.3	0.0	360	N/A
LC-FS	32	34.5	71	179.2	84.1	61.2	180	180
LC-FS	32	69.0	154	715.5	182.9	123.9	180	180
LC-FS	64	34.5	41	128.3	46.5	46.5	180	180
LC-FS	64	69.0	98	525.2	122.4	59.0	180	180
LC-FS	96	34.5	36	120.2	39.8	39.8	180	180
LC-FS	96	69.0	85	433.0	107.0	67.9	180	180
LC-RO	32	34.5	89	171.1	80.4	80.4	243	117
LC-RO	32	69.0	170	828.3	227.2	115.8	180	180
LC-RO	64	34.5	94	168.2	98.1	98.1	243	117
LC-RO	64	69.0	129	672.7	129.1	129.1	180	180
LC-RO	96	34.5	93	166.0	59.0	49.4	243	117
LC-RO	96	69.0	97	510.4	119.5	60.5	180	180
LC-TV	32	34.5	19	174.8	20.7	1.5	224	136
LC-TV	32	69.0	66	752.4	73.0	13.3	233	127
LC-TV	64	34.5	22	149.0	28.0	-3.0	200	160
LC-TV	64	69.0	56	563.5	67.1	0.0	360	N/A
LC-TV	96	34.5	18	129.1	47.9	24.3	210	150
LC-TV	96	69.0	45	478.0	68.6	0.0	360	N/A

Table 1 Summary of data analysis (Cont'd)

Friction Pair	Speed, km/h	Pressure, kN/m^2	Sensitivity, S	Mean, Nm	Amplitude1, Nm	Amplitude2, Nm	$\alpha 1$, Deg.	$\alpha 2$, Deg.
HC-BL	32	34.5	16	115.1	13.3	0.0	360	N/A
HC-BL	32	69.0	35	660.9	52.4	0.0	360	N/A
HC-BL	64	34.5	12	113.6	14.0	0.0	360	N/A
HC-BL	64	69.0	31	661.6	47.9	0.0	360	N/A
HC-BL	96	34.5	18	132.8	20.7	0.0	360	N/A
HC-BL	96	69.0	30	614.4	47.2	0.0	360	N/A
HC-FS	32	34.5	107	186.6	148.3	96.6	180	180
HC-FS	32	69.0	184	795.9	205.1	146.8	180	180
HC-FS	64	34.5	102	183.7	124.7	90.7	180	180
HC-FS	64	69.0	151	759.0	168.9	104.0	180	180
HC-FS	96	34.5	90	177.0	115.1	81.1	180	180
HC-FS	96	69.0	138	742.0	87.0	87.0	180	180
HC-RO	32	34.5	40	141.6	60.5	25.8	195	165
HC-RO	32	69.0	79	687.4	81.1	56.1	204	156
HC-RO	64	34.5	31	119.5	50.2	19.2	200	160
HC-RO	64	69.0	77	645.4	53.8	53.8	200	160
HC-RO	96	34.5	38	123.9	42.8	14.0	180	180
HC-RO	96	69.0	67	539.9	62.0	62.0	210	150
HC-TV	32	34.5	10	146.0	2.2	0.0	360	N/A
HC-TV	32	69.0	22	755.3	22.1	22.1	180	180
HC-TV	64	34.5	9	158.6	3.7	0.0	360	N/A
HC-TV	64	69.0	36	712.5	39.1	-7.4	180	180
HC-TV	96	34.5	11	171.9	14.0	0.7	180	180
HC-TV	96	69.0	37	666.1	38.4	-14.8	210	150

In Table 1, the degrees of drum rotation, $\alpha 1$ and $\alpha 2$, indicate contact points along the brake path. For instance, a value for $\alpha 2$ of 117 degrees indicates heel-toe contact and that of 180 degrees indicates leading-trailing shoe contact. This is due to the fact that the nominal lining arc is 117 degrees. $\alpha 2$ values in between 117 and 180 require further investigation due to the possible interference of multiple heel-toe and leading-trailing shoe contacts. The sum of $\alpha 1$ and $\alpha 2$ equals 360 degrees.

Based on the amplitudes in Table 1, the values of amplitude 1 indicate that the HC material pairs (HC-FS) are more sensitive than the LC material (LC-FS) to drum flat spots. However, since the runout in the flat spots was not equal, further investigation is warranted. The negative values for amplitude 2 are only observed for TV friction pairs indicating that the second peak is lower than the mean torque value. This has the effect of lowering the sensitivity values indicating lower torque fluctuation.

Based on the $\alpha 2$ values in Table 1, simple heel-toe contact is indicated in case of LC-RO friction pairs at lower pressure, which is eliminated at higher pressures. However, the LC-TV and HC-RO friction pairs show interference peaks. Also, based on mean torque values, drum irregularities in LC friction pairs are observed to cause a greater decrease in performance than HC friction pairs. The HC friction pairs also have lower speed sensitivity than the LC pairs.

Figure 2 is a graphical representation of the sensitivity values, S, as a function of drum geometry for LC friction pairs at the indicated speeds and pressures.

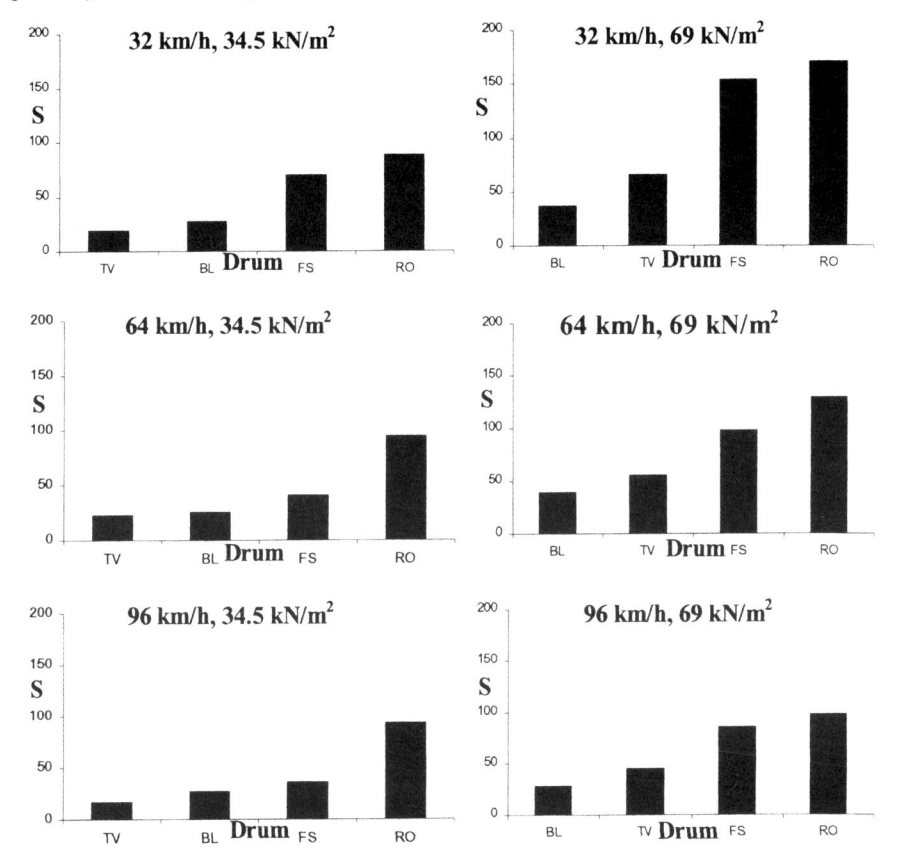

Fig. 2   Sensitivity of LC friction pairs at the indicated speed and pressure.

From Figure 2, the RO drum has the highest sensitivity value in all cases. The speed effects are greater for FS drum than the RO drum at lower pressure. In case of FS drums, the sensitivity decreases with speed. In all cases, the sensitivity increases with pressure and pressure has a greater effect on it than speed. At lower pressure, TV drums have the lowest sensitivity values.

Figure 3 is a graphical representation of the sensitivity values, S, as a function of drum geometry for HC friction pairs at the indicated speeds and pressures.

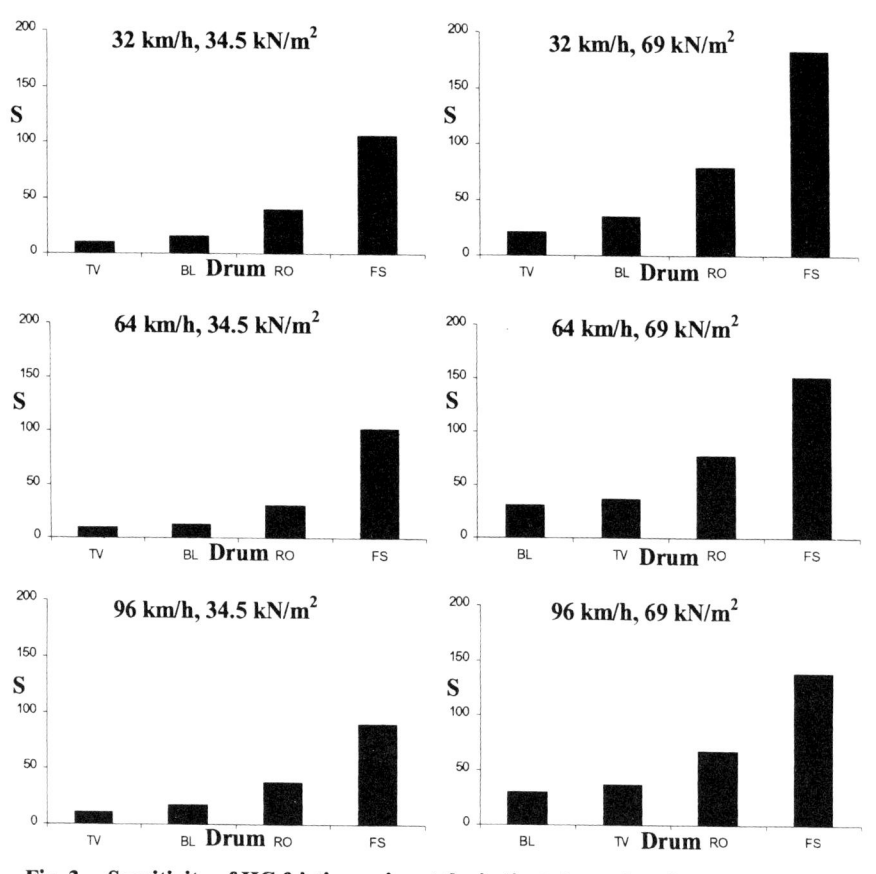

Fig. 3 Sensitivity of HC friction pairs at the indicated speed and pressure.

From Figure 3, FS has the most sensitivity. In all cases the sensitivity increases with pressure and pressure has a greater effect on it than speed. At lower pressure, TV drums have the lowest sensitivity values. A marginal increase in sensitivity is observed with increasing speed at higher pressure

5 CONCLUSIONS AND FUTURE WORK

(1) A decrease in performance occurs in case of the LC friction pairs with drum irregularities.
(2) Drum thickness variation along the brake path has a positive effect on the torque fluctuation in friction pairs.
(3) An HC material is more forgiving than a LC material to variations caused due to machining operations.
(4) Future work will focus on drum ID runout effects on friction pairs.

ACKNOWLEDGEMENTS

The authors wish to thank the Staff of Carlisle, Motion Control Industries, Inc. for their permission to publish this work. A personal thanks also goes to Devin McPherson, Dave Sarginger, Dave Wunderlich and the Dynamometer crew for their valuable input and help towards the completion of this work.

REFERENCES

(1) Flick, M. A. An overview of Heavy Vehicle Brake System Test Methods, SAE Technical Paper # 962215, 1996.
(2) Lang, A. M. PhD Thesis, Loughborough University of Technology, May 1994.
(3) Hulten, J. "Drum Brake Squeal - A Self-Exciting Mechanism with Constant Friction", SAE Technical Paper # 932965, 1993.
(4) Newcomb, T. P., and Spurr, R. T. Braking of Road Vehicles, Chapman and Hall Ltd., 1967.
(5) Day, A. J., and Harding, P. R. J. "Performance Variation of Cam Operated Drum Brakes", IMechE, 1983.
(6) Lang, A. M, and Smales, H. "An approach to the solution of disc brake vibration problems", C37/83, ImechE, 1983.
(7) Aronov, V., D'Souza, A. F., Kalpakjian, S. and Shareef, I. "Interactions Among Friction, Wear, and System Stiffness – Effect of Normal Load and Stiffness", Transactions of ASME, 106, p. 54, 1984.
(8) Ishihara, N., Nishiwaki, M., and Shimizu, H. Experimental Analysis of Low-Frequency Brake Squeal Noise, SAE Technical Paper 962128, 1997.
(9) Blau, P. J. Friction and Wear Transitions of Materials, Noyes Publications, 1989.

APPENDIX

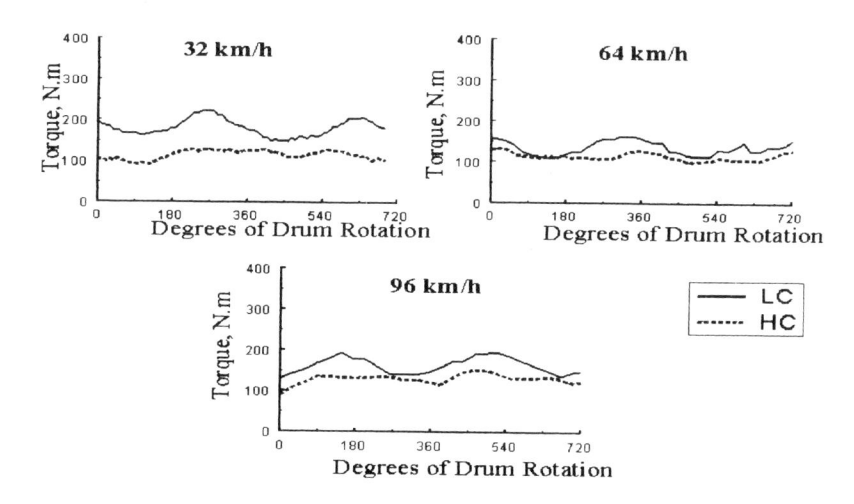

Fig. 4 Torque variation at 34.5 kN/m^2 as a function of drum rotation for BL friction pairs

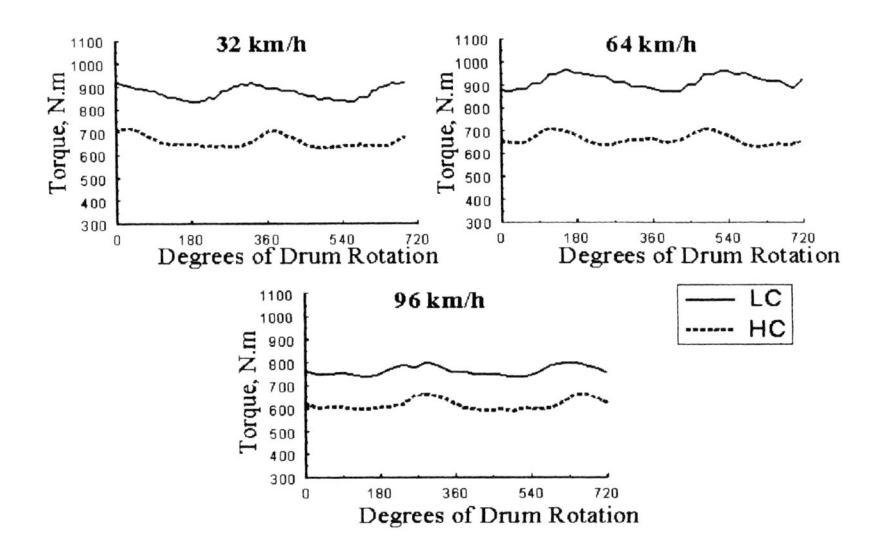

Fig. 5 Torque variation at 69 kN/m^2 as a function of drum rotation for BL friction pairs

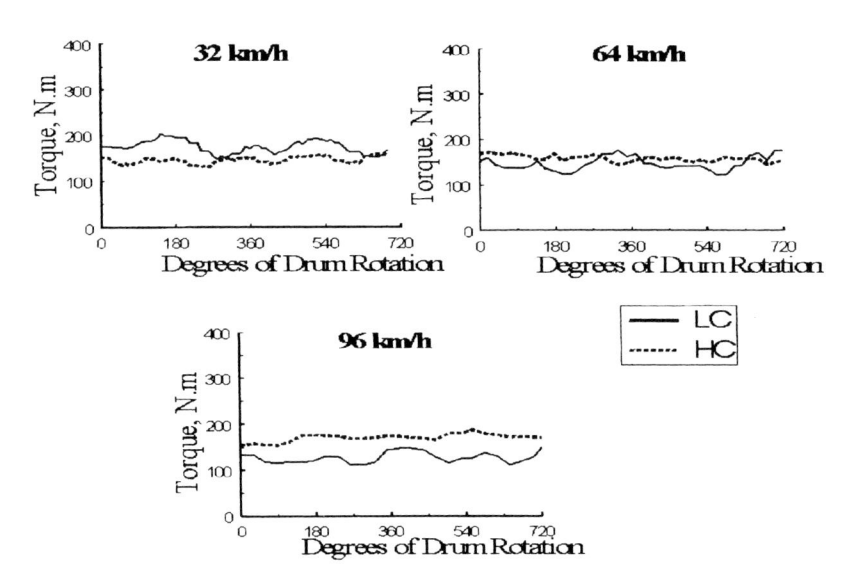

Fig. 6 Torque variation at 34.5 kN/m^2 as a function of drum rotation for TV friction pairs

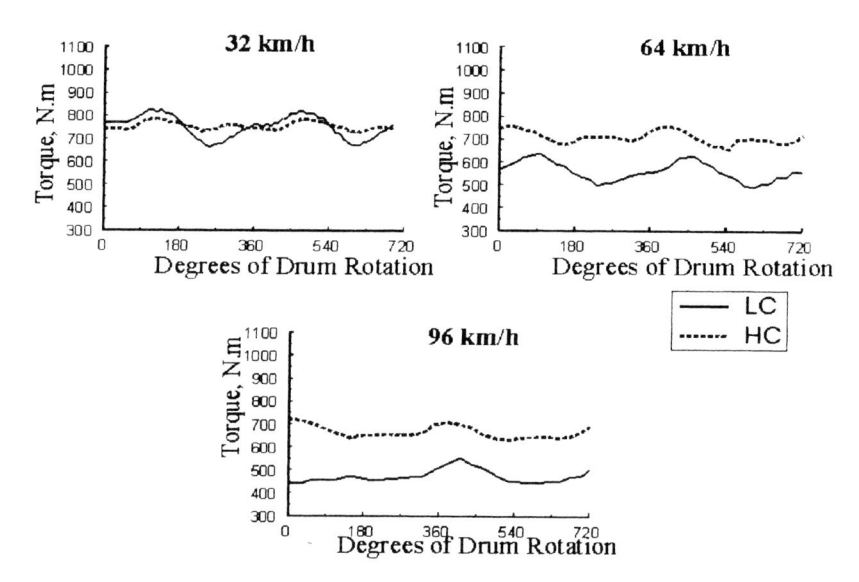

Fig. 7 Torque variation at 69 kN/m² as a function of drum rotation for TV friction pairs

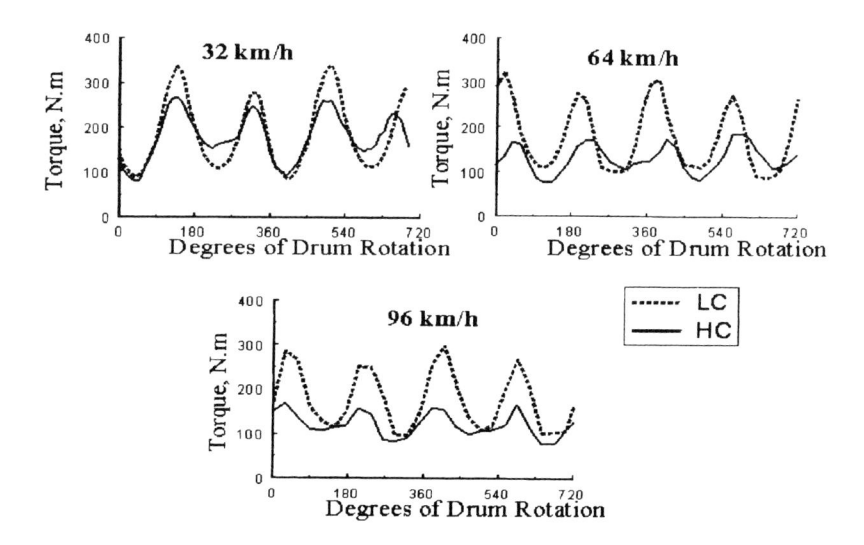

Fig. 8 Torque variation at 34.5 kN/m² as a function of drum rotation for FS friction pairs

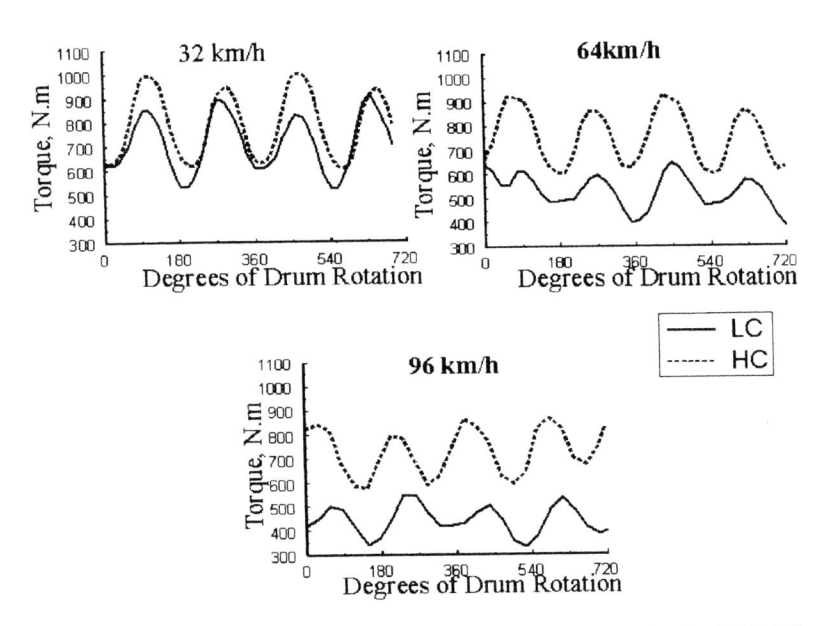

Fig. 9 Torque variation at 69 kN/m² as a function of drum rotation for FS friction pairs

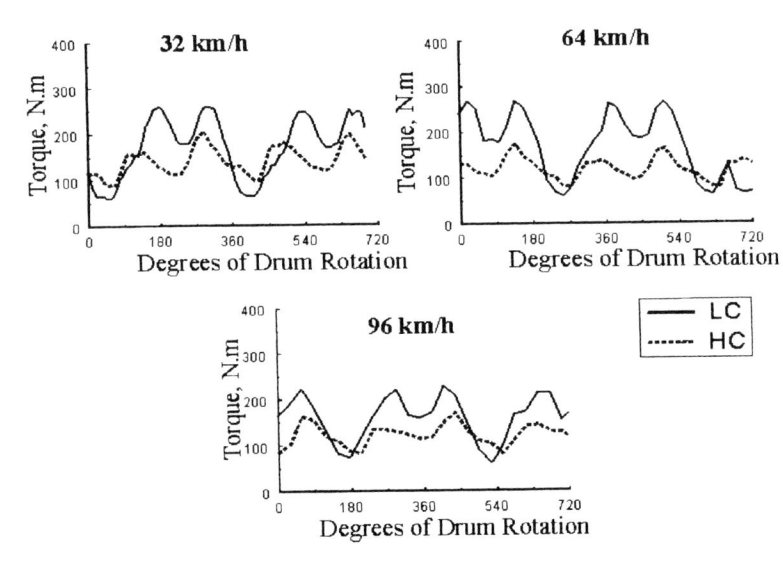

Fig. 10 Torque variation at 34.5 kN/m² as a function of drum rotation for RO friction pairs

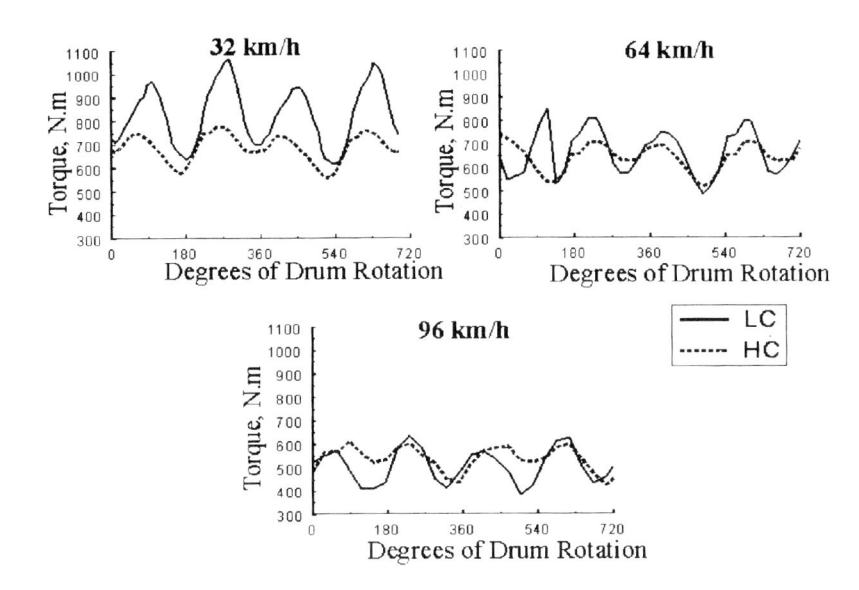

Fig. 11 Torque variation at 69 kN/m^2 as a function of drum rotation for RO friction pairs

Authors' Index

A
Abo, J 99–112

B
Bartholomew, R 49–66
Brookfield, D J 3–14
Buonfico, P 77–84

C
Cartmell, M P 3–14

D
Dalka, T M 113–126
Daudi, A R 127–144
Denholm, M J 205–214
Dickerson, W E 127–144
Douarre, A 231–264

E
Ellis, K 191–204
Evans, C E 215–230

F
Fash, J W 113–126
Fieldhouse, J D 27–48

G
Greening Jr, C W 87–98

H
Hartsock, D L 113–126
Hecht, R L 113–126
Heppes, P 15–26

J
James, S 3–14
Janevic, J 67–76

K
Kao, T K 231–264
Karthik, R 113–126
Krosnar, J G 177–190
Kühne, V 265–278
Kuroda, T 99–112

L
Lemon, T H 159–174

M
Macnaughtan, M P 177–190
Malosh, J 67–76
McLellan, R 279–292
Mottershead, J E 3–14

N
Narain, M 127–144

O
Ouyang, H 3–14

R
Rennison, M 27–48
Richmond, J W 231–264
Riesland, D 67–76
Ryan, T E 145–158

S
Stringham, W 67–76

T
Tron, B 77–84

V
Varma, T 279–292
Vikulov, K 77–84